Optics and Photonics:
An Introduction

F. Graham Smith and Terry A. King

John Wiley & Sons, Ltd

CHICHESTER · NEW YORK · WEINHEIM · BRISBANE · SINGAPORE · TORONTO

Other Wiley Editorial Offices

John Wiley & Sons, Inc., 605 Third Avenue,
New York, NY 10158-0012, USA

WILEY-VCH Verlag GmbH, Pappelallee 3,
D-69469 Weinheim, Germany

Jacaranda Wiley Ltd, 33 Park Road, Milton,
Queensland 4064, Australia

John Wiley & Sons (Asia) Pte Ltd, 2 Clementi Loop # 02-01,
Jin Xing Distripark, Singapore 129809

John Wiley & Sons (Canada) Ltd, 22 Worcester Road,
Rexdale, Ontario M9W IL1, Canada

Library of Congress Cataloging-in-Publication Data

Graham-Smith, Francis, Sir, 1923–
    Introduction to optics & photonics / F. Graham-Smith & T. A. King.
        p. cm.
    Includes bibliographical references and index.
    ISBN 0–471–48924–7 (hd : alk. paper) — ISBN 0–471–48925–5 (pbk. : alk. paper)
    1. Optics. 2. Photonics. I. Title: Introduction to optics and photonics. II. King, T. A.
III. Title.
    QC446.2.G73 2000
    535–dc21
                                                                    99–087078

British Library Cataloguing in Publication Data:

A catalogue record for this book is available from the British Library

ISBN 0 471 48924 7 (ppc)
    0 471 48925 5 (paper)

Typeset in 10/12.5pt Times by Kolam Information Services Pvt. Ltd, Pondicherry, India
Printed and bound in Great Britain by Bookcraft (Bath) Ltd
This book is printed on acid-free paper responsibly manufactured from sustainable forestry,
in which at least two trees are planted for each one used for paper production.

Contents

Authors' preface

My Design in this book is not to explain the Properties of Light by Hypothesis, but to propose and prove them by Reason and Experiments; in order to which I shall premise the following Definitions and Axioms.

The opening sentence of Newton's *Opticks*, 1717.

There has been a revolution in the field of optics over the last 30 years. It has assumed increasing importance in physics, other sciences, engineering and industry and in everyday life.

The dominant factor in this has been the discovery and development of many forms of lasers. The remarkable properties of laser radiation have led to a wealth of new techniques such as non-linear optics, atom trapping and cooling, femtosecond dynamics and electro-optics. It has also engendered a deeper understanding of light involving coherence and quantum optics, and optical coherence techniques have made a major impact in atomic physics. Not only physics but also many other areas of science have been enhanced by the use of laser-based methods in chemistry, biology, engineering and medicine. This has led to a wonderful range of new applications such as holography, optical communications, picosecond and femtosecond probes, optoelectronics, medical imaging and optical coherence tomography. Myriad applications have become prominent in industry and everyday life.

A modern optics course must now place equal emphasis on the photon and the wave aspects of light. The new term 'photonics' in the title reflects the importance of both aspects in understanding the new developments that the laser has brought to the subject, such as the development of optical fibres and semiconductor technology for optical sources and detectors. The field of optics is pervasive and continues to hold high potential for exploitation. This book presents an introduction to the essential elements of optics and photonics, suitable for an undergraduate lecture course but including an exposition of key modern developments. Optics is increasingly extending beyond physics into chemistry, biology, engineering and medicine; the book may also be used as a reference text for optics-related courses in those areas. The approach is to emphasize the basic concepts with the objective of developing student understanding. Mathematical content is sufficient to aid the physics description but without undue complication. Extensive sets of problems are included, devised to develop understanding and provide experience in the use of the equations as well as being thought provoking.

The chapters are grouped to provide an ordered progression of the concepts. The first six chapters lay foundations in waves and photons, geometric optics and optical instruments, periodic and non-periodic waves, electromagnetic waves, concepts of Fourier series and transforms, and polarization. Chapters 7 to 10 describe the fundamental optics and properties of interference and diffraction. Applications of optics in optical instruments, prism and grating spectrometers, Fabry–Perot and Michelson interferometers and spectroscopy (including a succinct description of Fourier transform spectroscopy) are grouped in Chapters 11 to 13. Aspects of spectroscopy not usually found in much detail in many optics texts are included here because of their importance. Chapter 14 brings together the concepts of coherence and correlation that unite these applications of interferometry.

Lasers and the properties of laser light are grouped in Chapters 15 to 17, including an introduction to the semiconductor physics that is basic to many types of lasers and detectors. Chapter 18 shows how fibre optic communications depend on many of the concepts introduced in this book as well as on lasers. Chapter 19 is devoted to holography. Chapter 20 on the interaction of light with matter is concerned with radiation mechanisms, absorption and scattering. Chapter 21 concentrates on the detection of light, with particular attention to semiconductors.

Finally, Chapter 22 on optics and photonics in nature describes the many instances of optical aspects that occur in the living world, including eyes with graded index lenses, polarization detectors, image dissectors, reflectors used for colouration and camouflage and in eyes, and photodetectors in plants, insects and animals.

This book replaces the earlier text, *Optics* (Smith and Thomson) in the Manchester Physics Series, from which some material has been drawn. The authors acknowledge invaluable help from Mike Wilson of Royal Holloway and from our editor, David Sandiford.

F. Graham Smith
Terry A. King
November 1999

Physical constants
Conversion factors

Velocity of light *in vacuo* $(c) = 2.99792458(\text{exactly}) \times 10^8 \, \text{m s}^{-1}$
Permittivity of free space $(\epsilon_0) = 8.854 \times 10^{-12}$ farad metre^{-1} (Fm^{-1})
Permeability of free space $(\mu_0) = 4\pi \times 10^{-7}$ henry metre^{-1} (Hm^{-1})
Planck's Constant $(h) = 6.626 \times 10^{-34} \, \text{Js}$
Boltzmann's constant $(k) = 1.381 \times 10^{-23} \, \text{JK}^{-1}$
Mass of electron $(m_e) = 9.109 \times 10^{-31} \, \text{kg}$
Mass of proton $(m_p) = 1.673 \times 10^{-27} \, \text{kg}$
Electronic charge $(e) = 1.602 \times 10^{-19} \, \text{C}$
Avogadro's number $(N_0) = 6.022 \times 10^{23} \, \text{mol}^{-1}$
Stefan-Boltzmann constant $(\sigma) = 5.671 \times 10^{-8} \text{Wm}^{-2} \text{K}^{-4}$
Wavelength of non-relativistic electron with energy V (in electron volts)
$= 1.23 \times V^{-1/2} \, \text{nm}$
Compton wavelength of an electron $(\lambda_c = h/mc) = 2.47631 \times 10^{-12} \, \text{m}$

Micron $(\mu\text{m}) = 10^{-6} \, \text{m}$
Nanometre (nm) $= 10^{-9} \, \text{m}$
Angstrom unit (Å) $= 10^{-10} \, \text{m}$
Picosecond (ps) $= 10^{-12} \, \text{s}$
Femtosecond (fs) $10^{-15} \, \text{s}$
Electron-volt (eV) $= 1.602 \times 10^{-19} \, \text{J} = 8066 \, \text{cm}^{-1}$

Acknowledgements

We gratefully acknowledge illustrations provided by Robert Alcock (Fig. 3.13), Paul Treadwell (Fig. 8.3), Andrew Kennaugh (Fig. 11.6), Stuart Jackson (Plate 15.1), and Tom Muxlow (Plate 14.1), all of the University of Manchester; William Livingston (Plate 22.1), Paul Neiman (Plate 22.2), from *Color and Light in Nature*, Cambridge University Press; Steve Bagley (Fig. 3.1), and Terry Allen (Plate 14.2) of the Paterson Institute, Manchester; Peter Vikusic and Gavin Wakeley (Plate 15.1) of the University of Exeter; Miles Padgett, University of Glasgow (Fig. 15.11), Rodney Loudon, University of Essex (Fig. 17.2), Hans Bjelkhagen, De Montford University (Fig. 19.6), Michael Land, University of Sussex (Fig 22.6 and other material); Bernard Schutz (Plate 11.1); Minolta Cameras (Plate 3.2); the Royal Society Library (Fig. 22.3), and Professor Philip Russell, University of Bath (Fig. 18.19).

1

Light as waves, rays and photons

Light is an *electromagnetic wave*: light is also a stream of *photons*, discrete particles carrying packets of energy and momentum. How can these two statements be reconciled? Similarly, while light is a wave, it nevertheless travels along straight lines or rays, allowing us to analyse lenses and mirrors in terms of geometric optics. Can we use these descriptions of waves, rays and photons interchangeably, and how should we choose between them? These problems, and their solutions, recur throughout this book, and it is useful to start by recalling how they have been approached as the theory of light has evolved over the last three centuries.

1.1 THE NATURE OF LIGHT

In his famous book *Opticks*, published in 1704, Isaac Newton described light as a stream of particles or corpuscles. This satisfactorily explained rectilinear propagation, and allowed him to develop theories of reflection and refraction, including his experimental demonstration of the splitting of sunlight into a spectrum of colours by using a prism. The particles in rays of different colours were supposed to have different qualities, possibly of mass, or size or velocity. White light was made up of a compound of coloured rays, and the colours of transparent materials were due to selective absorption. It was, however, more difficult for him to explain the coloured interference patterns in thin films, which we now call Newton's rings (see Chapter 9). For this, and for the partial reflection of light at a glass surface, he suggested a kind of periodic motion induced by his corpuscles, which reacted on the particles to give 'fits of easy reflection and transmission'. Newton also realized that double refraction in a calcite crystal (Iceland spar) was best explained by attributing a rectangular cross-section to light rays, which we would now describe as *polarization* (see Chapter 6). He nevertheless argued vehemently against an actual wave theory, on the grounds that waves would spread in angle rather than travel as rays, and also that there was no medium to carry light waves from distant celestial bodies.

The idea that light was propagated as some sort of wave was published by Descartes in *La Dioptrique* (1637); he thought of it as a pressure wave in an elastic medium. Christiaan Huygens, a Dutch contemporary of Newton, developed the wave theory; his explanation of rectilinear propagation is now known as 'Huygens' construction'. He correctly explained refraction in terms of a lower velocity in a denser medium. Huygens' construction is still a useful concept, and we use it later in this chapter.

It was not, however, until 100 years after Newton's *Opticks* that the wave theory was firmly established and the wavelength of light was found to be small enough to explain rectilinear propagation. In Thomas Young's double slit experiment (see Chapter 7), monochromatic light from a small source passed through two separate slits in an opaque screen, creating interference fringes where the two beams overlapped; this effect could only explained in terms of waves. Augustin Fresnel, in 1821, then showed that the wave must be a *transverse* oscillation, as contrasted with the longitudinal oscillation of a sound wave; following Newton's ideas of rays with 'sides', this was required by the observed polarization of light as in double refraction. Fresnel also developed the theories of partial reflection and transmission (Chapter 5), and of diffraction at shadow edges (Chapter 8). The final vindication of the wave theory came with James Clerk Maxwell, who synthesized the basic physics of electricity and magnetism into the four Maxwell equations, and deduced that an electromagnetic wave would propagate at a speed that equalled that of light.

The end of the nineteenth century therefore saw the wave theory on an apparently unassailable foundation. Difficulties only remained with understanding the interaction of light with matter, and, in particular, the 'black-body spectrum' of thermal radiation. This was, however, the point at which the corpuscular theory came back to life. In 1900, Max Planck showed that the form of the black-body spectrum could be explained by postulating that the walls of the body consisted of harmonic oscillators with a range of frequencies, and that the energies of those with frequency ν were restricted to integral multiples of the quantity $h\nu$. Each oscillator therefore had a fundamental energy quantum

$$E = h\nu,$$ (1.1)

where h became known as *Planck's constant*. In 1905, Albert Einstein explained the photoelectric effect by postulating that electromagnetic radiation was itself quantized, so that electrons are emitted from a metal surface when radiation is absorbed in discrete quanta. Newton was right after all! Light was again to be understood as a stream of particles, later to become known as photons.

If light is a wave that has properties usually associated with particles, could particles correspondingly have wave-like properties? This was proposed by Louis de Broglie in 1924, and confirmed experimentally three years later in two classical experiments by George Thomson and by Clinton Davisson and Lester Germer. Both showed that a beam of particles, like a light ray encountering an obstacle, could be diffracted, behaving as a wave rather than a geometric ray. The diffraction pattern formed by the spreading of an electron beam passing through a hole in a metal sheet, for example, was the same as the diffraction pattern in light that we explore in Chapter 7. Furthermore, the wavelength λ involved was simply related to the momentum p of the electrons by

$$\lambda = \frac{h}{p}.$$ (1.2)

The constant h was again Planck's constant, as in the theory of quanta in electromagnetic radiation; for material waves λ is the *de Broglie* wavelength. A general wave theory of the behaviour of matter, *wave mechanics*, was developed in 1926 by Schrödinger following de Broglie's ideas. Wave mechanics revolutionized our understanding of how microscopic particles were described and placed limitations on the extent of information one could have about such systems – the famous Heisenberg uncertainty relationship.

Photons and electrons evidently behave both as particles and as waves; which description best describes their behaviour depends on the circumstances. Although photons and electrons have very similar wave-like characteristics, there are several fundamental differences in their behaviour. Unlike electrons, photons are not conserved and can be created or destroyed in encounters with material particles. Again, their statistical behaviour is different in situations where many photons or electrons can interact, as, for example, for electrons in a metal or photons in a laser. No two electrons in such a system can be in exactly the same state, while there is no such restriction for photons: this is the difference between Fermi–Dirac statistics for electrons and Bose–Einstein statistics for photons.

In the first two-thirds of this book we shall be able to treat light mainly as a wave phenomenon, returning to the concept of photons when we consider the absorption and emission of electromagnetic waves.

1.2   WAVES AND RAYS

We now return to the question: how can light be represented by a ray? Huygens' solution was to postulate that light is propagated as a wavefront, and that at any instant every point on the wavefront is the source of a secondary wave that propagates outward as a spherical wave (Fig. 1.1).

The secondary waves then combine to form a new wavefront, so that successive positions of the wavefront can be found by a step-by-step process; a ray is then the normal to a wavefront. This simple Huygens' wavefront concept allows us to understand the rectilinear propagation of light along ray paths, and the basic geometric laws of reflection and refraction. There are obvious limitations: for example, what happens at the edge of a portion of the wavefront, as in Fig. 1.1, and why is there no wave reradiated backwards? We return to these questions when we consider diffraction theory in Chapters 7 and 8.

Reflection of a plane wavefront W_1 reaching a totally reflecting surface is understood according to Huygens in terms of secondary wavelets set up successively along the surface as the wavefront reaches it (Fig. 1.2(a)). These secondary wavelets propagate outwards and combine to form the reflected wavefront W_2. The rays are normal to the incident and reflected wavefronts. Light has travelled along each ray from W_1 to W_2 in the same time, so all path lengths from W_1 to W_2 via the mirror must be equal. The basic law of reflection

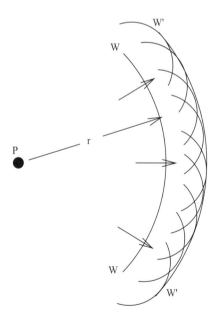

Fig. 1.1   Huygens' secondary wavelets. A spherical wavefront W has originated at P and after a time t has a radius $r = ct$, where c is the speed of light. Huygens' secondary wavelets originating on W at time t combine to form a new wavefront W' at time t', when the radii of the wavelets are $c(t' - t)$.

follows: the incident and reflected rays lie in the same plane and the angles of incidence, i, and reflection, r, are equal.

Figure 1.2(b) shows the same reflection in terms of rays. Here we may find the same law of reflection as an example of *Fermat's Principle of Least Time*, which states that the *time* of propagation is a minimum (or more strictly either a maximum or a minimum) along a ray path.[1] It is easy to see that the path of a light ray between the two points A, B (Fig. 1.2(b)) is a minimum if the angles i, r are equal. The proof is simple: construct the mirror image A′ of A in the reflecting surface, when the line A′B must be straight for a minimum distance. Any other path AP′B is longer.

Why are these two approaches essentially the same? Fermat tells us that the time of travel is the same along all paths close to an actual ray. In terms of waves this means that waves along these paths all arrive together, and add as in Huygens' construction. When we consider periodic waves, we will express this by saying that they are *in phase*.

The basic law of refraction (Snell's law) may be found by applying either Huygens' or Fermat's principles to a boundary between two media in which the velocities of propagation v_1, v_2 are different; as Huygens realized, his secondary waves must travel more slowly in an optically denser medium. The *refractive indices* are defined as $n_1 = c/v_1, n_2 = c/v_2$ where c is the velocity of light in free space. As we now show, the Fermat approach shown in Fig. 1.3 leads to Snell's law via some simple trigonometry.

The Fermat condition is that the travel time $n_1 \text{AP}c^{-1} + n_2 \text{PB}c^{-1}$ is a minimum; this gives the shortest time of travel between the two fixed points A and B.

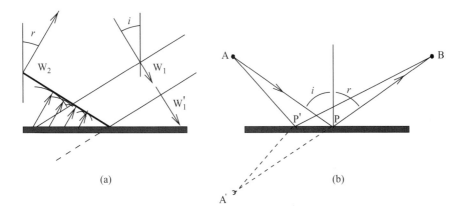

(a)   (b)

Fig. 1.2 Reflection at a plane surface: (a) Huygens' wave construction. The reflected wave W_2 is made up of wavelets generated as successive points on the incident plane wave W_1 reach the surface; (b) Fermat's principle. The law of reflection is found by making the path of a reflected light ray between the points A and B a minimum.

---

[1] This explanation of the basic law of reflection was first given by Hero of Alexandria (first century AD).

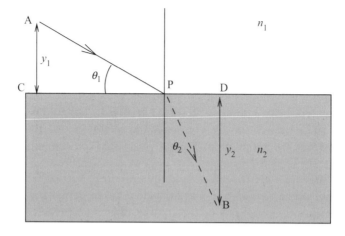

Fig. 1.3 Refraction at a plane surface between transparent media with refractive indices n_1 and n_2. Snell's law corresponds to a minimum of the optical path $n_1 \mathrm{AP} + n_2 \mathrm{PB}$.

The distance $n_1 \mathrm{AP} + n_2 \mathrm{PB}$ is called the *optical path*. Setting the perpendicular distances $\mathrm{AC} = y_1$ and $\mathrm{DB} = y_2$, the optical path is $n_1 y_1 \sec \theta_1 + n_2 y_2 \sec \theta_2$. The point P for minimum optical path is found by differentiation, giving

$$n_1 y_1 \sec^2 \theta_1 \sin \theta_1 \, d\theta_1 + n_2 y_2 \sec^2 \theta_2 \sin \theta_2 \, d\theta_2 = 0. \qquad (1.3)$$

The distance $\mathrm{CD} = y_1 \tan \theta_1 + y_2 \tan \theta_2$ is constant, so that its differential is zero:

$$y_1 \sec^2 \theta_1 \, d\theta_1 + y_2 \sec^2 \theta_2 \, d\theta_2 = 0. \qquad (1.4)$$

Solution of these simultaneous equations gives Snell's law of refraction

$$\boxed{n_1 \sin \theta_1 = n_2 \sin \theta_2.} \qquad (1.5)$$

In Chapter 5 we show how the laws of reflection and refraction may be derived from electromagnetic wave theory.

1.3 TOTAL INTERNAL REFLECTION

Referring again to Fig. 1.3, and noting that the geometry is the same if the ray direction is reversed, we consider what happens if a ray inside the refracting medium meets the surface at a large angle of incidence θ_2, so that $\sin \theta_2$ is greater than n_1/n_2 and Eq. (1.5) would give $\sin \theta_1 > 1$. There can then be no ray

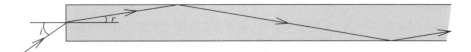

Fig. 1.4   The light pipe. Rays entering at one end are totally internally reflected, and can be conducted along long paths that may include gentle curves.

above the surface, and there is *total internal reflection*. The internally reflected ray is at the same angle of incidence to the normal as the incident ray.

The phenomenon of total internal reflection is put to good use in the light pipe (Fig. 1.4), in which light entering the end of a glass cylinder is reflected repeatedly and eventually emerges at the far end. The same principle is applicable to the transmission of light down thin optical fibres, but here the effect of the relation of the wavelength of light to the fibre diameter must be taken into account (Chapter 18).

1.4   THE LIGHT WAVE

We now consider in more detail the description of the light wave, starting with a simple expression for a plane wave of any quantity ψ, travelling in the positive direction z with velocity v:

$$\psi = f(z - vt). \tag{1.6}$$

The function $f(z)$ describes the shape of ψ at the moment $t = 0$, and the equation states that the shape of ψ is unchanged at any later time t, with only a movement of the origin by a distance vt along the z axis (Fig. 1.5). The minus sign in $(z - vt)$ indicates motion in the $+z$ direction; a plus sign would correspond to motion in the $-z$ direction.

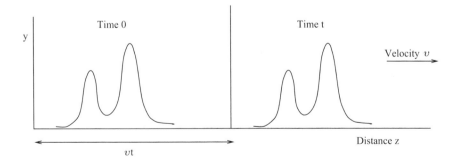

Fig. 1.5   A wave travelling in the z direction with unchanging shape and with velocity v. At time $t = 0$ the waveform is $\psi = f(z)$, and at time t it is $\psi = f(z - vt)$.

The variable quantity ψ may be a scalar, for example the pressure in a sound wave, or it may be a vector. If it is a vector, it may be *transverse*, i.e. perpendicular to the direction of propagation, as are the waves in a stretched string, or the electric and magnetic fields in the electromagnetic waves that are our main concern. For most of optics it is sufficient to consider only the transverse electric field; indeed, as we shall see later, the results of scalar wave theory are sufficiently general that for many purposes we may just think of the magnitude of the electric field and forget about its vector nature.

At any one time the variation of ψ with z, i.e. the slope of the graph in Fig. 1.5, is $\partial \psi / \partial z$, and at any one place the rate of change of ψ is $\partial \psi / \partial t$. Changing to the variable $z' = (z - vt)$ and using the chain rule for partial differentiation:

$$\frac{\partial \psi}{\partial z} = \frac{\partial \psi}{\partial z'} \frac{\partial z'}{\partial z} = \frac{\partial \psi}{\partial z'}, \tag{1.7}$$

$$\frac{\partial \psi}{\partial t} = \frac{\partial \psi}{\partial z'} \frac{\partial z'}{\partial t} = -v \frac{\partial \psi}{\partial z'}. \tag{1.8}$$

Similarly, the second differential of ψ with respect to z, i.e. $\partial^2 \psi / \partial z^2$, which is the curvature of the graph in Fig. 1.5, is related to the second differential with respect to time, i.e. the acceleration of ψ, by

$$\frac{\partial^2 \psi}{\partial t^2} = v^2 \frac{\partial^2 \psi}{\partial z^2}. \tag{1.9}$$

This so-called one-dimensional wave equation applies to any wave propagating in the z direction with uniform velocity and without change of form.

The wave equation (1.9) may be extended to three dimensions, giving:

$$\frac{\partial^2 \psi}{\partial x^2} + \frac{\partial^2 \psi}{\partial y^2} + \frac{\partial^2 \psi}{\partial z^2} = \frac{1}{v^2} \frac{\partial^2 \psi}{\partial t^2}, \tag{1.10}$$

or in a more general and concise notation[2]

$$\boxed{\nabla^2 \psi = \frac{1}{v^2} \frac{\partial^2 \psi}{\partial t^2}.} \tag{1.11}$$

The form of the wave $f(z - vt)$ may be any continuous function, but it is convenient to analyse such behaviour in terms of *harmonic* waves, taking the simple form of a sine or cosine. (In Chapter 4 we show that any continuous function can be synthesized from the superposition of harmonic waves.) At any point such a wave varies sinusoidally with time t, and at any time the wave varies sinusoidally with distance z. The waveform is seen in Fig. 1.6, which

---

[2] Recall that ∇^2 is the *Laplacian operator*: $\nabla^2 = \partial^2 / \partial x^2 + \partial^2 / \partial y^2 + \partial^2 / \partial z^2$.

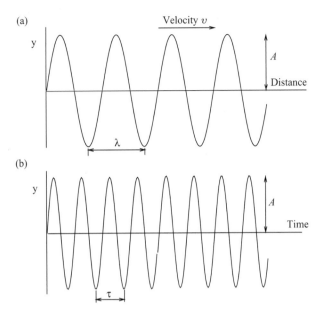

Fig. 1.6   A progressive sine wave: (a) the wave at a fixed time; (b) the oscillation at a fixed point.

introduces the *wavelength* λ and *period* τ. At any point there is an oscillation with *amplitude A*. Equation (1.6) then becomes

$$\psi = A \sin\left[2\pi\left(\frac{z}{\lambda} - \frac{t}{\tau}\right)\right], \tag{1.12}$$

which is easily seen to be a solution of the general wave equation (1.9).

The *frequency* of oscillation is $\nu = 1/\tau$. It is often convenient to use an *angular frequency* $\omega = 2\pi\nu$, and a *propagation constant* or *wave number*[3] $k = 2\pi/\lambda$. Equation (1.12) may then be written in terms of k and ω as

$$\psi = A \sin(kz - \omega t). \tag{1.13}$$

The vector quantity $\mathbf{k} = (2\pi/\lambda)\hat{\mathbf{n}}$, where $\hat{\mathbf{n}}$ is the unit vector in the direction of $\mathbf{k}$, is also termed the *wave vector*. In Eq. (1.13) the plane wave is moving in the direction $+z$, so $\mathbf{k}$ is pointing in the $+z$ direction.

Another powerful way of writing harmonic plane wave solutions of Eq. (1.11) is in terms of complex exponentials

$$\psi = A \exp[i(kz - \omega t)]. \tag{1.14}$$

---

[3] Beware: the term *wave number* is also used in spectroscopy for $1/\lambda$, without the factor 2π.

Because complex numbers consist of pairs of real numbers, the advantage of this type of solution is that it can vastly simplify the process of combining waves of different amplitudes and phases, as we shall see in Chapter 4.

1.5   ELECTROMAGNETIC WAVES

Although the idea that light was propagated as a combination of electric and magnetic fields was developed qualitatively by Michael Faraday, it required a mathematical formulation by Maxwell before the process could be clearly understood. In Chapter 5 we derive the electromagnetic wave equation from Maxwell's equations, and show that light and all other electromagnetic waves travel with the same velocity in free space: here we are concerned only with the nature of the variables. There are two variables in an electromagnetic wave, the electric and magnetic fields E and B; both are vector quantities, but each can be represented by the variable ψ in the wave equation (1.11). As shown in Chapter 5, they are both transverse to the direction of propagation, and mutually perpendicular. In free space their magnitudes[4] are related by

$$E = cB,$$

(1.15)

where c is the velocity of light in free space. Since the electric and magnetic fields are mutually perpendicular and their magnitudes are in a fixed ratio, only one need be specified, and the magnitude and direction of the other follows, but note that the velocity v in a dielectric such as glass is less than c; the *refractive index* n of the medium is

$$n = \frac{c}{v}.$$

(1.16)

As Huygens realized, light travels more slowly in dense media than in a vacuum.

In a transverse wave moving along a direction z the variable quantity is a vector that may be in any direction in the orthogonal plane x, y. The relevant variable for electromagnetic waves is conventionally chosen as the electric field E. The *polarization* of the wave is the description of the behaviour of the vector E in the plane x, y. The plane of polarization is defined as the plane containing the ray, i.e. the z-axis, and the electric field vector. If the vector E remains in a fixed direction, the wave is *linearly* or *plane* polarized; if the direction changes randomly with time, the wave is *randomly* polarized, or *unpolarized*. The vector E can also rotate uniformly at the wave frequency, as observed at a fixed point on the ray; the polarization is then *circular*, either right- or left-handed, depending on the direction of rotation.

---

[4] We use the SI system of electromagnetic units throughout.

Polarization plays an important part in the interaction of electromagnetic waves with matter, and Chapter 6 is devoted to a more detailed analysis.

1.6   THE ELECTROMAGNETIC SPECTRUM

The wavelength range of visible light covers about one octave of the electromagnetic spectrum, approximately from 400 nm to 800 nm. The electromagnetic spectrum covers a vast range, stretching many decades through infrared light to radio waves and many more decades through ultraviolet light and X-rays to gamma rays (Fig. 1.7). The differences in behaviour across the electromagnetic spectrum are very large. Frequencies, ν, and wavelengths, λ, are related to the velocity of light c, by $\lambda\nu = c$. The frequencies vary from 10^4 Hz for long radio waves (1 Hz equals one cycle per second), to more than 10^{21} Hz for commonly encountered gamma rays; the highest energy cosmic gamma rays so far detected reach to 10^{35} Hz (4×10^{20} eV). It is unusual to encounter a

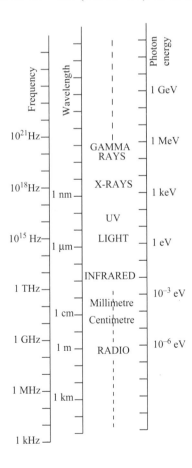

Fig. 1.7   The electromagnetic spectrum.

quantum process in the radio frequency spectrum, and even more unusual to hear a physicist refer to the frequency of a gamma ray, instead of the energy and the momentum carried by a gamma ray photon.

Although wave aspects dominate the behaviour of the longest wavelengths, and photon aspects dominate the behaviour of short wavelength X-rays and gamma rays, the whole range is governed by the same basic laws. It is in the optical range that we most usually encounter the 'wave-particle duality' that requires a familiarity with both concepts.

The propagation of light is determined by its wave nature, and its interaction with matter is determined by quantum physics. The relation of the energy of the photon to common levels of energy in matter determines the relative importance of the quantum at different parts of the spectrum: cosmic gamma rays, with a high photon energy and a high photon momentum, can act on matter explosively or like a high-velocity billiard ball, while long infrared or radio waves, with low photon energies, usually only interact with matter through classical electric and magnetic induction. (See numerical examples 1.1 to 1.4.)

The photon energy of light waves, ranging from 1.5 to 3 eV, is such that quantum effects dominate only some of the processes of emission and absorption or detection. The visible spectrum contains the marks of quantum processes in the profusion of colour from line emission and in line absorption; it can also display a continuum of emission over a wide range of wavelengths, giving 'white' light, whose actual colour is determined by the large-scale structure of the continuum spectrum rather than its fine detail.

1.7   WAVES AND PHOTONS

At the start of this chapter we remarked on the apparently complete understanding of optics at the beginning of the twentieth century. The wave nature of light was fully understood, stemming from the classical experiments of Young, Fresnel and Michelson, and substantiated by Maxwell's electromagnetic theory. Much of the content of our later chapters on interference and diffraction is derived directly from that era (with some refinements). Even Planck's bombshell announcement in 1900 that black-body radiation is emitted by quantized oscillators, and Einstein's demonstration in 1905 of the reality of photons through his explanation of the photoelectric effect completed rather than disturbed the picture; they had cleared up a mystery about the interchange of energy between matter and electromagnetic waves. Einstein's theory of that interaction, however, contained the seed of another revolution in optics, which germinated half a century later with the invention of the laser.

Einstein, in 1917, showed that there are three basic processes involved in the interchange of energy between a light wave and the discrete energy levels in an atom. All three involve a quantum jump of energy within the atom; typically in the visible region this is around 2 eV. Figure 1.8 illustrates the three basic photon processes; the processes are illustrated adopting a model with only

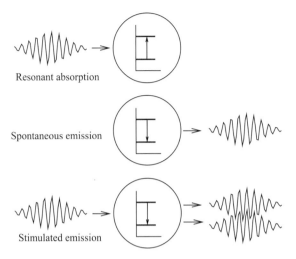

Resonant absorption

Spontaneous emission

Stimulated emission

Fig. 1.8   Three basic photon processes: absorption, spontaneous emission, and stimulated emission. For simplicity only two energy levels are shown.

two energy levels, although there are many more energy levels for the atom. As depicted in Fig. 1.8, the first is the *absorption* of a photon, which can occur when the quantum energy $h\nu$ equals the energy difference between the two levels (a *resonant* condition); the atom then gains a quantum of energy. The second is *spontaneous emission*, when an atom emits a photon, losing a quantum of energy in the process. The third is *stimulated emission*, in which the emission of a photon is triggered by the arrival of another, resonant photon. This third process was shown to be essential in the overall balance between emission and absorption. What emerged later was that the emitted photon is an exact copy of the incident photon; furthermore, each could then stimulate more photon emissions, leading to the build-up of a *coherent* wave of very great irradiance (or 'intensity', in old terminology).[5] This requires the number of atoms in the higher energy level to exceed the number in the lower level, a condition known as population inversion, so that the rate of stimulated emission exceeds the rate of absorption. The energy supply used to create the population inversion is often referred to as a *pump*, which in Fig. 1.9 is light absorbed between a ground level E_0 and level E_1. If the excitation of this level is short-lived, and it decays to a lower but longer-lived level E_2, the process leads to an accumulation and overpopulation of atoms in level E_2 compared with E_0. Stimulated emission, fed by energy from a pump, is the essential process in a *laser*. Prior to the laser, stimulated emission had been demonstrated in 1953 in the microwave region of the spectrum by Basov, Prokhorov and Townes, an achievement for which they were awarded the Nobel Prize. We describe the earliest laser, due to Maiman in 1960, in Chapter 15.

---

[5] See the Appendix for the definition of *irradiance* and other photometric terms.

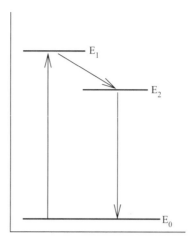

Fig. 1.9 Energy levels in the three-level laser. Energy is supplied to the atom by absorption from the ground level to the excited level E_1; spontaneous emission to the long-lived level E_2 then results in overpopulation of that level. Transitions from E_2 to ground are then the stimulated emission in the laser.

The process of stimulated emission in a laser builds up a stream of identical photons, which add coherently as the most nearly ideal monochromatic light, with very narrow frequency spread and correspondingly great *coherence length* (Chapter 14). Paradoxically, lasers, which depend fundamentally on quantum processes, produce the most nearly ideal waves. Lasers have allowed the classical experimental techniques of interferometry and spectroscopy to be extended into new domains, which we explore in Chapter 11 on the measurement of length and Chapter 13 on high-resolution spectrometry.

Largely as a result of the discovery and development of lasers, a new subject of *photonics* has developed from pre-laser studies of transmission and absorption in dielectrics. Coherent laser beams easily have an irradiance many orders of magnitude greater than that of any thermal source, leading to very large electric fields and *nonlinear* effects in dielectrics, such as harmonic generation and frequency conversion. There are many practical applications, some of which are more familiar in electronic communications, such as switching, modulation, and frequency mixing. The title of this book indicates the current importance of lasers and photonics; the materials involved, including those used in nonlinear optics, are included in Chapter 17 on laser light and Chapter 21 on detectors.

1.8   FURTHER READING

A. R. Hall, *All Was Light – an Introduction to Newton's Opticks*, Clarendon Press, Oxford, 1993.
V. Ronchi, *The Nature of Light*, Harvard University Press, 1971.

J. Simmons and M. Guttmann, *States, Waves and Photons: a Modern Introduction to Light*, Addison-Wesley, 1970.

NUMERICAL EXAMPLES 1

1.1 What would be the velocity of a tennis ball, mass 60 g, with the same energy as a 10^{20} eV cosmic gamma ray?

1.2 At what temperature would a molecule of hydrogen gas have the same energy as a photon of the 21 cm hydrogen spectral line?

1.3 What wavelength of electromagnetic radiation has the same photon energy as an electron accelerated to 100 eV?

1.4 An X-ray photon with wavelength 1.5×10^{-11} m arrives at a solid. How much energy (in electron-volts) can it give to the solid?

1.5 GaAs is an important semiconductor used in photoelectronic devices. It has a refractive index of 3.6. For a slab of GaAs of thickness 0.3 mm show that a point source of light within the GaAs on the bottom face will give rise to radiation outside the top face from within a circle of radius R centred immediately above the point source. Find R.

PROBLEMS 1

1.1 In the Pulfrich refractometer (Fig. 1.10), the refractive index n of a liquid is found by measuring the emergent angle e from the prism whose refractive index is N. Show that if i is nearly $90°$

$$n \approx (N^2 - \sin^2 e)^{1/2}.$$

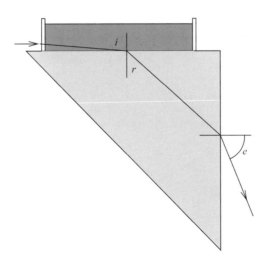

Fig. 1.10 Pulfrich refractometer.

1.2 The angular radius of a rainbow, measured from a point opposite to the Sun, may be found from the geometry of the ray in Fig. 1.11, which lies in the meridian plane of a spherical drop of water with refractive index n. The angular radius is a stationary value of the angle through which a ray from the Sun is deviated; show that it is given by

$$\cos i = \left(\frac{n^2 - 1}{3}\right)^{1/2}$$

Note that the internal reflection is near the Brewster angle (see Section 5.4), so that the rainbow light is polarized along the circumference of the bow.

1.3 Show that the apparent diameter of the bore of a glass capillary tube of refractive index n, as seen normally from the outside, is independent of the outer diameter, and is n times the actual diameter.

1.4 Show that the lateral displacement d of a ray passing through a plane-parallel plate of glass refractive index n, thickness t, is related to the angle of incidence θ by

$$d \approx t\theta\left(1 - \frac{1}{n}\right),$$

provided that θ is small.

1.5 If the refractive index n of a slab of material varies in a direction y, perpendicular to the x-axis, show by using Huygens' construction that a ray travelling nearly parallel to the x-axis will follow an arc with radius

$$n\left(\frac{dn}{dy}\right)^{-1}.$$

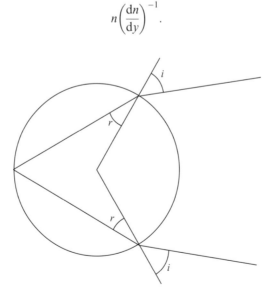

Fig. 1.11   A ray refracted in the meridian plane of a spherical raindrop.

(Consider a sector of wavefront δy across, and compare the distances travelled in time τ by secondary waves from each end of the sector.)

1.6 Show that the geometric distance of the horizon as seen by an observer at height h metres is approximately $3.5\,h^{1/2}$ km. The radius of the Earth ≈ 6000 km.

Use the result of Problem 1.5 to calculate how this is affected by atmospheric refraction, if this is due to pressure changes only with an exponential scale height of 10 km. The refractive index of air at ground level is approximately 1.00028.

1.7 The refractive index of solids at X-ray wavelengths is generally less than unity, so that a beam of X-rays incident at a glancing angle may be reflected, as in total internal reflection. If the refractive index is $n = 1 - \delta$, show that the largest glancing angle for reflection is $\simeq \sqrt{\delta}$. Evaluate this critical angle for silver at $\lambda = 0.07$ nm where $\delta = 5.8 \times 10^{-6}$.

2

Geometric optics

Optics is either very simple or else it is very complicated.
Feynman, *Lectures on Physics*, Addison-Wesley, 1963.

That y^e rays wch make blew are refracted more $y^n y^e$ rays wch make red appears from this experim^{nt}.
Isaac Newton, *Quaestiones.*

Light, which is propagated as an electromagnetic wave, may often conveniently be represented by rays, which are geometrical lines along which light energy flows; the term geometric optics is derived from this concept. Rays are lines perpendicular to the wavefronts of the electromagnetic wave. An alternative concept is to regard the action of the various components of optical systems, such as convex and concave mirrors and lenses, as modifying a wavefront by changing its direction of travel or its curvature. This wavefront concept is useful, but the precise geometry of *ray tracing* is nevertheless essential for the detailed design of optical instruments.

We start our exposition of geometric optics by analysing the action of a thin prism and a simple lens in terms both of waves and rays, and then develop the basic ray theory of imaging. Images are inevitably imperfect, apart from trivial cases such as images in plane mirrors; we analyse the imperfections as various types of *aberration*. The use of a lens as a simple magnifier, and the combination of optical components in systems such as the microscope and telescope, will be considered in the following chapter.

2.1 THE THIN PRISM

The wavefront concept is usefully applied to the bending of a light ray in a thin prism, with small apex angle α and refractive index n, assuming free space[1] outside the prism. We first calculate the angle of deviation θ by applying Snell's

---

[1] The optical properties of free space and air are nearly the same, and are taken as identical in this chapter.

law (Eq. (1.5)) to each surface in turn, and find a useful approximation for a thin prism at near normal incidence. We then show that the wavefront approach leads directly to this approximation.

The ray approach

In Fig. 2.1 (a) the ray is incident on the first surface at angle β_1. Following the ray through the prism we have for the two refracting surfaces

$$n \sin \beta_2 = \sin \beta_1,$$
$$n \sin \beta_3 = \sin \beta_4. \tag{2.1}$$

The total deviation is $\theta = \beta_1 - \beta_2 - \beta_3 + \beta_4$. In the triangle OAB we have $\alpha = \beta_2 + \beta_3$, so that

$$\theta = \beta_1 + \beta_4 - \alpha. \tag{2.2}$$

Figure 2.1(b) shows the results of a numerical solution of Eqs (2.1) and (2.2), giving θ for a prism with $\alpha = 10°$ and $n = 1.5$ with β between $0°$ and $20°$. There is a minimum deviation when the ray passes symmetrically through the prism, at $\beta_1 = 7.5°$. The angle of deviation varies only between $5.02°$ and $5.23°$ over the whole range in Fig. 2.1(b).

   If the analysis is restricted to small values of α and β, so that to a good approximation $\sin \beta \approx \beta$, Eq. (2.1) becomes

$$\beta_1 = n\beta_2 \quad \text{and} \quad n\beta_3 = \beta_4, \tag{2.3}$$

and Eq. (2.2) becomes

$$\boxed{\theta = (n - 1)\alpha.} \tag{2.4}$$

In the example above the simplified equation gives $\theta = (1.5 - 1) \times 10 = 5°$, close to the correct result $\theta = 5.02°$ at minimum deviation.

The wavefront approach

We now derive Eq. (2.4) from the wavefront approach. In Fig. 2.1(c), the incident wavefront is AB and the emergent wavefront A'B'. The prism is arranged symmetrically, for minimum deviation, but the same argument can be applied for wavefronts over a range of angles about this position.

   To calculate the angle of deviation we note that the optical paths AA' and BB' are equal. (Remember from Section 1.2 that this implies that the time of travel from A to A' is the same as from B to B'.) The refracting face length of

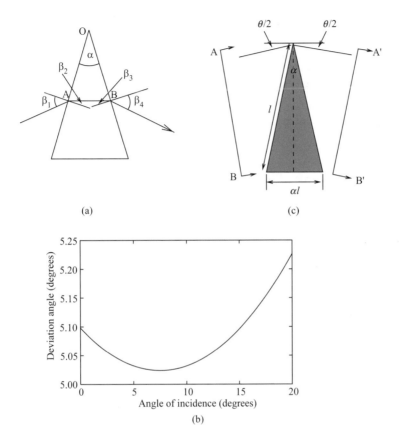

(a)

(b)

(c)

Fig. 2.1   A prism with small angle α refracts a wavefront through an angle that is nearly independent of the angle of incidence, provided it is near normal: (a) the ray approach; (b) deviation angle for a 10° prism over a range of angles of incidence; (c) the wavefront approach.

the prism is l. While the wavefront at B passes through a length $2l \sin \frac{1}{2}$ of the prism, the wavefront at A passes through a length $2l \sin \frac{1}{2}(\theta + \alpha)$ of air. The wave velocity is a factor n slower inside the prism, so that the two equal optical paths are $2nl \sin \frac{1}{2}\alpha$ and $2l \sin \frac{1}{2}(\theta + \alpha)$. At minimum deviation θ is therefore given by

$$\sin \frac{\theta + \alpha}{2} = n \sin \frac{\alpha}{2}. \tag{2.5}$$

As in the ray treatment, we approximate for the small angle prism by writing the sine of an angle as the angle itself (in radian measure), and the angle of deviation θ is given very simply by

$$\boxed{\theta = (n - 1)\alpha.} \tag{2.6}$$

as in Eq. (2.4).

2.2   THE LENS AS AN ASSEMBLY OF PRISMS

A convex lens, shown in section in Fig. 2.2, is familiar as a simple hand-held magnifying glass. The lens is also shown as a series of thin prisms with the apex angle increasing with distance y from the axis. As before, we assume all angles are small. If the radius of curvature of both surfaces is r, the prism angle at height y is $2y/r$, giving a wavefront deviation

$$\theta = (n - 1)2y/r. \tag{2.7}$$

As shown in Fig. 2.2, a plane wavefront passing through the lens will become curved, and will converge to a focal point at a distance $f = y/\theta = r/2(n - 1)$ from the lens. This is the *focal length* of the lens. The action of the convex lens is to add a curvature[2] $2(n - 1)/r$ to the plane wavefront. Within the approximation of small angular deviation the wavefront over the whole of the lens converges on a single focal point. Moreover, a wavefront arriving at a different angle will converge on a different point in the same focal plane, so that the lens gives an image of a distant scene.

Figure 2.3(a) shows the effect of a convex lens on a diverging wavefront originating from a point source P_1 at distance u from the lens; the diverging wavefront already has a negative curvature $-1/u$, the lens adds a positive curvature $2(n - 1)/r$, and the wavefront converges on the point P_2 at distance v. Similarly, a concave lens, as in Fig. 2.3(b), adds a negative curvature and the wavefront diverges. The addition of these curvatures gives the lens equation[3]

$$-\frac{1}{u} + \frac{2(n - 1)}{r} = \frac{1}{v}, \tag{2.10}$$

or

$$\boxed{\frac{1}{u} + \frac{1}{v} = \frac{1}{f}.} \tag{2.11}$$

This equation relates the image distance v to the object distance u and the focal length f of the lens. It also relates the focal length of the lens to the refractive index n and the radius of curvature r of the two equally curved surfaces.

---

[2] A spherical wavefront converging on a point at distance R is said to have a *curvature* $1/R$.
[3] Note that these equations will change when we introduce the Cartesian sign convention in Section 2.3 to become

$$-\frac{1}{v} + \frac{1}{u} = \frac{1}{f}, \tag{2.8}$$

and

$$\frac{1}{f} = (n - 1)\left(\frac{1}{r_1} - \frac{1}{r_2}\right). \tag{2.9}$$

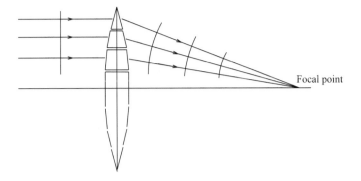

Fig. 2.2   A simple converging lens as an assembly of prisms.

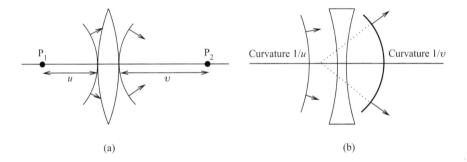

(a)   (b)

Fig. 2.3   Convex and concave lenses changing the curvature of a wavefront.

A simple extension of this analysis to a thin lens with different radii of curvature r_1, r_2 gives the focal length

$$\frac{1}{f} = (n-1)\left(\frac{1}{r_1} + \frac{1}{r_2}\right).$$

(2.12)

The *power* P of a lens is defined as the inverse of its focal length, so that $P = 1/f$; measuring f in metres, the power of a lens is specified in *dioptres*.

2.3   REFRACTION AT A SPHERICAL SURFACE

In the more formal analysis of lens systems it is essential to use a consistent sign convention for distances and curvatures. There are two conventions in general use, the Cartesian and the 'real-positive' systems.

The Cartesian system is to be preferred for the logical development of geometrical formulae, especially when several surfaces are involved, and we illustrate its use in the following examples. The 'real-positive' system, by

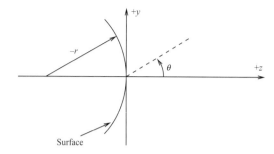

Fig. 2.4 The Cartesian coordinate system in geometric optics. The signs of distances z and y follow normal geometric convention, and anticlockwise angles are positive. A wavefront with centre of curvature to the left has a negative radius of curvature.

contrast, is based on the actual distances travelled by a light wave, which are all regarded as positive; it is therefore well suited to descriptions of the behaviour of a wavefront, as in the previous section. In the Cartesian system, the signs of distances are determined by the normal conventions of coordinate geometry. Usually, but not necessarily, rays are considered to travel from left to right. This sign convention is illustrated in Fig. 2.4. Distances along the optical axis z and along y above the axis are positive. Angles measured counterclockwise from the z-axis are positive. The radius of curvature of a surface is measured from the surface to the centre of curvature; it is negative if the centre of curvature is to the left of the surface, and positive if to the right.[4] Quantities referring to the left-hand space may be distinguished from those in the right-hand space by subscripts 1 and 2.

Refraction of a ray at a single spherical surface, radius r, bounding media with refractive indices n_1, n_2, is shown in Fig. 2.5. Note that the labelled angles should all be considered small, so that sines and tangents are approximated by the angle itself, and the point A is taken to be not far from the axis $P_1 C P_2$. This is the *paraxial* approximation, which applies to rays that are not far from parallel to the optical axis. Then we can take the object distance $P_1 A = u$ and the image distance $A P_2 = v$, as in our previous analysis of the lens. The relation between object distance u and image distance v is obtained by constructing perpendiculars $P_1 M_1$ and $P_2 M_2$ to the radial line through A, when the similar triangles $P_1 M_1 C$, $P_2 M_2 C$ give:

$$\frac{CM_1}{P_1 M_1} = \frac{CM_2}{P_2 M_2},$$

$$\frac{-u \cos \phi_1 + r}{-u \sin \phi_1} = \frac{-v \cos \phi_2 - r}{-v \sin \phi_2}. \tag{2.13}$$

---

[4] Numerical examples 2.1 and 2.2 may be useful as tests of a sign convention.

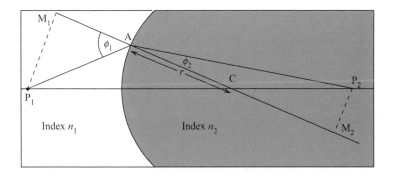

Fig. 2.5 Geometry of a ray refracted at a spherical surface between media of refractive indices n_1 and n_2. P_1 and P_2 are *conjugate points*. The surface as shown has positive power since $n_2 > n_1$.

Using the relation $n_1 \sin \phi_1 = n_2 \sin \phi_2$ this becomes

$$-\frac{n_1}{u} + \frac{n_2}{v} = \frac{n_2 \cos \phi_2 - n_1 \cos \phi_1}{r}. \tag{2.14}$$

In the paraxial approximation, the cosines are taken to be unity, and the relation is obtained

$$-\frac{n_1}{u} + \frac{n_2}{v} = \frac{n_2 - n_1}{r} = P, \tag{2.15}$$

where P is defined as the power of the surface.

2.4 TWO SURFACES: THE SIMPLE LENS

The simple thin lens in air, with two convex surfaces, is analysed by adding two equations of the form of (2.15) and assuming that the thickness of the lens is negligible. We give a negative sign to the second radius since the centre of curvature is to the left. For the first surface we set $n_1 = 1$ and $n_2 = n$, the refractive index of the glass, and find an image distance v_1, which becomes the object distance for the second surface. For object distance u from the lens we obtain for the first surface

$$-\frac{1}{u} + \frac{n}{v_1} = \frac{n - 1}{r_1}, \tag{2.16}$$

and for the second surface, refracting from glass to air

$$-\frac{n}{v_1} + \frac{1}{v} = \frac{1 - n}{-r_2}. \tag{2.17}$$

The sum of these gives the lens equation

$$\frac{1}{u} - \frac{1}{v} = (n-1)\left(\frac{1}{r_1} - \frac{1}{r_2}\right). \tag{2.18}$$

The focal length may be written, using Eq. (2.12) (but using the Cartesian sign convention)

$$\boxed{\frac{1}{f} = (n-1)\left(\frac{1}{r_1} - \frac{1}{r_2}\right).} \tag{2.19}$$

The power of a thin lens is the sum of the powers of the two surfaces. If the object is at infinity, v in Eq. (2.18) becomes the focal length f. The power is then $1/f$.

2.5 IMAGING IN SPHERICAL MIRRORS

Figure 2.6(a) shows the action of a spherical concave mirror M on a wavefront, showing the similarity with the action of a lens as shown in Fig. 2.3. Figure 2.6(b) shows the geometry of an axial ray P_1CV and a ray at a small angle to the axis. A ray from the object at P_1 is reflected at A on the mirror surface, and reaches the image point P_2 on the axis, which is defined by the line from P_1 through the centre of curvature C. The angles θ of incidence and reflection are equal, so that the angle P_1AP_2 is bisected by the line AC, giving the geometric relation

$$\frac{P_1C}{P_1A} = \frac{CP_2}{P_2A}. \tag{2.20}$$

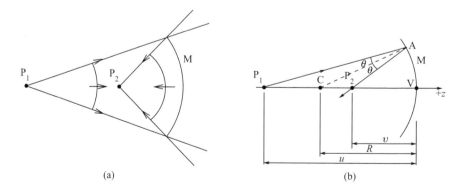

(a)   (b)

Fig. 2.6   A concave spherical mirror: (a) action of the mirror on a wavefront; (b) the geometry of a paraxial ray.

With object and image distances u, v and radius of curvature R, we find that Eq. (2.20) becomes

$$\frac{u - R}{u} = \frac{R - v}{v},\qquad(2.21)$$

where we have used a paraxial approximation by writing $P_1 A \approx P_1 V = u$ and $P_2 A \approx P_2 V = v$. We obtain the *mirror formula*

$$\boxed{\frac{1}{u} + \frac{1}{v} = \frac{2}{R}.}\qquad(2.22)$$

The same equation applies for a convex mirror, having due regard for the sign convention.

It is instructive to observe one's own image in convex and concave mirrors, especially noting the position and magnification of the image in a concave mirror as the object (the face!) is placed in front of, or behind, the centre of curvature. At the centre of curvature one's image is immediately in front of one's face, and so appears huge. Close to the mirror one sees a normal image, not much different from that in a plane mirror; outside the centre of curvature one sees an image not far behind the mirror, reduced in size and inverted.

2.6 GENERAL PROPERTIES OF IMAGING SYSTEMS

It is remarkable how well simple optical systems can work, despite the approximations that we have made in the lens theory. Even if the object point is at some distance from the axis of a simple lens, rays still converge on an off-axis image point found from the lens equation. A simple lens can therefore make an image of an extended object, in which the scale of the image is almost the same over a considerable area. The object and image planes containing an object and its image are called *conjugate planes*, and we now find the *magnification* of the image, which is the ratio between the sizes of the image and the object.

The geometric specification of a perfect optical system is that points and lines in the object space should correspond precisely to points and lines in the image space. Mathematically, the two spaces are linked by a projective transformation, and it can be shown that for this to be true there must be a linear relation between distances in the object and image spaces. Equation (2.18) is an example of such a relation involving axial distances only. There is also a linear relationship between perpendicular distances, giving the *transverse magnification* of the system. The magnification depends, of course, on the positions of the conjugate planes containing the object and image.

We have so far considered only the theory of a thin spherical lens, but the same concepts can be applied to a lens whose thickness cannot be neglected, and to a multiple lens system such as that used in camera lenses (see Chapter 3).

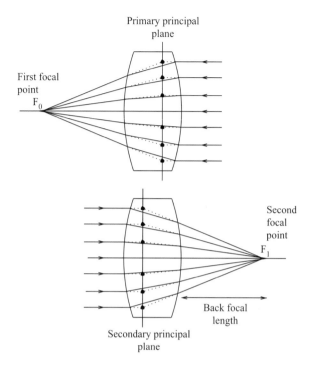

First focal
point
F_0

Primary principal
plane

Second
focal
point
F_1

Back focal
length

Secondary principal
plane

Fig. 2.7 A thick lens. Rays from infinity converge on the focal points F_0, F_1 from points of intersection on the two principal planes. The back focal length is measured from the surface of the lens.

The important concept is to define planes in the system from which the axial distances should be measured. Figure 2.7 shows the location of the *principal planes* in a thick lens. These are the planes on which rays from a focal point intersect corresponding rays from a point at infinity; the focal length is measured from the principal plane. (It may also be convenient, as, for example, in the design of a camera, to define a *back focal length* as the distance from the back of a lens system to the focal plane.)

   In general, the principal planes form a pair of conjugate planes between which the magnification is unity. The planes whose conjugates are at infinity are the *focal planes*. The magnification of an object on a focal plane is infinite and negative.

   The general linear relationship between distances in object and image planes becomes very simple when the system is axially symmetrical and when axial distances are measured from the focal planes, as in Fig. 2.8. Denoting axial and transverse distances by Z and Y, and using subscripts 1 and 2 for the object and image spaces, the relationship is

$$\frac{f_1}{Z_1} = \frac{Z_2}{f_2} = -\frac{Y_2}{Y_1}.$$

(2.23)

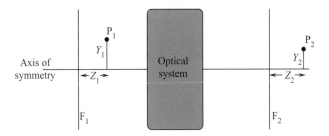

Fig. 2.8   Coordinate system for general axially symmetric optical system. P_1 and P_2 are conjugate points. Axial distances Z_1, Z_2 are measured from the focal planes F_1, F_2.

The *transverse magnification* is defined as Y_2/Y_1. Equation (2.23) can be verified by substituting $Z_1 = f_1$, obtaining unity magnification corresponding to a principal plane, and by substituting $Z_1 = 0$, obtaining infinite magnification. The constants f_1 and f_2 are the *focal lengths* of the system.

Equation (2.23) contains Newton's equation

$$Z_1 Z_2 = f_1 f_2.$$ (2.24)

A *longitudinal magnification* can be found by differentiating Newton's equation:

$$M_L = \frac{dZ_2}{dZ_1} = -\frac{f_1 f_2}{Z_1^2}.$$ (2.25)

This indicates, for example, the amount of refocusing required when an object moves closer to a camera lens.

An *angular magnification* M_A is defined by the ratio of angles at which a ray cuts the axis in the image and object planes. It is given by the ratio of the transverse and longitudinal magnifications:

$$M_A = \frac{Z_1}{Z_2} \frac{Y_2}{Y_1}.$$ (2.26)

2.7   SEPARATED THIN LENSES IN AIR

Many optical systems use components that are themselves made up of two or more lenses, as, for example, in a telescope eyepiece. The analysis of such systems by the repeated use of the simple lens formula (2.18) soon leads to tedious algebra, and it is more usual to follow a ray-tracing procedure. We now analyse the separated pair of Fig. 2.9 in this way, following an incident ray parallel to the axis as it is deviated by each lens.

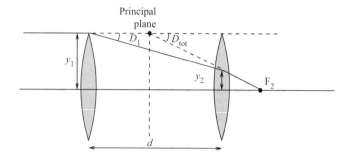

Fig. 2.9 Ray tracing in a separated lens system. A ray parallel to the axis is deviated in both lenses, and crosses the axis at the focus F_1.

At some distance from the axis the lens acts like a thin prism, as in Section 2.1. From Eq. (2.7) the angular deviation D of the ray at a distance y from the axis of a thin lens of power P is

$$D = (n-1)y\left(\frac{1}{r_1} - \frac{1}{r_2}\right) = yP. \tag{2.27}$$

The ray in Fig. 2.9, which meets the first lens at a distance y_1 from the axis, and then the second at distance y_2 from the axis, has a total angular deviation given by the sum

$$D_{tot} = D_1 + D_2 = y_1 P_1 + y_2 P_2. \tag{2.28}$$

As can be seen in Fig. 2.9, $y_2 = y_1 - dD_1$. Equation (2.28) therefore becomes

$$D_{tot} = y_1(P_1 + P_2 - dP_1P_2). \tag{2.29}$$

The power P_{tot} of the combination is $P_{tot} = D_{tot}/y_1$. The power of the pair of lenses separated by distance d is therefore

$$P_{tot} = (P_1 + P_2 - dP_1P_2). \tag{2.30}$$

The focal point can also be found geometrically from the figure, without recourse to tedious algebra.

Note that the power of the combination is less than the sum of their individual powers, unless they are in contact, when the powers add directly. For an assembly of n thin lenses in contact the total power P is

$$P = P_1 + P_2 + \ldots P_n. \tag{2.31}$$

2.8   PERFECT IMAGING

An ideal, or perfect, optical system would be one in which every point in an object space corresponds precisely to a point in an image space, being connected to it by rays passing through all points of the optical system. The optical path from any object point to its image is then the same along all rays. There is a fundamental reason, first formulated by Maxwell, why this cannot be achieved in any but the most elementary optical system. He showed that a perfect optical system can only give a magnification equal to the ratio of the refractive indices in the object and image spaces (Fig. 2.10). For example, if object $A_1B_1C_1$ and image $A_2B_2C_2$ are both in air, the magnification can only be unity, which may not be very useful. A plane mirror may be perfect, but a magnifying lens cannot be.

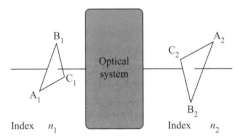

Fig. 2.10   Maxwell's theorem for a 'perfect' system. Optical path lengths must be equal for corresponding parts of the object and image, so that, for example, $n_1A_1B_1 = n_2A_2B_2$.

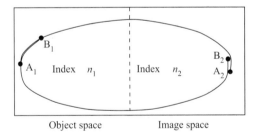

Fig. 2.11   Lenz's proof of Maxwell's theorem. In a perfect system the optical paths A_1B_1 and A_2B_2 are equal.

The following demonstration of Maxwell's theorem is due to Lenz. The theorem states effectively that if two object points A_1, B_1 in a medium of refractive index n_1 give rise to image points A_2, B_2 where the refractive index is n_2, the optical paths A_1B_1 and A_2B_2 must be equal. Suppose in Fig. 2.11 the rays A_1B_1 and B_1A_1 can both pass through the optical system. They must then pass through B_2A_2 and A_2B_2, respectively. Since both optical paths from A_1 to A_2 must have the same length, and also both optical paths from B_1 to B_2, it follows immediately that the optical paths $n_1A_1B_1$ and $n_2A_2B_2$ must be the same. This gives Maxwell's theorem

$$\frac{A_2B_2}{A_1B_1} = \frac{n_1}{n_2}. \qquad (2.32)$$

This proof appears at first sight to be very limited, since the rays AB and BA can hardly be expected both to pass through the optical system. It may, however, be generalized by constructing a curve similar to that in Fig. 2.11 but which is made up of many segments of actual rays, and integrating the whole path between the intersections of these segments.

The simplest example of a perfect optical system is a plane mirror. A plane refracting surface, in contrast, only approaches perfection for rays that are nearly normal to its surface; away from the normal, a bundle of rays from a single point does not form a point, or *stigmatic* image. A theoretical example of a perfect refracting system, known as the 'fish-eye' lens,[5] was invented by Maxwell; this uses an infinite spherical lens with a refractive index varying with the radius in such a way that all rays diverging from any point would converge on another point.

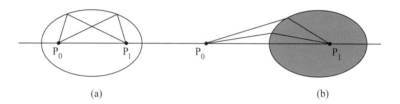

(a)   (b)

Fig. 2.12 Stigmatic imaging: (a) in an ellipsoidal mirror; (b) in a refracting Cartesian oval.

If in a more restricted system a single object point and its image point are specified, they can be connected by stigmatic rays in the optical systems of Fig. 2.12. The ellipsoidal mirror has the two points as its two foci; if one point is infinitely distant, then the reflector becomes the familiar paraboloid of revolution used in reflecting telescopes and car headlights. The refracting surface is the more complicated Cartesian oval, named after Descartes.

2.9   PERFECT IMAGING OF SURFACES

The severe restriction of the 'perfect' optical system, in which magnification can only be equal to the ratio of refractive indices in the object and image spaces, does not apply if the object points are restricted to lie on a single definite surface. This surface need not be plane, but the corresponding image points must lie on another conjugate surface if all points are to have sharp, or 'stigmatic', images.

---

[5] See M. Born and E. Wolf, *Principles of Optics*, 6th edn, Pergamon Press, Oxford, 1980, p. 147.

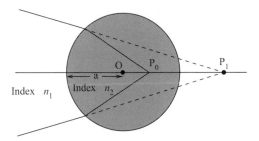

Fig. 2.13 A spherical lens. All rays diverging from the point P_0 appear to diverge from the point P_1 when $OP_0 = a/n$ and $OP_1 = na$. The points P_0 and P_1 lie on spherical conjugate surfaces.

An example of a curved, but truly stigmatic, imaging surface is provided by a spherical lens. Microscope objectives commonly use such a spherical lens, but with a flat face. Figure 2.13 shows a homogeneous spherical lens, centre O, with radius a and refractive index n. A point source P_0 inside the lens is imaged at P_1 outside the lens. All rays leaving P_0 towards the left appear to diverge from a single point P_1. This is only possible when $OP_0 = a/n$ and $OP_1 = na$; the conjugate surfaces are therefore spherical.

It is, of course, not always convenient to restrict object and image surfaces to a special curve such as a sphere, but if it is required that either or both should be plane it will be necessary to abandon the requirement that the images should be strictly stigmatic. We therefore turn in the next section to the description and control of imperfections in optical images.

2.10 RAY AND WAVE ABERRATIONS

We have noted that a useful optical instrument can ideally only give stigmatic images of points on a single surface, while even under this restriction the lenses or mirrors in the instrument cannot have simple spherical surfaces unless only a small bundle of paraxial rays is used to form the image. In spite of this it is evident that many very useful optical instruments exist that do not conform strictly to these conditions. The quality of their images may not be ideal, but the departures from perfection, known as *aberrations*, may be tolerable for their purpose. The design of an optical system is mainly concerned with the calculation of the various aberrations, and with their suppression below a tolerable level.

Aberration may be specified for any ray that contributes to the formation of a point image. The distance between an ideal image point and the intersection of the ray with the image plane is called the *ray aberration*. The total effect on the image is found by tracing sufficient rays from an object point so that the spread of intensity across the image can be found. Ray aberration is therefore a measure of the size of an ideal image point; its importance may be judged in

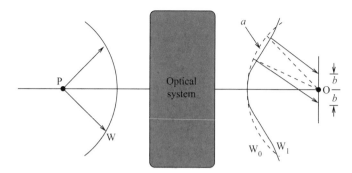

Fig. 2.14   Ray and wave aberrations. A spherical wave W leaving the point P is focused by the optical system into a converging wave W_1, which departs from the ideal spherical shape W_0, centred on O. Wave aberrations are shown as a, and ray aberrations as b, for two different rays.

relation to the size of the diffraction patch, which is the lower limit to the size of the image of a point object below which even an ideal instrument cannot go (see Chapter 7). Alternatively, the point image may be considered as the centre of a convergent wave, ideally spherical but in practice departing from sphericity; the departures are known as *wave aberrations*. The relation between ray and wave aberrations is seen in Fig. 2.14.

Wave aberrations for an object on axis may amount to some tens of wavelengths in a good camera lens, but usually it is less than one wavelength in an astronomical telescope. The corresponding ray aberrations may be found by geometric ray tracing rather than by analysis of wavefronts; there is, however, no need to draw a sharp distinction since both approaches lead to similar analytic results. The wave aberrations offer a clearer physical picture, as set out in the next section. Ray aberrations may be found from any pattern of wavefront aberrations by drawing ray normals from the wavefront, as in Fig. 2.14. The intensity at the nominal image point is best found from the wave aberrations, since these give directly the pattern of waves that must be added to give the amplitude at the image point. The efficiency with which light is concentrated into the image point increases with decreasing wave aberration until the aberration becomes small compared with $\lambda/2\pi$.

2.11   WAVE ABERRATION ON AXIS – SPHERICAL ABERRATION

As soon as it is admitted that a particular optical instrument, such as a camera, cannot meet the ideal of producing stigmatic images over the whole of an image, the possible range of optical designs at once becomes infinite, as does the variety of aberration patterns over the image. A simple pattern does, however, emerge from refraction or reflection at a spherical surface. Here, the pattern of aberrations separates into parts that depend on the angular spread of rays from a single on-axis object, and on the width of the field containing the object (see Chapter 3 on the design of cameras).

Following Fig. 2.15, the difference a in the optical path between the axial ray and the ray intersecting the surface at a distance y from the axis is given by

$$a = n_1 \text{AP}_1 + n_2 \text{AP}_2 - n_1 u - n_2 v. \tag{2.33}$$

The distance y is taken for convenience as the chord CA, so that the following geometrical relations hold:

$$\cos \psi = y/2r \text{ (from the isosceles triangle AOC)}$$

$$\text{AP}_1 = \left(u^2 + y^2 + 2uy \cos \psi\right)^{1/2} = u\left(1 + \frac{y^2}{u^2} + \frac{y^2}{ur}\right)^{1/2}$$

$$\text{AP}_2 = \left(v^2 + y^2 - 2vy \cos \psi\right)^{1/2} = v\left(1 + \frac{y^2}{v^2} - \frac{y^2}{vr}\right)^{1/2}. \tag{2.34}$$

The wave aberration, or difference in optical path, is then found by expanding Eq. (2.34) as a power series in y^2, which depends on the off-axis distance y as

$$a = \frac{y^2}{2}\left\{n_1\left(\frac{1}{u} + \frac{1}{r}\right) + n_2\left(\frac{1}{v} - \frac{1}{r}\right)\right\} - \frac{y^4}{8}\left\{\frac{n_1}{u}\left(\frac{1}{u} + \frac{1}{r}\right)^2 + \frac{n_2}{v}\left(\frac{1}{v} - \frac{1}{r}\right)^2\right\}$$

$$+ \text{ terms of higher order in } y. \tag{2.35}$$

As the radius of the aperture increases, so must the approximation in Eq. (2.35) be taken to higher orders in y. For paraxial rays where y is small, only the term in y^2 need be considered, and a is zero when the first half of Eq. (2.35) is zero. This gives the simple formula for refraction at a single surface (Eq. (2.15), which is the relation between conjugate points). The second half of Eq. (2.35) is then the *spherical aberration* expressed as a wave aberration. The magnitude of spherical aberration increases as the fourth power of the aperture of a spherical refractor.

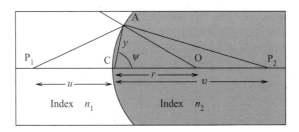

Fig. 2.15   Spherical aberration at a single spherical surface. The wave aberration for a ray at a distance y from the axis is found from the small difference between the optical paths $P_1\text{AP}_2$ and $P_1\text{CP}_2$.

The ray passing through A is normal to the wavefront, so that its direction departs from the correct direction AP_2 by the angle between the ideal wavefront and the actual wavefront. This angle is found from the rate of variation of wavefront aberration a with increasing y; the angular deviation of a ray at P_2 therefore varies as da/dy; that is, as y^3 rather than y^4. The ray aberration increases as the cube of the aperture.

Correction of spherical aberration is achieved very simply by changing the shape of the refracting surface. This can be made exactly correct for any chosen pair of conjugate points. Even if the surfaces are for simplicity constrained to be spherical, a lens may be corrected very well for spherical aberration by the 'bending' illustrated in Fig. 2.16, where the surfaces are still spherical but have different radii of curvature. An exact correction requires the use of aspheric surfaces, which are frequently used to correct image distortion in optical systems. It is important to note that any correction can only apply exactly to one particular object distance, and that objects at a different distance will still suffer from spherical aberration.

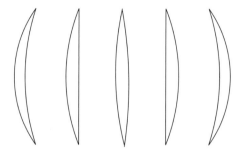

Fig. 2.16 Lenses with spherical surfaces, and with the same focal lengths, but 'bent' by different amounts. Spherical aberration is minimized by using a lens shaped so that the refraction is shared roughly equally between the two surfaces; the plane or concave surface should therefore be closest to the nearer of the object and image points.

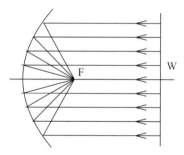

Fig. 2.17 A section through a paraboloidal reflector telescope, showing rays from a distant object converging on the focus F. All optical paths from the wavefront W to the focus are exactly equal, so that there is no spherical aberration for waves from a distant object.

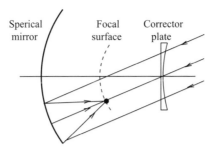

Fig. 2.18 The Schmidt corrector plate retards the wave in the outer parts of the aperture, removing spherical aberration. It is placed at the centre of curvature of a spherical mirror so that its effect is nearly independent of the ray inclination.

Reflecting telescopes, and particularly the large reflector radio telescopes, commonly use apertures with diameters of the same order as their focal lengths. It is usual to remove the spherical aberration by making the surface a paraboloid of revolution (Fig. 2.17), when the spherical aberration for an object at infinity and on the axis is exactly zero. A paraboloid of revolution does not, however, form a perfect image for objects off the axis, and if it is intended to use an extended field of view in an optical or radio telescope it will be necessary to consider the off-axis aberrations, which grow more rapidly with angle for a paraboloid than for a spherical reflector. A system using a spherical mirror that avoids spherical aberration and still produces good off-axis images is used in the Schmidt telescope (Fig. 2.18). Here, a thin corrector plate located at the centre of curvature introduces a correction to the wavefront that compensates for spherical aberration. The location at the centre of curvature provides good compensation for a wide angle off axis, although the focal surface is necessarily curved.

2.12 OFF-AXIS ABERRATIONS

The analysis in Section 2.11 of spherical aberration on axis may be extended to off-axis aberrations. The results are quoted below in terms of *ray aberrations*, which are the deviations of rays from the correct focal point. These deviations are found from the slope of the wavefront as it leaves the refracting surface (see Fig. 2.14).

Figure 2.19 shows a section of wavefront meeting a refracting surface centred on A, at a distance y from the axis at O. For each point B of the wavefront, with polar coordinates $\rho \sin \theta$, $\rho \cos \theta$, we require the wave aberration a and its slope. The wave aberration a is not circularly symmetrical: we must specify x and y components of the ray aberration b, respectively perpendicular and parallel to the plane containing the ray and the lens axis.

$$b_x = B\rho^3 \sin \theta - Fy\rho^2 \sin 2\theta + Dy^2\rho \sin \theta$$
$$b_y = B\rho^3 \cos \theta - Fy\rho^2 (1 + 2\cos^2 \theta) + (2C + D)y^2\rho \cos \theta - Ey^3. \qquad (2.36)$$

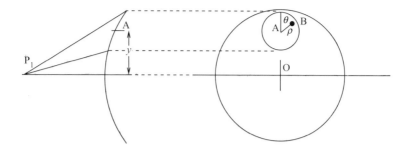

Fig. 2.19 Off-axis aberration. A pencil of rays from P_1 crosses the refracting surface at a distance y from the axis P_1O, showing the position ρ, θ of a ray within the pencil as it crosses the surface.

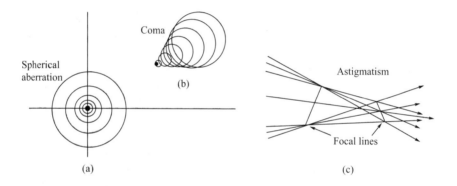

Fig. 2.20 The effects of (a) spherical aberration, (b) coma and (c) astigmatism.

Here, the coefficients B, C, D, E, F depend on the configuration of the optical system. This formulation is used in the traditional developments of aberration theory,[6] in which the aberrations associated with each coefficient are known as the *Seidel aberrations*. The first term $B\rho^3 \sin\theta$ is the spherical aberration already described (Fig. 2.20(a)).

Coma is a wavefront distortion additional to spherical aberration which only appears for object points off the axis (terms with coefficient F). Rays intersect the image plane in a comet-like spread image, whose length increases as the square of the distance off axis (Fig. 2.20(b)). Rays from a line across the aperture, at $\theta = 90°$, do not contribute to coma, but the rays along the line at $\theta = 0°$ cannot focus at a point and spread out beyond the true image. The typical comatic image consists of superposed circular images, successively shifted further from the axis and focused less sharply.

---

[6] See, for example, M. Born and E. Wolf, *Principles of Optics*, 6th edn, Pergamon Press, Oxford, 1980, p. 211 *et seq.*

Astigmatism (Fig. 2.20(c)), related to coefficients C and D, is the result of a cylindrical wavefront aberration, which increases as the square of the distance off axis and ρ. The effect is unfortunately familiar in many human eyes, which show astigmatism even for objects on axis. The focus, shown in Fig. 2.20(c), consists of two concentrations of rays known as the focal lines, with a blurred circular region between representing the best approximation to a point focus. This is called the *circle of least confusion*.

The *curvature* term, also related to C and D, in which the wavefront has an added curvature proportional to y^2, shows that the focal length of the lens changes for off-axis points. A flat object plane will then give a curved image surface. It is usual to find curvature still present in a lens that is corrected for astigmatism; this remaining curvature is referred to as the Petzval curvature.

Distortion, related to E in Eq. (2.36), represents an angular deviation of the wavefront, increasing as y^3. This spreads or contracts the image, destroying the linear relation between dimensions in object and image.

Since we know from the start that all aberrations cannot be eliminated from a useful optical system, it becomes a matter of choice which aberrations are the most nuisance and which can most easily be tolerated. For example, a photograph with distortion may be more displeasing to the eye than one with some blurring due to spherical aberration or coma. An astronomical photograph might, on the other hand, be required to show small symmetrical point images over the whole of a plate covering a large solid angle, while it might be less important to minimize the distortion of angular scale near the edges of the plate.

We can now appreciate Feynman's remark that optics is either very simple (as in paraxial approximations) or very complicated (when a compromise must be made between conflicting aberrations). The difficult part is made easier by automatic methods of ray tracing, which can rapidly demonstrate the performance of any optical system, however complex. Many modern camera lenses use components with non-spherical surfaces, derived from computation programs that optimize performance. Such computational methods nevertheless require a performance specification and an outline solution, which can only be provided with a knowledge and understanding of the basic aberration theory.

2.13   THE INFLUENCE OF APERTURE STOPS

The amount of spherical aberration introduced by an uncorrected lens or reflector system varies as the cube of the lens aperture. If a large aperture is necessary, the aberration must either be tolerated or corrected, but an improvement in images can obviously be made by restricting the aperture by means of a stop. For the single purpose of restricting spherical aberration in a lens the stop would be placed against the lens itself, but the other aberrations are also affected by the stop in ways that depend on the separation of the stop from the lens. This is demonstrated in Fig. 2.21 (a,b), which shows a pencil of rays from an off-axis point passing through an aperture stop in front of a lens.

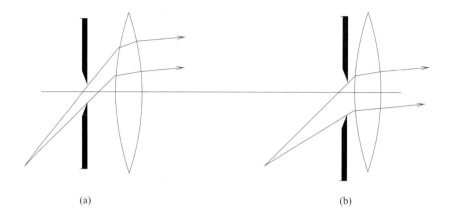

(a)   (b)

Fig. 2.21 The positioning of an aperture-stop. In (a) the stop is spaced away from the lens, so that off-axis points are focused by the outer part of the lens. The shape of the lens may then be changed so that aberrations are reduced. In (b) the same part of the lens is used for all ray inclinations. Aberrations are then less controllable, although they will be smaller for spherical surfaces.

When the aperture stop is separated from the lens the rays from an off-axis point are constrained to pass through the outer part of the lens, as in (a). Depending on the shape of the lens, this may reduce or increase the off-axis aberrations. An important difference between (a) and (b), where the stop allows rays to reach the lens by a shorter path, is that the magnification of off-axis points will be greater in (b) than (a), since the object distance is less in (b). Distortion can therefore be controlled by the correct positioning of the aperture stop.

2.14   THE CORRECTION OF CHROMATIC ABERRATION

The power of a spherical refracting surface, radius r, is given by Eq. (2.15) as $(n_2 - n_1)/r$; where $n_2 - n_1$ is the difference of refractive index across the surface. So far no account has been taken of the need to focus light of a wide range of colour by the same optical system; since the refractive index inevitably varies with the wavelength of the light, any optical system that depends on refraction rather than reflection will behave differently for different colours. *Chromatic aberration* is a measure of the spread of an image point over a range of colours. It may be represented either as a longitudinal movement of an image plane, or as a change in magnification, but basically it is a result of the dependence of the power of a refracting surface on wavelength. It may be compensated for by combining lenses made of different materials.

The power of a single thin lens used in air may be written as $P = (n - 1)/R$, where $1/R = 1/r_1 + 1/r_2$, the sum of the curvature of the two surfaces. A small change of refractive index δn therefore changes the power by δP where

$$\delta P = P \frac{\delta n}{n - 1}. \tag{2.37}$$

Varieties of optical glass differ quite widely in the way in which n varies with wavelength, so that it is possible to combine two lenses with power P_1 and P_2 in such a way that $\delta P_1 + \delta P_2 = 0$ without at the same time making the total power $P_1 + P_2 = 0$. Since $\delta n/(n - 1)$ has the same sign for all glasses, this means that P_1 and P_2 must have the opposite sign, and the combined lens system has a lower power than either component. Two colours separated in wavelength by $\delta\lambda$ will be focused together when

$$P_1 \frac{\delta n_1}{n_1 - 1} + P_2 \frac{\delta n_2}{n_2 - 1} = 0. \tag{2.38}$$

The powers must therefore be of opposite sign and inversely proportional to the value of $(\delta n/\delta\lambda)(n - 1)^{-1}$ for the two glasses. The two lenses may be in contact if two surfaces have the same radii of curvature. It is advantageous to reduce the number of interfaces between glass and air, since light is lost by partial reflection at each step in the refractive index. The step between the two kinds of glass is smaller than for interfaces between glass and air, but the advantage is lost unless the two lenses are cemented together using a transparent glue with a refractive index approximately the same as that for glass. Both the power and the focal plane of such an *achromatic doublet* can be made the same over a range of wavelengths, or at any two widely separated wavelengths. Outside these wavelengths, however, it will generally still suffer from chromatic aberration.

The *dispersive power* of glass is often quoted in terms of the refractive index at specific wavelengths, which have traditionally been those of the three *Fraunhofer lines* F, D and C. (These are prominent absorption lines in the solar spectrum.) Table 2.1 shows the refractive indices for crown and flint glass.

Table 2.1 The refractive indices for crown and flint glass

Designation	Wavelength (nm)	Crown glass	Flint glass
F blue	486	1.5286	1.7328
D yellow	589	1.5230	1.7205
C red	656	1.5205	1.7076

Dispersive power Δ is defined as

$$\Delta = \frac{n_F - n_C}{n_D - 1}, \tag{2.39}$$

so that the dispersive powers of crown and flint glass are respectively 1/65 and 1/29.

2.15   ACHROMATISM IN SEPARATED LENS SYSTEMS

The provision of a focal length that does not vary with wavelength is not a sufficient condition to provide completely achromatic images, since it provides only a constant transverse magnification. The position of the focal plane can still vary with wavelength, giving a longitudinal chromatic aberration. This is often less important than the reduction of transverse chromatic aberration, especially in the eyepieces of microscopes and telescopes, so that it is common to find in these a very simple system for achromatism, using two identical lenses separated by the focal length of one lens. This provides a constant transverse magnification, giving a great improvement over an uncorrected lens at very little cost. At one particular wavelength the power of two lenses, separated by a distance d, is given by Eq. (2.30) as

$$P = P_1 + P_2 - dP_1P_2. \tag{2.40}$$

At a different wavelength the net change in total power is given by

$$\delta P = \delta P_1 + \delta P_2 - d(P_1 \delta P_2 + P_2 \delta P_1). \tag{2.41}$$

The change in total power is zero when

$$\frac{\delta P_1}{P_1 P_2} + \frac{\delta P_2}{P_1 P_2} = d\left(\frac{\delta P_2}{P_2} + \frac{\delta P_1}{P_1}\right). \tag{2.42}$$

If the lenses are made of the same glass, then $\delta P_1/P_1 = \delta P_2/P_2$ for all wavelengths, and the achromatic condition becomes

$$\frac{1}{P_1} + \frac{1}{P_2} = 2d \quad \text{or} \quad d = \frac{f_1 + f_2}{2}. \tag{2.43}$$

The doublet, therefore, is achromatic when the lenses are separated by half the sum of their focal lengths. This configuration is used in the Huygens and Ramsden eyepieces of microscopes and telescopes (see Chapter 3).

2.16   ADAPTIVE OPTICS

The angular resolution of large optical telescopes is usually limited by turbulence in the atmosphere, which causes random fluctuations in the refractive index. Ideally, the wavefront reaching the telescope from a distant point-like source is plane over the whole aperture. Turbulence disturbs the wavefront, so that it can only behave as a plane wave over a small distance d instead of the whole aperture diameter D. The width of the effective telescope aperture determines the angular resolution (see Chapter 7), so that instead of angular resolution $\sim \lambda/D$ we have the larger angle $\sim \lambda/d$. Typically, $d \sim 0.3$ m, giving a resolution limited to ~ 1 arcsec, even for the largest telescope apertures. It may seem impossible to improve on this limit, apart from observing from a telescope in space, such as the Hubble Space Telescope. Only if the atmospheric distortion can be known instantaneously, and corrected for, can the full resolution be restored. The wavefront distortions change rapidly, typically in less than 100 ms, so that the measurement and correction have to be completed and repeated within this short time. How can this be achieved?

The form of the wavefront distortion can be found by a simultaneous observation of a nearby bright star, whose image will be distorted in the same way as that of the target object. For example, the wavefront across the whole aperture may be tilted, so that both objects appear to change position. The image movement can be detected if the bright star image falls on an array detector (see Chapter 21). Such a wavefront tilt can be compensated by tilting a small mirror in the optical path, near the detector. A mirror with small mass can be controlled very rapidly by a piezoelectric actuator, holding the images of both the reference star and the target object steady.

The correction of wavefront tilt is the simplest example of *adaptive optics*. Further improvements can be made by dissecting the wavefront from the reference star into a number of separate segments, and correcting each individually for tilt, using a dissected compensating mirror. Rapid measurement and computation are essential to such a scheme. Obviously, such a technique is only applicable to fields of view containing a sufficiently bright reference star. An artificial star can, however, be created by shining a laser beam up through the atmosphere, when the back scattered light from the upper atmosphere simulates a point source. Laser light tuned to sodium atoms is used, since it is scattered from sodium atoms in the upper atmosphere. A powerful laser beam can be pulsed on, and the wavefront distortion measured, in about 1 ms, well below the 100 ms in which correction and normal observation must be achieved.

2.17   FURTHER READING

M. Born and E. Wolf, *Principles of Optics*, 6th edn, Pergamon Press, Oxford, 1980.
P. Mourculis and J. Macdonald, *Geometrical Optics and Optical Design*, Oxford University Press, 1997.
D. G. O'Shea, *Elements of Modern Optical Design*, Wiley, 1985.
F. Roddier, *Adaptive Optics in Astronomy*, Cambridge University Press, 1999.
W. T. Welford, *Aberrations of Optical Systems*, Hilger, 1986.

NUMERICAL EXAMPLES 2

2.1 A simple 35 mm camera has a lens with focal length 2 cm. How far from the film must the lens be for a person 2 m from the camera to be in focus?

2.2 A shaving mirror has a concave surface on one side with a radius of curvature of 40 cm, and a plane mirror on the other side. When looking at oneself imaged in the plane side, how far from the mirror should one's face be for the image to be 25 cm away from the real face? Obviously, 12.5 cm, since $-v = u$. Now repeat for the concave side of the mirror.

If the pupil of one's eye is 2 mm in diameter, what is the angular width of the image of the pupil as seen by the eye in both cases. For the plane mirror this is $0.2/25 = 0.008$ rad $= 27.5$ arcmin. For the concave mirror we shall need the magnification.

2.3 A plastic sphere of radius 1 cm and refractive index 1.4 has a small light bulb 10 cm from its centre. Where is the image of the bulb?

PROBLEMS 2

2.1 A thin mirror that is part of a spherical surface is silvered on both sides. An object O on the concave side is reflected in it as a virtual image at O′, as a check on your sign convention show that an object at O′ will be imaged correctly as a virtual image at O.

2.2 Two identical planoconvex lenses are each silvered on one face only, one on the plane face and the other on the convex face. Find the ratio of their focal lengths for light incident on the unsilvered side.

2.3 A lens with refractive index 1.52 is submerged in carbon disulphide, which has refractive index 1.63. What happens to its focal length?

2.4 A small fish swims along the diameter of a spherical gold fish bowl, directly towards an observer. Find how the fish's apparent position varies in terms of the bowl radius R and liquid refractive index n. Can the image be inverted?

2.5 A thick lens consists of two spherical surfaces with curvature R_1, R_2 separated by a thickness d of material of refractive index n. Show that the power P of the thick lens is given by

$$P = P_1 + P_2 - \frac{d}{n} P_1 P_2,$$

where P_1, P_2 are the powers of the two surfaces. (Follow the argument of Section 2.7, noting the factor n in Eq. (2.16).)

2.6 In a plane-parallel circular disc of refracting material, the refractive index $n(r)$ is made to be a function only of the distance from the axis of the disc. Show that the radius of curvature of a ray nearly parallel to the axis is $(1/n \, dn/dr)^{-1}$. Find the form of $n(r)$ that will make the disc act as a concave lens.

Design a microwave lens with focal length 5 m from such a disc with radius 1 m and thickness 30 cm, allowing the refractive index to vary between 1.1 and 1.4.

2.7 A reflecting surface giving stigmatic images of two conjugate points is a paraboloid of revolution when one of the conjugate points is on the axis and at infinity. Show that a single refracting surface between refractive indices n_1 and n_2 is similarly free from spherical aberration for an object at infinity when it is (i) an ellipsoid of revolution, (ii) a hyperboloid of revolution.

2.8 The magnification between the object and image at the foci of the elliptical mirror (Fig. 2.12) might be calculated from paraxial rays reflected either to the left or to the right of the foci. These evidently give different magnifications, while symmetry apparently demands unit magnification. What is wrong with this analysis?

2.9 If a thin glass filter, thickness d and refractive index n, is inserted between a camera lens and the photographic plate, show that the plate must be moved a distance $((n-1)/n)d$ away from the lens for focus to be maintained.

2.10 The distance between an object and its image formed by a thin lens is D. The same distance is found in a second position of the lens when it is moved a distance x. Show that the focal length of the lens is

$$f = \frac{D^2 - x^2}{4D}.$$ (2.44)

Show incidentally that the minimum distance between an object and its image is $4f$.

2.11 Consider a glass sphere of radius r and a narrow pencil of light parallel to the axis but off the axis by a distance αr chosen so that the refracted ray meets the opposite side of the sphere exactly on the axis. Find α in terms of the refractive index n.

2.12 A thin equiconvex lens with radii of curvature 220 mm is made of crown glass with refractive indices 1.515 and 1.508 for blue and red light, respectively. Find the focal length of the lens and the axial chromatic aberration. A thin plano-concave lens of flint glass is to be used to compensate this chromatic aberration, with the concave face towards the first lens. The refractive indices of flint glass for blue and red light are 1.632 and 1.615, respectively. Find the required radius of curvature of the concave surface and the focal length of the combination.

2.13 A point source of light is at distance u from a concave spherical mirror, radius of curvature r, aperture $2h$. Following the method and approximations of Section 2.11, show that
  (i) the wave aberration a at the edge of the mirror is given by

$$a = \frac{h^2}{4}\frac{1}{r}\left(\frac{1}{u}-\frac{1}{r}\right)^2$$ (2.45)

  (ii) the transverse ray aberration b is given by

$$b = v\frac{da}{dh} = h^3\left(\frac{1}{u}-\frac{1}{r}\right)^2\frac{1}{r}\left(\frac{1}{2r}-\frac{1}{u}\right)^{-1}$$ (2.46)

  (iii) the longitudinal ray aberration c is given by

$$c = \frac{v}{h}b = h^2\left(\frac{1}{u}-\frac{1}{r}\right)^2\frac{1}{r}\left(\frac{1}{2r}-\frac{1}{u}\right)^{-2}.$$ (2.47)

2.14 Find the transverse and the longitudinal spherical aberration for a concave spherical mirror 1 m diameter, focal length 10 m, for a distant source of light such as a star.

2.15 Find the difference in thickness across a Schmidt corrector plate (Fig. 2.18), refractive index 1.4, used to correct the spherical aberration of the previous example without moving its focal plane.

2.16 Plane-parallel light is incident normally on the vertex of a glass hemisphere with radius 70 mm. If the refractive indices for red and blue light are 1.61 and 1.63, respectively, find the axial chromatic aberration.

3

Optical instruments

I knew a man who, failing as a farmer, / Burned down his farmhouse for the fire insurance / And spent the proceeds on a telescope / To satisfy a life-long curiosity / About our place in the stars. / And how was that for otherworldiness?

Robert Frost (1875–1963), 'The Star Splitter'.

And besides the observations of the Moon I have observed the following in the other stars. First, that many fixed stars are seen with the spyglass that are not discerned without it; and only this evening I have seen Jupiter accompanied by three fixed stars, totally invisible because of their small mass.

Galileo Galilei, 7 January 1610.

Optical imaging systems, of which the most important is the human eye, obtain information about an object or a scene in three basic ways. In the eye, a complete image is formed on the retina where an array of detectors works simultaneously to send information to the brain; the action of a conventional photographic film camera is in the same category. In another group, the object may be dissected and scanned by a single detector, as in a television camera; or there may be an array of such independent detectors, each with a simple lens, as in the multiple eyes of many insects. Finally, the light from an object may be analysed to obtain its spatial Fourier components, followed by a reconstruction either mathematically or optically, as in a hologram; this will be the subject of Chapter 19.

In this chapter we deal with the typical image-forming instruments: the eye, the telescope, the microscope and the camera.

3.1 THE HUMAN EYE

The human eye is a miracle of evolution, with many parts subtly adapted to their individual purposes. The essential elements are shown in Fig. 3.1. The eye is nearly spherical, about 25 mm in diameter. The transparent front portion is more sharply curved and is covered with a tough membrane, the *cornea*. Between the cornea and the *lens* is a liquid, the *aqueous humour*. Behind the lens is the thin jelly-like *vitreous humour*, filling the volume in front of the *retina*,

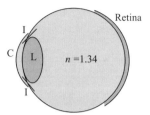

Fig. 3.1   The focusing system of the human eye. Most of the refraction occurs at the front surface of the cornea C, which has a refractive index 1.38. The lens L has a refractive index graded from 1.41 to 1.39, and the refractive index of the main volume is 1.34. The focal length of the lens is adjusted by tension in the surrounding ciliary muscles. The iris I adjusts the aperture according to the available illumination.

on which the image is focused. The network of nerves from the sensitive cells of the retina is on the front surface of the retina; it is gathered into the *optic nerve*, which passes through the retina, and the *sclera*, which is the outer case of the eye. The hole in the retina may be detected as a blind spot[1] in the field of view. The *iris*, which gives individual eyes their distinctive pattern and colour, is located in front of the lens; it expands and contracts in response to the light intensity.

The eye analyses light by focusing wavefronts from different directions on to different parts of the retina. The wavefronts are very nearly plane; by adjusting the eye to slightly diverging wavefronts a correct focusing can be obtained for objects as close as a limiting distance D_{near}, known as the nearest distance of distinct vision. The ability of the eye to change its effective focal length to image objects over a range of distances is known as *accommodation*. In the human eye there are two focusing elements: the cornea (Fig. 3.1) has a fixed power of about 40 dioptres,[2] while the lens, which is adjustable by the surrounding ciliary muscles, brings the total power to around 60 dioptres. (In fish the adjustment is achieved by moving the lens, and in some birds it is achieved by changing the surface of the cornea.) The back focal length for the human eye is about 17 mm.

The lens of the human eye operates at a relatively small focal ratio $(f/D \approx 5)$, but is remarkably free from spherical aberration. This is partly due to an outward gradient of the refractive index from 1.41 to 1.39 within the lens while the surrounding aqueous fluid has an index of 1.34. Defects are, however, common, as is evident by the number of wearers of spectacles.

The common defects of short sight (*myopia*) and long sight (*hyperopia* or *hypermetropia*) are illustrated in Fig. 3.2. Without correction the cornea and lens of the myopic eye bring rays from a distant object to a focus in front of the retina (myopic eyes do, on the other hand, have the advantage that they can focus on objects closer than D_{near}, allowing them to resolve more detail.) The hyperopic eye cannot focus on close objects, and often not even on distant ones.

---

[1] With one eye closed, concentrate on one of a pair of spots about 6 cm apart on a card 20 cm away. Using the left eye, the left-hand spot will disappear if the gaze is fixed on the right-hand spot.
[2] Power in dioptres $= 1/f$, where the focal length f is in metres.

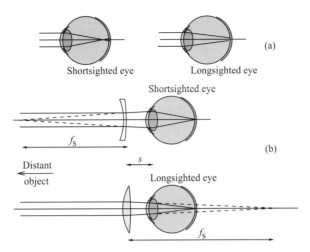

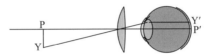

Fig. 3.2   (a) Shortsighted (myopic) and longsighted (hyperopic) eyes; (b) their correction by diverging and converging spectacle lenses, respectively.

Fig. 3.3   A spectacle lens located at the first focal point of the eye does not affect the magnification. The central ray from the object PY is undeviated by the lens, and forms an image P'Y' as for a perfect eye; all other rays from Y also reach Y'.

The power of the corneal surface may be corrected by a contact lens, which must add negative power for myopia, and positive power for hyperopia. The power of the combination is the sum of the powers of the surface and the lens. The contact lens brings the focal point on to the retina, but it also changes the magnification of the image. Fortunately, the brain is able to compensate for a small change in magnification, and it is only in severe cases needing correction in excess of about 8 dioptres that the effect is important.

More commonly, a lens is used, spaced at some distance from the eye, as in Fig. 3.2. The spacing has an advantage: if the lens is located near the first focal point of the eye, about 16 mm in front of the cornea, it does not affect the magnification, giving the same scale of image with and without the lens. In Fig. 3.3 the central ray from the off-axis object point Y passes without deviation through the spectacle lens at the focal point of the eye lens, and traverses the eye parallel to the axis, reaching the retina at Y'. The position of Y' is unaffected by the spectacle lens, provided it is near the front focal point.

To compare the power P_S of a spectacle lens at a distance s from the eye with the power P_C of an equivalent contact lens, we use the ray diagram of Fig. 3.2 for a distant object. Treating the eye as a simple lens, the combination must bring rays to a focus at a distance b (the back focal length) behind the eye lens.

The focal length of the uncorrected eye is f_e, and $f_S = 1/P_S$. Then the simple lens formula gives

$$\frac{1}{b} - \frac{1}{f_S - s} = \frac{1}{f_e}. \tag{3.1}$$

Noting that the power of the equivalent contact lens is

$$P_C = \frac{1}{b} - P_e \tag{3.2}$$

we obtain

$$P_C = \frac{P_S}{1 - P_S s}. \tag{3.3}$$

It is also common to find *astigmatism* in on-axis images, resulting from uneven curvature of the cornea. This can usually be corrected with *anamorphic* lenses, which have different powers in two perpendicular meridians.

3.2   THE SIMPLE LENS MAGNIFIER

The angular resolution of the eye is determined by the focal length and the separation of sensitive elements on the retina. This matches well the limit of angular resolution set by diffraction at the iris, the aperture of the main part of the eye lens. In the centre part of the retina, known as the *macula*, the sensitive elements are *cones* spaced about 3 μm apart, matching the angular resolution $\sim 5'$ expected from an iris diameter of ~ 5 mm (see Chapter 7).

Since the angular resolution is very nearly unchangeable, it follows that the linear resolution of the unaided eye is greatest for objects as close as possible, i.e. at the near distance D_{near}; this is generally taken to be 25 cm. Closer objects are out of focus, but if the eye is aided by a convex lens, an object at a very small distance can be focused, with a corresponding increase in linear resolution (Fig. 3.4). In wavefront terms, the lens assists the eye by converting a wavefront that is diverging sharply into the nearly plane wave that the eye can focus unaided.

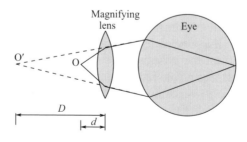

Fig. 3.4 A simple lens used as a magnifier. An object at O close to the eye can be focused by the eye as though it were at a more distant point O'.

The magnification of the image depends on the position of the lens and the eye; a very large (but distorted) image can be seen if the object is near the focal plane of the lens and the eye is some distance from the lens. Normally, the lens is close to the eye and the image is at the near point $D_{\text{near}} = 25\,\text{cm}$. The magnifying power M of a simple lens used in this way is given by the ratio of the apparent size of an object, seen as an image at the near point, to its actual size. This is the ratio, D/d, where D is the image distance (25 cm) and d is the object distance. From the simple lens equation, if the lens has power P:

$$P = \frac{1}{d} - \frac{1}{D_{\text{near}}}, \qquad (3.4)$$

and therefore

$$\boxed{M = 1 + PD_{\text{near}}.} \qquad (3.5)$$

A typical commercial magnifying glass has a power of 12 dioptres, so giving

$$M = 1 + 12 \times 0.25 = 4. \qquad (3.6)$$

Often, $M \gg 1$; it is then convenient to approximate by putting $M \simeq PD_{\text{near}}$.

Practical values of M are limited by the difficulty of making a single lens with very high power and free of aberration. Some improvement is obtained from double lens systems, which are used in the eyepieces of microscopes and telescopes, but for magnifications greater than about 10 or 20 the simple magnifier is replaced by the compound microscope, which is considered in Section 3.6.

3.3   THE TELESCOPE

When the eye attempts to distinguish details of a distant object, it is attempting to separate nearly plane waves that are inclined at small angles to each other. The limit of resolution can now only be improved by using an instrument that increases the angular separations of a range of plane waves. This is the action of a telescope.

If a plane wave at a small angle θ_1 to the axis of a telescope is to emerge as a plane wave at a larger angle θ_2, the refractive index at both ends being the same, then it may be shown that the width of the wavefront is reduced in the ratio θ_1/θ_2. Consider the wavefronts entering and leaving a telescope, as shown in Fig. 3.5. This shows the simple *astronomical* telescope, using two convex lenses with long and short focal lengths f_1 and f_2; these are called the *objective* and the *eyepiece*. As in a compound microscope, the eyepiece can be regarded as a lens magnifier, which is used to view an image formed by the objective. The magnification is the ratio f_1/f_2.

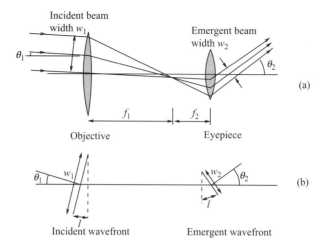

Fig. 3.5   The action of a simple telescope, with convex objective and eyepiece lenses, focal lengths f_1 and f_2: (a) a pencil of parallel rays from a distant source enters at angle θ_1 and emerges at angle θ_2. The magnification of the telescope is $\theta_2/\theta_1 = f_1/f_2$; (b) the widths of the wavefront as it enters and emerges are w_1 and w_2. The optical paths l are identical; the angles are small in practice, so that $l = w_1\theta_1 = w_2\theta_2$.

The wavefront enters the telescope with width w_1. It is at an angle θ_1 to the axis of the telescope, so that the difference in optical path l across the wavefront in the diagram is $l = w_1\theta_1$ (where a small angle approximation may be used). This path difference l is preserved as the wavefronts traverse the telescope, so that the difference in angle as the wavefronts leave the telescope is determined by l and the new width w_2 of the wavefront. The angular magnification is therefore:

$$\frac{\theta_2}{\theta_1} = \frac{w_1}{w_2} = \frac{f_1}{f_2}.$$

(3.7)

Practical arrangements for telescopes giving angular magnification are shown in Fig. 3.6, which includes many of the conventional varieties of telescope. For each, the figure shows the reduction of the width of a plane wavefront. (In some optical systems, notably in laser optics, a telescope may be used in reverse to expand rather than contract the area of a wavefront.) The emerging wavefront may be observed directly by eye, or it may be focused by a camera on to a photographic film or an array detector; the eyepiece may then become part of the camera. The angular magnification of all these arrangements is given by the ratio of the widths of the plane wavefronts entering and leaving the telescope; this is numerically equal to the ratio of the focal lengths of the two optical elements, either lenses or mirrors, which form the objective and eyepiece elements of the system.

A telescope also has the advantage over the eye that it can gather a larger area of plane wavefront, so that a point source of light becomes more easily

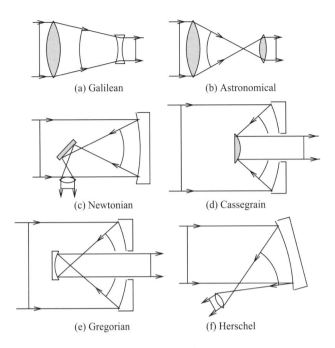

(a) Galilean   (b) Astronomical

(c) Newtonian   (d) Cassegrain

(e) Gregorian   (f) Herschel

Fig. 3.6 The reduction in width of a wavefront in various types of telescope. The telescopes are all adjusted for direct viewing of the emergent beam; the emergent wave could instead be made convergent, focusing an image on a photographic plate.

visible. It is most important not to confuse this increase in sensitivity with the question of the visibility of a uniformly bright object with a finite size: the surface of the moon is no brighter as seen through a telescope, while stars that are effectively point sources of light may easily be seen through a telescope even if they are invisible to the naked eye. If the object is already resolved in angle by the eye, then its visibility is related to its *brightness*, which is the light emitted per unit area into unit solid angle. The *apparent brightness* as seen through the telescope cannot be greater than the original, since the apparent size of the object increases in proportion to the increase in total light collected. There can in practice only be a loss of brightness in a telescope, due to partial reflection at lens surfaces or to incomplete reflection at a mirror surface.

3.4   ADVANTAGES OF THE VARIOUS TYPES OF TELESCOPE

The types of telescope in Fig. 3.6 are distinguished mainly as reflecting or refracting by the use either of mirrors or of lenses for the objective,[3] and by

---

[3] A telescope system that uses only lenses, such as the Galilean, is referred to as *dioptric*, and with mirrors only as *catoptric*; a combination of lenses and mirrors, as in the Schmidt telescope (Fig. 2.18) is a *catadioptric* system.

the use of second elements with positive or negative power. Systems such as the Galilean telescope and the Cassegrain telescope have the advantage that they are shorter than the corresponding instruments using second elements with positive power (astronomical and Gregorian). They may, however, be unsuitable for terrestrial survey and position measurement because they have no real image point at which a graticule can be placed to use as a reference mark or scale. Most reflector telescope systems are axially symmetric, and consequently the secondary mirror tends to obstruct part of the aperture. The Herschel system uses the whole of the aperture without obstruction, but at the cost of using the primary off the axis; it is easier to control aberrations with the more symmetrical mirror arrangements (c), (d) and (e).

The control of aberrations has already been discussed in Chapter 2, but the detailed application to a full telescope system becomes complicated. Modern astronomical telescopes commonly use a Cassegrain system, and the control of aberrations may entail the addition of a Schmidt corrector plate in the beam as it enters the telescope, or a corresponding asphericity of both primary and secondary mirror.

A lens has the great advantage over a mirror that a small distortion due to gravity or uneven temperature has no first-order effect on the optical path through it, whereas if part of a mirror bends forward by an amount z it shortens the optical path by $2z$. If nearly perfect images are required, in which all optical paths are near equal, this means that the mounting of the mirror must be considered much more carefully than that of a lens. On the other hand, a mirror has no inherent chromatic aberration, and very little light is lost at a reflection. The largest telescopes, and some of the best survey theodolites, use mirror systems. A mirror must, of course, be used for wavelengths where no good lens can be made, as at infrared or radio wavelengths.

Most mirrors for optical telescopes use a thin silver or aluminium film evaporated on to glass, or preferably a ceramic with near-zero thermal coefficient of expansion. Mirrors for astronomical telescopes must be as large as possible to obtain sufficient light-gathering power; some of the largest have been cast in a single piece up to 8 m in diameter, as for the Gemini telescopes in Hawaii and Chile. The mirrors of the Keck telescopes on Hawaii are even larger, with diameters of 10 m; these are built up of hexagonal elements mounted to produce an almost complete single mirror. The Hubble Space Telescope is necessarily smaller, with a diameter of 2.3 m; it has, of course, the tremendous advantage of avoiding the effects of absorption and random refraction in the atmosphere. Larger space telescopes must use multi-element mirrors assembled in space. For all these large telescope mirrors the surface profile must be accurate to a small fraction of a wavelength; this is extremely demanding both in manufacture and in the support systems that maintain the shape in use. Errors may be hard to rectify; the Hubble Space Telescope was launched with serious spherical aberration due to a faulty test procedure, and the wavefront entering the cameras and spectrometers now needs correction by special optical systems.

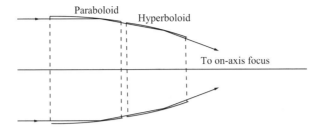

Fig. 3.7 The Wolter X-ray telescope. The grazing incidence reflecting elements are sections of a paraboloid followed by a section of a hyperboloid.

Radio telescopes use a simple metal surface, fabricated or polished so that the surface profile is correct within a small fraction of a wavelength. X-ray telescopes present a different problem; the only efficient reflector is a polished metal surface at a grazing angle of incidence. The Wolter telescope of Fig. 3.7 uses a section of a paraboloid that is only slightly tapered, followed by a second reflector element that is part of a hyperboloid (the combination reduces off-axis aberrations, giving a wider field of view). X-ray telescopes in spacecraft use Wolter telescopes, often with several concentric reflector systems so as to increase the effective collecting area. The aperture is typically 1 m in diameter, and the focal point is several metres beyond the reflector system. Electronic detector arrays, such as the charge coupled detector arrays described in Chapter 21, are used to obtain remarkably detailed images of the X-ray-emission of energetic astronomical objects such as active galactic nuclei.

The attainment of the theoretical angular resolving power of telescopes (given approximately by the ratio wavelength/diameter, see Chapter 7) depends on a number of factors. For infrared and longer wavelengths the full resolution may be achieved, even for the largest astronomical telescopes. In the optical region, however, atmospheric effects limit the resolution to around 0.3 arcsec, while for X-ray telescopes the limitation is the accuracy of the mirror surfaces.

3.5 BINOCULARS

The binocular telescope, or 'binoculars', as used by bird-watchers and amateur astronomers, must be one of the most widely used forms of telescope; for many people, using both eyes gives a considerable improvement in the perception even of diffuse objects. Binoculars comprise a pair of refracting telescopes, with objective lenses some centimetres across and with eyepieces allowing normal vision of the magnified scene. Each telescope is basically an astronomical telescope, with internally mounted prisms to correct the inversion of the image. Let us imagine that we are to design a binocular telescope for general use.

We know already that the magnification of a telescope focused for object and image at infinity is given by the ratio f_o/f_e between the focal lengths of objective and eyepiece. A hand-held instrument does not usually have a magnification greater than about ×8 or ×10, since otherwise the image could not be held sufficiently steady without a tripod mounting. A magnification of ×8 is common for binoculars. We also know that the total amount of light entering the instrument is determined by the aperture of the objective, and that this affects the visibility of point sources of light. A large-diameter objective is therefore important. We now discuss the factors that determine the field of view of the binoculars, what sorts of lenses we must use, and what determines the diameter of the eyepiece.

The eye is especially sensitive to chromatic aberration, which has the effect of colouring the edges of objects away from the axis. Binocular objectives must therefore be carefully corrected; they are therefore made as cemented achromatic doublets (Section 2.14). In the eyepiece a single cemented achromatic pair is insufficient to control other aberrations over a wide field of view; practical eyepieces usually consist of a separated pair, one of which is itself a cemented doublet. Figure 3.8 shows only a simple pair in the form known as the Ramsden eyepiece.

An important part of eyepiece design concerns the position of the eye. The rays from a point source in Fig. 3.8 cross the axis beyond the eyepiece; at this point they fill the *exit pupil* of the system. The eye is placed at the exit pupil, which is separated from the eye lens by a distance known as the *eye relief*. The optimum size of the exit pupil is determined by the size of the pupil of the eye. If the exit pupil is smaller than the eye pupil, then the *eye* is used inefficiently, since only part of the eye pupil is illuminated. If the exit pupil is larger than the eye pupil, then some light is wasted and the *telescope* is being used inefficiently. In practice, the exit pupil should be somewhat larger than the eye pupil so that the exact position of the eye pupil is not too critical; the binoculars are then easier to use.

The focal plane of eyepieces of the Ramsden type is close to the first lens. A real image of any object at infinity exists at this point, so that in this case the angular width of the field of view is determined by the aperture of the first lens of the eyepiece. An aperture that limits the field in this way is known as a *field stop*; the first lens is therefore often called the field lens. The angular width of the field of view is the diameter of the field lens divided by the focal length of the objective. A long objective focal length f_o therefore gives a large magnification but a small field of view. We might therefore expect to see large magnifications

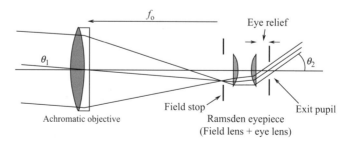

Fig. 3.8   Astronomical telescope system, as used in binoculars.

obtained instead by using a small eyepiece focal length f_e. However, as we will see, the diameter of the eyepiece is fixed by other considerations, and reducing f_e becomes difficult without introducing aberrations.

Let us assume that the magnification is fixed at the comfortable limit of ×10, and find the diameters and focal lengths that must be used for eyepiece and objective. We shall find that all of these depend on our requirements for the *field of view*. Consider again the rays entering the eye at the exit pupil in Fig. 3.8. The angular spread of rays at this point is the angular width of the field of view multiplied by the magnification; it is therefore a large angle, often about 50°. The eye lens must be larger than the exit pupil to accommodate these rays, and the field lens must be somewhat larger again. The field lens must therefore be at least 15 mm in diameter; the size of the real image at this point must be the same size, since this lens constitutes the field stop. The field of view, which in this example would be 50° divided by the magnification, determines the objective focal length f_o; this must therefore be about 17 cm. The focal length of the eyepiece is therefore one-tenth of this, i.e. 1.7 cm.

Finally, the diameter D of the objective must, for a magnification M, be M times the diameter of the exit pupil, which is usually about 4 mm to match the pupil of the eye. The objective is therefore 40 mm in diameter. We have reached the specification in the form usually quoted: these binoculars would be specified as 10×40.

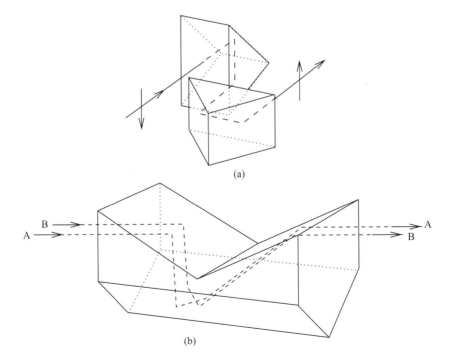

(a)

(b)

Fig. 3.9(a) Inverting prisms: (a) a pair of prisms, as used in prismatic binoculars; (b) the roof prism, which performs the same function and is more compact. Note the reversal of the rays AA, BB.

Two remaining problems are solved simultaneously by the use of a pair of prisms, as seen in Fig. 3.9(a), called a Porro prism. These invert the image, so that it appears upright, and they fold the telescope so that the total length is much less than $f_o + f_e$. A more compact form is the Abbe roof prism shown in Fig. 3.9(b). A Galilean arrangement would, of course, provide an upright image without prisms, but without folding, the telescope length is too great for anything more than the small magnification used in opera glasses.

3.6   THE COMPOUND MICROSCOPE

The simple lens magnifier of Section 3.2 converts a diverging wavefront into a nearly plane wavefront, so that separate points of an image give rise to separate components of an angular range of plane wavefronts. If further linear resolution is required, then the angular spread of these wavefronts can be increased as it is in a telescope. In Fig. 3.10 the objective lens, which has the short focal length required of the simple magnifier, is placed so that the object is just beyond the focal plane. The rest of the instrument is like the astronomical telescope.

The magnification of the compound microscope is calculated in two stages. First, the objective lens forms a real image; this is then magnified further by the eyepiece (Fig. 3.10). If the real image is formed at a distance g beyond the focus F of the objective, whose power is P_o, then the magnification is $P_o g$. (See Section 2.5 for a general derivation; alternatively, this may be shown to be equal to v/u for a simple lens.) The magnification of the eyepiece is $(1 + P_e D)$ (see Section 3.2), giving the overall magnification as $P_o g(1 + P_e D)$. The length g is known as the optical tube length of the microscope, since it accounts for most of the length of the instrument.

The diffraction theory of the microscope, dealt with in Chapter 14, shows that a high resolving power requires the objective lens to collect the spherical wavefront emerging from an object point over as wide an angle as possible. The design of the objective lens is crucial to the success of a microscope, since aberrations are easily introduced when rays traverse the lens at large angles. The highest magnification is obtained with an oil-immersion lens (Fig. 3.11(a)), in which the oil film has the same refractive index as the glass. The property of

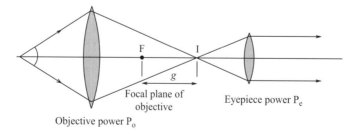

Fig. 3.10 Basic optics of the compound microscope. A wide-angle divergent wavefront is converted into a narrow, nearly parallel, wavefront emerging from the eyepiece.

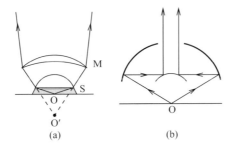

Fig. 3.11   Microscope objectives: (a) oil-immersion objective in which the object O and virtual objective O′ are on the aplanatic surfaces of a sphere S. The wavefront curvature is again reduced by a series of meniscus lenses M; (b) reflecting objective. This is a close relation of the Cassegrain telescope.

aplanatic surfaces (Section 2.9) in which all rays focus correctly is used to form a virtual image of O′ and successive meniscus lenses M are then used to reduce the curvature of the wavefront. The requirement to collect rays over a wide angle has also been met in the reflecting microscope objective of Fig. 3.11(b).

The performance of the objective in collecting light over a large angle is measured by its *numerical aperture* $n \sin \theta$ where n is the refractive index of the oil and θ is the half angle of the light cone. Even with a large numerical aperture, a microscope can only resolve detail at a scale greater than a wavelength of the illuminating light. The electron microscope, which uses beams of electrons in place of rays of light, is similarly restricted in resolving power by the equivalent wavelength of the electrons and by the numerical aperture of the system.

3.7   THE CONFOCAL SCANNING MICROSCOPE

A high-power conventional microscope is at a disadvantage when examining three-dimensional objects, when well-focused parts of an object are seen overlaid by confusing out-of-focus images of other parts at different depths. This becomes more confusing at larger magnifications and for larger numerical apertures, when the depth of focus becomes smaller. This disadvantage is overcome in the confocal microscope shown in Fig. 3.12.

In this instrument only one point at a time is illuminated and focused to a single small electronic detector element situated behind a pinhole stop. The signals from the detector are stored and used later for a reconstruction of the image. The image at a particular depth is scanned either by moving the whole microscope with its light source detector, or more simply by moving the object, in a raster scan as in television. The scanning can be extended to different depths by refocusing. This process may appear to be elementary and slow, but the results are spectacular. As shown by the broken rays in Fig. 3.12, light from planes away from the required focal plane is mainly spread outside the pinhole detector, avoiding the confusion inherent in the conventional instrument.

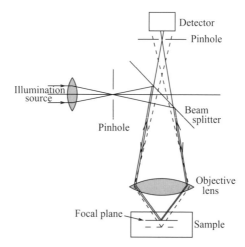

Fig. 3.12 The confocal scanning microscope. The beam splitter allows the same objective lens system to be used for illumination and for focusing light on to the pinhole detector. Only light from the focal plane enters the detector; the broken lines show rays from a different depth in the sample.

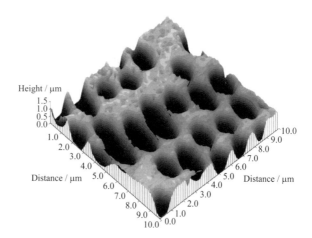

Fig. 3.13 Microphotograph of a portion of a compact recording disc (CD), with pits 0.5 μm across in tracks 1.6 μm apart. A scanning confocal microscope was focused at a series of depths to produce the layers of images that can be seen within the pits. (Robert Alcock, University of Manchester.)

Individual sections of the object are scanned in sequence and combined. These separate sections can be seen in Fig. 3.13, which shows a confocal microscope scan of a portion of a compact recording disc (CD). In this microphotograph the pits that form the digital recording are 0.5 μm across, in tracks 1.6 μm apart. Many spectacular microphotographs of biological subjects, such as the network of cells in Plate 3.1, have been scanned in sections and recombined in this way.

3.8   THE CAMERA

Astronomical research is seldom conducted by looking through a telescope: instead an image is formed on a photographic plate or detector array (Chapter 21), or it may be focused on the slit of a spectrograph. The telescope then becomes a camera, which is an artificial eye; the photographic plate is the retina and the lens of the eye is the primary lens or mirror of the telescope. A camera is usually focused on an object at a distance that is large compared with the focal length of the lens; the linear size of the image is then given directly by the product of the focal length of the lens and the angular size of the object.

Similarly, a camera may be arranged to focus on very near objects, when it becomes a photomicroscope. Photographic and television cameras are often provided with interchangeable sets of lenses with a range of focal lengths, so that the scale of a picture may be selected according to the required angular resolution; alternatively, a 'zoom' lens may be used, which is an adjustable compound lens whose focal length can be varied over a range that may be as large as five to one. Small cameras commonly have a lens with focal length about 40 mm; an astronomical telescope may have a focal length of 10 m or more, so as to provide a sufficiently large linear scale on the photographic plate. Even with a focal length of 10 m, an angle of 1 arcsec corresponds to only 0.05 mm at the focal plane; since diffraction images smaller than 1 arcsec are obtainable in large telescopes, a stellar image usually has a microscopic scale on the image plane. The effective focal length may be adjusted by any of the devices of Fig. 3.6, since we have already seen how these affect the angular scale of a pattern of plane waves. It is in fact only necessary to change the position of the secondary lens or mirror to obtain a real image at any desired distance. For example, the Galilean telescope may be converted into the telephoto lens (Fig. 3.14) by moving the secondary away from the primary. The advantage over the use of a single objective lens is that a long focal length is available without a corresponding and inconveniently long distance between the first lens and the photographic plate.

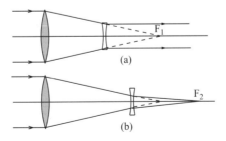

(a)

(b)

Fig. 3.14   Comparison of Galilean telescope (a) and telephoto lens (b). In the telescope the diverging secondary lens is placed so that the foci of the two lenses coincide at F_1; in the telephoto lens the secondary is moved so that a real focus F_2 is located on a photographic plate.

The objective lens of a camera with a wide field of view, which is required of most modern cameras, usually consists of at least five elements. A high-quality image can only be obtained by careful control of aberrations (see Chapter 2). This is achieved by combining several optical elements with aberrations of opposite sign, so that the aberrations of one element are corrected by other elements. A minimum of three lens elements (a triplet) is required to correct all seven third-order aberrations. The typical triplet lens has a negative lens element between two positive elements; a stop is often placed just before the second positive lens. An example of a six-element lens is shown in the single lens reflex camera of Fig. 3.15; this Biotar design, and also the four-element Tessar, are widely used. The two main elements are cemented pairs designed to correct for achromatism, while the outer lenses provide correction for geometrical aberrations. In this camera the viewfinder uses the same lens, viewing the field via a mirror that hinges out of the light path when the film is exposed. The image on a translucent screen is seen upright through a reversing prism.

In a so-called *compact* camera there is no room to place a folding mirror, and the lens system is placed closer to the film. Many such cameras have a zoom lens, with a range of three or more in magnification: it is then necessary to incorporate a separate zoom system in the viewfinder, coupled mechanically to the main lens system. Control of aberrations over the full range requires further sophistication in design; high-quality compact cameras accordingly use aspheric surfaces in some of their lens elements.

A modern compact automated camera (Plate 3.2) conceals from the user many other sophisticated design features. All automatic focusing and exposure metering is done through the viewfinder. Luminance is measured by a photometer, followed by automatic focusing and aperture adjustments, exposure and automatic film winding. Electronic array detectors such as the CCD (Chapter 21) may also be used, with their own complex circuitry, offering possibilities of enhanced sensitivity and spectral range but as yet not as good image detail as the photographic plate.

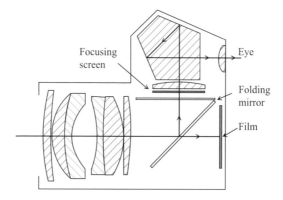

Fig. 3.15   A single lens reflex (SLR) camera, showing the multiple element lens, and the viewfinder arrangement.

A telescope or camera photographing an extended object produces an image in which we need to know the amount of light energy per unit area, i.e. the illumination. This depends on the light energy leaving unit area of the source in the direction of the observer, i.e. on the *luminance* of the source. The illumination of the image is proportional to the aperture area, but it also varies inversely as the area of the image, which is itself proportional to the square of the focal length. The intensity on the photographic plate therefore varies as the ratio $(f/D)^{-2}$, where f/D is the familiar 'focal ratio', or F-number, of a camera lens; it is the ratio of focal length f to aperture diameter D. Many compact cameras incorporate a *zoom* lens, which will adjust the focal length over a range of two or three; the effect on the field of view is presented to the photographer by adjusting the viewfinder in synchronism.

The depth of focus, which is the range of object distance over which the image is effectively in focus, depends on the F-number. Consider first a camera focused on infinity. In Fig. 3.16 a distant point object forms a point image at P, distance $v = f$ from the lens. A closer point object at distance u forms an image at $v + \delta v$. The converging rays form a blurred image at P with diameter d; if this is small enough the object is still effectively in focus. The lens diameter is D, so by simple proportion the image diameter on the focal plane is given by

$$\frac{d}{\delta v} = \frac{D}{f}.$$

(3.8)

The case for a camera focused on infinity is dealt with in Problem 3.5. More generally, and provided that δv is small, we can find the depth of focus δu by differentiating the lens equation

$$\frac{1}{u} + \frac{1}{v} = \frac{1}{f}$$

(3.9)

$$\frac{\delta u}{u^2} = -\frac{\delta v}{v^2}.$$

(3.10)

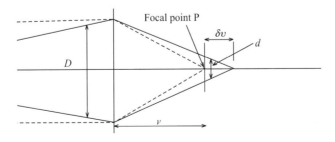

Fig. 3.16   Geometrical construction for the depth of focus.

After some algebra this gives

$$\delta u = \frac{F(u-f)}{f^2}d, \tag{3.11}$$

where $F = f/D$. For example, a camera with $F = 2.5$ and focal length 5 cm focused on an object at 2 m distance, using film with an acceptable blurring diameter of 50 μm will be in focus for objects 10 cm in front of or behind the 2 m position.

3.9   ILLUMINATION IN OPTICAL INSTRUMENTS

The discussion of the compound microscope started by assuming that wave-fronts left the object over a wide range of angles. The object may, of course, be illuminated naturally by diffuse light, but this is often insufficient. Extra illumination must be provided, and for efficiency and good angular resolution the light must be encouraged to leave the object in the right range of directions. This is achieved for transparent objects by the use of a *condenser*, which may be a concave mirror or a lens system, as in Fig. 3.17. No great optical quality is required, since only a rough image of a diffuse source of light need be formed on or near the objective plane of the microscope. The total light entering the microscope depends on the solid angle Ω over which the condenser collects the light; the condenser must therefore have a short focal length both for this reason and so that the object plane is well illuminated by light traversing it over a wide range of angles.

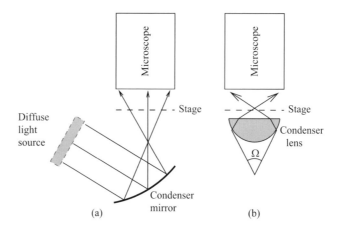

Fig. 3.17   Condenser systems for a microscope: (a) concave mirror; (b) lens system. A short focal length is needed to collect light over a large solid angle Ω and to cover a wide range of angles as it enters the microscope.

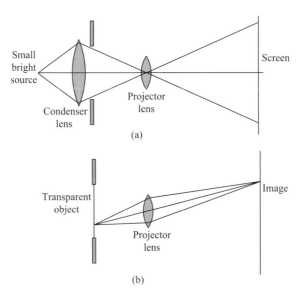

Fig. 3.18   Illumination in a projector system: (a) the action of the condenser lens is to collect light emerging from a small bright source over a wide angle, providing an even illumination over the transparent object, and concentrating the light as it passes through the projector lens; (b) the projector lens forms an image of each part of the object in a narrow pencil of rays determined by the illumination.

A similar problem is encountered in projection systems and enlargers; there the requirement is to obtain as much light as possible through a system consisting of a transparency, a projection lens and a screen. The illumination of the transparent object must be even, but there is no requirement for illumination over a wide range of angles. Fig. 3.18(a) shows the way in which light from a small source traverses a projection system: it is important not to confuse this diagram with the more conventional ray diagram of Fig. 3.18(b), which is concerned with the image on the screen of a point on the transparent object. This image is formed by a narrow pencil of rays within the light paths of Fig. 3.18(a).

The condenser lens of a projector need not be an accurate high-quality component. For the familiar 'overhead' projector a stepped lens is used (Fig. 3.19); this is a thin sheet of glass or plastic with an embossed array of prisms.[4] The equivalent simple lens that it replaces would be an impossibly thick and massive piece of glass, while the imperfections of the stepped lens are unimportant.

----

[4] The stepped lens is known as the Fresnel lens after its inventor, who was the first to use the principle in lighthouse lenses.

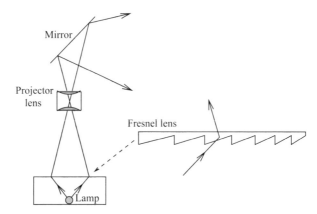

Fig. 3.19 Overhead projector. The light from the lamp is concentrated into the projector lens by the stepped lens plate, known as a Fresnel lens.

3.10 FURTHER READING

L. Levi, *Applied Optics*, Wiley, 1968.
D. Malacara, *Geometrical and Instrumental Optics*, Academic Press, 1988.
C. J. R. Sheppard, D. M. Hotton and D. Shotton, *Confocal Laser Scanning Microscopy*, Springer-Verlag, 1995.
W. J. Smith, *Modern Lens Design*, McGraw-Hill, 1992.

NUMERICAL EXAMPLES 3

3.1 A simple astronomical telescope has an objective with diameter 40 mm and focal length $f_1 = 30$ cm. What should be the focal length f_2 and diameter D of the second lens to give a magnification of 15 and an angular field of view 5° in diameter?

3.2 Light from a star makes an angle of 0.01 rad with the axis of a simple telescope whose objective has a focal length of 50 cm and an eyepiece of focal length 2 cm. Calculate the distance D beyond the eyepiece at which the ray of light from the star crosses the axis of the telescope. (This is the eye relief.)

PROBLEMS 3

3.1 The microscopist Antoni van Leeuwenhoek (1632–1723) used a single lens to obtain magnifications up to ×200 or more. Find the diameter of a spherical glass bead that would give such a magnification. (Leeuwenhoek apparently fabricated biconvex lenses of such a small diameter.)

3.2 Show that the magnification of the astronomical telescope in Fig. 3.5 can be measured by placing a scale across the objective lens and measuring the magnification of this scale in the image formed by the eyepiece.

3.3 The immersion technique, in which a liquid fills the space between a microscope objective and a slide cover glass, can give improved image brightness by allowing more light to enter the objective. Calculate the improvement that can be obtained when the objective in air accepts a cone of half-angle 30° and the liquid and glass both have refractive index 1.5.

3.4 Compare the fields of view of the Gregorian and astronomical (Keplerian) telescopes of Fig. 3.6, in the following example. The objective diameters are both 2 cm, with focal lengths 20 cm, and the eyepiece diameters are both 1 cm, with focal lengths −10 and +10 cm, respectively. Show that the magnifications are +2 and −2, respectively, and the fields of view are 1/10 and 1/30 rad, respectively. Show that the exit pupil of the Gregorian telescope is located between the objective and the eyepiece lenses, whereas for the astronomical telescope it is beyond the eye lens.

3.5 A camera focused on infinity has a depth of focus depending on the focal ratio F, the focal length f and the acceptable image diameter d. Show that objects beyond a distance u_1 are in focus, where $u_1 = f^2/Fd$. Find this distance for $F = 2.5, f = 5$ cm, $d = 50$ μm.

If the camera is focused on u_1, what is the nearest object in focus? (Use the approximate analysis of Section 3.8.)

3.6 For binoculars specified as 8×40, with objective focal length 15 cm, what is (i) the magnification of a distant object, (ii) the focal length of the eyepiece, (iii) the diameter of the exit pupil, (iv) the angular field of view?

4

Periodic and non-periodic waves

Fourier, Jean Baptiste Joseph (1768–1830), French mathematician... born Auxerre... son of a tailor... soon distinguished himself as a student, and made rapid progress, delighting most of all, but not exclusively, in mathematics.

Encyclopaedia Brittanica, 9th Edn, 1898.

The propagation of light, and, in particular, its behaviour in interference and diffraction, is determined by its wave nature. Although the plane wave $\psi = f(z - vt)$, progressing in the positive z-direction with velocity v, may have any wave shape provided that it keeps the same shape as it progresses, it is both convenient and physically meaningful to concentrate on the simple harmonic waveform or sinusoidal wave introduced in Chapter 1:

$$\psi = A \cos k(z - vt). \tag{4.1}$$

Using the angular frequency ω and adding an arbitrary phase ϕ the wave becomes

$$\boxed{\psi = A \cos(kz - \omega t + \phi).} \tag{4.2}$$

Note that when $\phi = -\pi/2$ the wave becomes the sinusoidal wave of Eq. (1.13). We will also write the same wave as an exponential function

$$\boxed{\psi = A \exp i(kz - \omega t + \phi),} \tag{4.3}$$

which lends itself well to analytical work, although it represents less obviously the same wave and will require explanation in this chapter.

However it is expressed, the simple harmonic wave is a type of wave that is easily recognizable, as, for example, in a sound wave composed of a single pure note, or the monochromatic light from a laser. Familiarity with the various ways of representing and visualizing simple harmonic waves is essential to understanding the behaviour of light.

In this chapter we introduce the representation of a simple harmonic wave mathematically by complex exponential functions and graphically by a rotating vector known as a *phasor*. We show how simple harmonic waves are added, taking account of phase; this is at the heart of interference and diffraction phenomena in optics. We then show how any waveform can be built up by the addition of simple harmonic motions, using Fourier synthesis, or separated into component parts by Fourier analysis.

4.1   SIMPLE HARMONIC WAVES

The constants in Eq. (4.2) are the *angular frequency* ω (which is $2\pi\nu$ where ν is the *frequency*), the *wave number* k, and the *phase* ϕ. A cycle of oscillation occurs at time intervals of one *period* $= 2\pi/\omega$, and at distance intervals of one *wavelength* $\lambda = 2\pi/k$.

The velocity of the wave is $v = \omega/k$ and its form at any time is a simple sine or cosine wave along the z axis. The phase term ϕ determines the position of the cosine wave at $t = 0$ (Fig. 4.1). Adding $-\pi/2$ to ϕ makes the cosine wave into a sine wave, moving it along the z-axis by a quarter wavelength. A wave with an arbitrary phase can be expressed as the sum of cosine and sine components; for example, the cosine wave, Eq. (4.2) with phase ϕ, can be expanded[1] to give

$$\psi = A\cos\phi\cos(kz - \omega t) - A\sin\phi\sin(kz - \omega t). \qquad (4.4)$$

A great simplification in the treatment of periodic oscillations is achieved by introducing the idea of complex exponential frequency terms, allowing the sine and cosine terms in Eq. (4.4) to be included in one expression. We recall that a complex quantity like $A\exp i\phi$ has real and imaginary parts

$$A\exp i\phi = A\cos\phi + iA\sin\phi. \qquad (4.5)$$

We may therefore think of the cosine and sine components of a simple wave at any one frequency as a *complex amplitude*. The complex amplitude contains both the actual amplitude and the phase of the wave; it therefore behaves like a vector, which is often referred to as a *phasor*. Fig 4.2(a) shows a general phasor of length A at an angle $(kz - \omega t + \phi)$ to the axis. This phasor rotates at angular velocity ω in a clockwise direction. The component of the phasor along the horizontal axis is A $\cos(kz - \omega t + \phi)$, i.e. Eq. (4.2), and this is the physical quantity represented. Often phasors are drawn at $t = 0$ and $z = 0$, as in the example of Fig. 4.2(b).

---

[1] As a reminder: $\cos(\alpha \pm \beta) = \cos\alpha\cos\beta \mp \sin\alpha\sin\beta$, $\sin(\alpha \pm \beta) = \sin\alpha\cos\beta \pm \cos\alpha\sin\beta$.

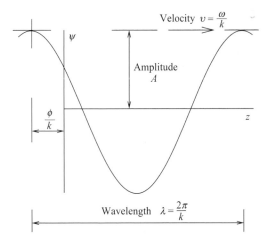

Fig. 4.1   The progressive cosine wave $\psi = A\cos(kz - \omega t + \phi)$.

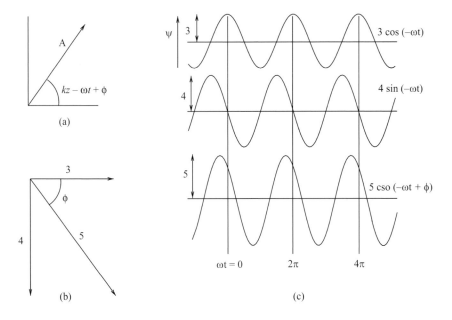

Fig. 4.2   (a) A general phasor of amplitude A at an angle of $(kz - \omega t + \phi)$ to the horizontal axis; (b) the addition of phasors at $z = 0$ and $t = 0$ representing $3\cos(kz - \omega t)$ and $4\sin(kz - \omega t)$ give a resultant phasor of length 5 at angle ϕ, representing $5\cos(kz - \omega t + \phi)$ where $\phi = -\tan^{-1}(4/3)$; (c) the two waves and their resultant as a function of ωt for $z = 0$.

Note that a phasor can represent any periodically varying quantity, which may itself be a scalar, such as pressure in a sound wave, or a vector, such as the electric and magnetic fields **E** and **B** in an electromagnetic wave. The phasor is a representation of the amplitude and phase of a harmonic wave. When two or more waves at the same frequency are superposed, each with its own amplitude and phase, the way they add depends on their relative phases: the *interference* between them may give an increased or decreased amplitude. This is done algebraically by adding the complex amplitudes, but it may be pictured by adding the corresponding phasors as vectors to produce a single phasor representing the combination in amplitude and phase.

To summarize this important concept: we can represent a harmonic wave in amplitude and phase by a complex number, the complex amplitude; summing the complex amplitudes for a combination of waves gives the complex amplitude for the combined wave.

It may seem perverse to represent a real oscillation or a wave by a complex mathematical quantity, but it turns out to be very convenient to think of amplitude and phase in this way. For example, to increase the phase of a wave by α is to rotate its phasor by angle α; this is achieved by multiplying the exponential by $\exp(i\alpha)$. Thus, for a wave with initial phase ϕ we have

$$
\begin{aligned}
A_{\text{rotated}} &= A\exp(i\phi)\exp(i\alpha) \\
&= A\exp i(\phi + \alpha).
\end{aligned}
\tag{4.6}
$$

A similar operation via the trigonometric formulation of Eq. (4.4) is not to be recommended.

The intensity of the wave is the square of the amplitude. It is obtained from the complex amplitude by removing the phase term and squaring the modulus, or by multiplying by the *complex conjugate* $A^* = A\exp(-i\phi)$, so that

$$
I = |A|^2 = AA^*.
\tag{4.7}
$$

We will see in Section 4.10 that any continuous waveform can be represented as the sum of simple cosine and sine waves. If the waveform is *periodic*, repeating at equal intervals of time with basic frequency ν, these will be a series of *harmonics* at frequencies that are integral multiples of ν. If the waveform is not truly periodic, but changes with time (and therefore with distance), then it can be constructed from a continuous spectrum of sine waves. The mathematical link between a waveform and its components is by way of Fourier analysis.

4.2   POSITIVE AND NEGATIVE FREQUENCIES

The phasors, and the complex numbers, of the previous section are concerned only with the amplitude and phase of a harmonic wave. Adding the frequency term is now simple: we multiply by the exponential $\exp(i\omega t)$, which rotates[2] the phasor at angular frequency ω. The result is a complex number $A\exp\{i(\omega t + \phi)\}$ representing an oscillation with amplitude A, frequency $\nu = \omega/2\pi$ and phase ϕ. For a travelling wave we have the equivalent of Eq. (4.2):

---

[2] Conventionally a positive rotation is anticlockwise.

$$\psi = A \exp\{i(kz - \omega t + \phi)\}. \tag{4.8}$$

The *exponential frequency term* exp $(-i\omega t)$ represents a point at unit distance from the origin in the complex plane, rotating clockwise around the origin $\omega/2\pi = \nu$ times per second. Similarly, an exponential frequency term in which the frequency is a *positive* quantity rotates the phasor in the opposite direction, continually adding phase rather than subtracting from it. Both signs are equally good mathematical representations of the same oscillation cos ωt, which is the real part both of exp $(i\omega t)$ and of exp$(-i\omega t)$. Why then do we bother with negative frequencies, when positive frequencies are equally useful? If we want to represent *any* function $f(x)$ by exponential components, we need frequencies with both positive and negative signs. A simple example is a cosine wave, where

$$\cos \omega t = \tfrac{1}{2}\{\exp(i\omega t) + \exp(-i\omega t)\}, \tag{4.9}$$

while a sine wave is represented by

$$\sin \omega t = \tfrac{1}{2i}\{\exp(i\omega t) - \exp(-i\omega t)\}. \tag{4.10}$$

A wave with intermediate phase, i.e. with both cosine and sine components, can be represented by a sum such as

$$\tfrac{1}{2}(a + ib)\exp(i\omega t) + \tfrac{1}{2}(a - ib)\exp(-i\omega t).$$

The two components now have *complex amplitudes* that are complex conjugates.

Extending this to a spectrum with a range of frequency components, a function $f(t)$ may be written as

$$f(t) = \tfrac{1}{2}a_0 + \int_{-\infty}^{+\infty} A(\omega)\exp(i\omega t)d\omega, \tag{4.11}$$

where $A(\omega)$ is the complex amplitude at frequency ω; a_0 is a constant (the amplitude at zero frequency).

Note that for $f(t)$ to be real, $A(-\omega) = A^*(\omega)$. The real and imaginary parts of the complex $A(\omega)$ together form the *complex spectrum* of the function $f(t)$, and may be plotted on two graphs, one for the real and one for the imaginary part. Fig. 4.3 shows the spectra of three periodic waves in this way. Notice that there is no need always to plot the negative frequency half of these spectra, because of the conjugate property.

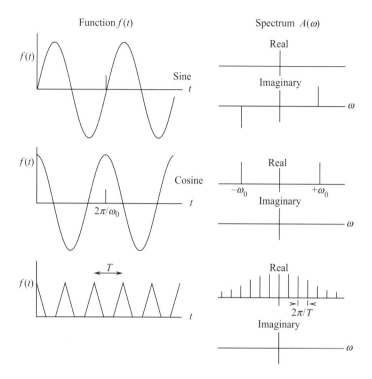

Fig. 4.3 Functions and their spectra. The functions themselves are real, but their spectra have in general both real and imaginary parts. The cosine has only real components, and the sine has only imaginary components. The chopped triangular function is an even function, like a cosine, so its spectrum is real, made up of lines spaced at 1/T. As its average is not zero it has a zero frequency component.

4.3   STANDING WAVES

The simplest example of two waves with the same frequency adding with varying phase is given by two cosine waves travelling in opposite directions adding to give an interference pattern of *standing waves*. This may be seen in water waves reflected from a pond wall, or heard in sound waves; it is often conspicuous in VHF radio (the FM band) where standing wave patterns inside a room may be explored by moving a portable receiver. Two waves with equal amplitudes travelling in the directions $+z$ and $-z$ add as

$$\psi = A\cos(kz - \omega t) + A\cos(-kz - \omega t). \tag{4.12}$$

It is a useful exercise to write this as the sum of waves represented by the exponential terms $A\exp{-i(\omega t \pm kz)}$, obtaining

$$\psi = A\exp(-i\omega t)\{\exp{-ikz} + \exp{ikz}\}$$
$$= 2A\cos kz \exp{i\omega t}. \tag{4.13}$$

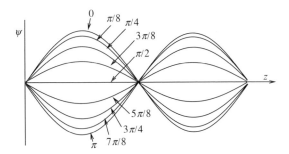

Fig. 4.4   The envelope of the oscillation in a standing wave pattern, with successive plots at intervals of one-sixteenth of the period.

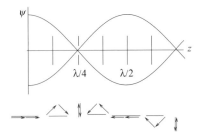

Fig. 4.5   Phasor diagrams for the standing wave pattern formed by two waves of equal amplitudes travelling in opposite directions. The phasor diagrams show the waves in phase at $z = 0$, with the resultant amplitude depending on the relative phase as z increases.

This is a wave with the same phase everywhere, but with an amplitude varying with distance z. Fig. 4.4 shows the envelope pattern of the standing wave, with the actual displacement ψ at intervals of one-sixteenth of the period, i.e. at phase intervals of $\pi/8$. The amplitude of the superposition varies along the z-axis as the relative phase of the two component waves changes. If the oscillations are in phase at $z = 0$, the phase of one wave increases and the phase of the other decreases as kz, as indicated in Fig. 4.5. The phase reference is the phase of the oscillation at $z = 0$. Equation (4.13) shows that the sum of the two waves gives a maximum at $z = 0$, falling as $\cos kz$ to zero at $z = \lambda/4$ and increasing to a maximum at $z = \lambda/2$. The successive minima and maxima are called *nodes* and *antinodes*.

The pattern of standing waves provides a simple example of the phasor representation of amplitude and phase. Figure 4.5 shows the phasors for the two waves at intervals of $\lambda/8$ along the z-axis. Starting at an antinode with two waves in phase at $z = 0$ and moving to larger values of z, the phase difference increases in steps of $\pi/4$. The sum of the two vectors decreases to zero to give the first node, and then increases to give the next antinode at $z = \lambda/2$; here, the

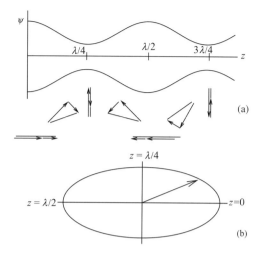

Fig. 4.6 The envelope of the standing wave pattern, with corresponding phasor diagrams (a), for waves of unequal amplitude. The resultant phasor traces out an ellipse (b).

phasor is seen to be rotated through angle π compared with the first antinode, i.e. there is a phase change of π.

The standing wave pattern for waves of unequal amplitude does not have zero amplitude at the nodes. Figure 4.6 shows the envelope of the standing wave pattern, with the phasor diagrams at intervals of $\lambda/8$ along the z-axis. The phasor representing the standing wave then traces an ellipse as it varies along the z-axis. The major and minor axes represent amplitudes at antinodes and nodes; for equal amplitudes the ellipse degenerates into a straight line.

4.4   BEATS BETWEEN OSCILLATIONS

The addition of two oscillations with slightly different frequencies gives the effect of beating, which is familiar in sound waves. This is closely analogous to the standing wave patterns of the previous section, with the relative phases of two oscillations varying with time rather than with distance. The addition of two sinusoids, one of which is smaller in amplitude and which has a slowly increasing phase, is shown in Fig. 4.7(a). Here, the phase of the larger oscillation is taken as the reference phase, so that the phasor representing the smaller oscillation rotates; the tip of the phasor representing the sum oscillation traces a small circle as the relative phase changes. Although the amplitude of the sum varies sinusoidally, the phase does not, as shown in Figs 4.7(b) and (c).

The phase reference has been taken as one of the two oscillations, with the phase of the other increasing uniformly with time. One oscillation then has angular frequency ω, and the other has $\omega + \Delta\omega$, and the beat frequency is $\Delta\omega/2\pi$. A more convenient phase reference may be taken as that of an

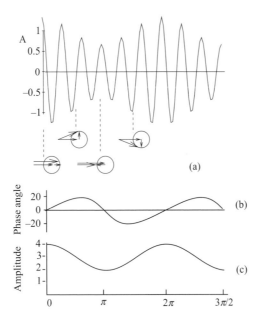

Fig. 4.7   (a) The sum of two sinusoidal oscillations with different amplitudes and with a slowly changing relative phase corresponding to slightly different frequencies, showing phasor diagrams; (b) the phase variations relative to the phase of the larger oscillation, shown together with (c), the amplitude variations.

oscillator with a frequency halfway between these two, so that we are adding angular frequencies $\omega - \Delta\omega/2$ and $\omega + \Delta\omega/2$. The time variation of the phasors now looks like the spatial variation in the standing wave pattern of Fig. 4.5. For equal amplitudes the phase of the resultant is now constant for half a period, reversing at the instants of zero amplitude.

4.5   SIMILARITIES BETWEEN BEATS AND STANDING WAVE PATTERNS

The common use of the phasor diagram to illustrate the phenomena of beats and of standing waves demonstrates their underlying similarity. Beats are variations of amplitude with time at one point, whilst standing waves are variations of amplitude at different positions at one time. Both phenomena can, in fact, be produced simultaneously in very simple circumstances. In Fig. 4.8 two sources of sinusoidal waves S_1, S_2 are separated by a distance of several wavelengths, so that the distances S_1X, S_2X to a point X may differ by anything from zero to several wavelengths. Such a situation may occur, for example, with two sources of sound, or with two radio transmitters. At X the two waves add, with a phase relation that depends on the relative phase of the sources S_1 and S_2, and on the difference $S_1X - S_2X$ expressed in terms of the wavelength λ.

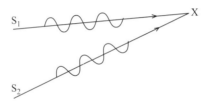

Fig. 4.8 Waves from the two sources S_1, S_2 reach X by different path lengths. As X moves it explores an interference pattern: alternatively, if S_1 and S_2 transmit different frequencies, beats will be heard at X.

Beats can now be produced at X by keeping the geometric arrangement fixed, and transmitting two different frequencies from S_1 and S_2. Alternatively, the two transmitters can be set to the same frequency, and arranged to transmit exactly in phase. Then, if $(S_1X - S_2X) = n\lambda$, the waves will arrive in phase at X, while if at another point X' the path difference $(S_1X' - S_2X') = (n + \frac{1}{2})\lambda$, the waves will arrive there out of phase. There is, therefore, a pattern of waves resulting from the interference of the waves, with maxima and minima of amplitude following a simple geometric pattern.

Now let the phase of S_1 change slowly with respect to S_2. The result can be described in two ways: either the interference pattern is moving, or at each point there is a beat between the two transmitters. The physical situation can be reversed so that X represents a single source or transmitter and S_1 and S_2 represent two receivers connected together in such a way that the relative phase of the two waves they receive determines the sum of their signals. This situation occurs in optical and radio interferometers, such as the Michelson stellar interferometer. The radio interferometer, as used in radio astronomy, collects waves from a single radio source in two separate antennas, adding them in a single receiver. As the Earth rotates, a celestial source moves across the interferometer, the path difference changes, and the receiver output varies. Alternatively, in the addition of the two waves an extra phase difference can be inserted deliberately, and if this increases steadily with time, then the rate of variation can be adjusted to compensate for the rotation of the Earth.

The basic equivalence in these examples is that an addition or subtraction of phase linearly with time in a sinusoidal oscillation is equivalent to a change of frequency.

4.6   STANDING WAVES AT A REFLECTOR

Standing wave patterns can most easily be demonstrated by arranging for the total reflection of a plane wave. A classic demonstration of the standing waves of light reflected by a mirror was carried out in 1891 by Lippmann. He used a photographic plate with a very fine grain, backed with a layer of mercury to act

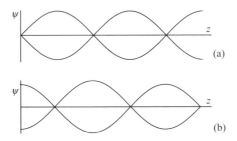

Fig. 4.9 Standing wave patterns due to reflections at different types of boundary: (a) zero amplitude at the boundary; (b) maximum amplitude at the boundary.

as a smooth reflector. A plane wave of monochromatic light, falling on the plate, formed a standing wave that could be seen in the developed film by cutting a section at a very shallow angle. Lippmann also showed that a plate exposed and developed in this way could be used for colour photography, since, as explained below, light was reflected selectively according to its wavelength from the planes of silver left in the emulsion. Coloured holographic images, which we describe in Chapter 19, are based on the same concept of selective reflection from a three-dimensional negative, made by interference between light beams within a photographic emulsion.

In an emulsion of 20 μm thickness, since the interference planes are separated by $\lambda/2$, some 40 interference planes are formed, depending on the wavelength. This array acts as a selective reflective filter for white light. A similar phenomenon of selective reflection is found in X-ray diffraction at planes of atoms within a crystal (Chapter 10).

The Lippmann demonstration can now be repeated much more easily by using radio waves with wavelengths of a few centimetres; but it had a particular historical importance in that it showed not merely the wave pattern but the way in which the pattern was related to the reflecting surface. The two standing wave patterns in Fig. 4.9 show the patterns obtained at two different kinds of boundaries. This difference is well known in wind instruments: the resonant oscillations in an organ pipe have a node or antinode at the end of the pipe for closed and open pipes, respectively. For electromagnetic waves the relevant boundary conditions are determined by the dielectric and magnetic properties of the material at the boundary. In 1891 it was still interesting to prove that a metal surface, being an excellent conductor, would determine that there would be an essentially zero electric field at the surface, giving the pattern of Fig. 4.9(a). The Lippmann films showed this clearly, giving layers of silver starting a one-quarter wavelength above the surface. The pattern of Fig. 4.9(a) is produced by two waves out of phase at the surface, while in Fig. 4.9(b) the two waves are in phase at the surface. The properties of the surface determine the relation between the incident and the reflected waves, i.e. the reflection coefficient, as in Chapter 5.

The concepts of interference in space and in time are well illustrated in the Doppler radar system used for measuring the speed of moving aeroplanes or automobiles, which we discuss after setting out the theory of the Doppler effect itself.

4.7   THE DOPPLER EFFECT

The Doppler effect is familiar as a change in pitch of a sound as the source or observer moves. It applies over the whole of the electromagnetic spectrum, but for light, in particular, it is important on physical scales from atomic to cosmic. On the atomic scale we shall be concerned with the spread in frequency of spectral lines due to the thermal velocities of atoms or molecules in a gas (Chapter 13), and on the cosmic scale we observe the *redshift* of spectral lines from galaxies receding with velocities up to at least $0.95c$. We show how the simple theory of the Doppler effect can be refined to take account of such large velocities by incorporating the theory of special relativity.

From the point of view of a stationary observer, a source emitting ν_0 waves per second and moving towards the observer with velocity v will compress the ν_0 waves into a distance $v - u$, where u is the velocity of the waves, as in Fig. 4.10. The frequency will therefore be seen by the observer as

$$\nu_{obs} = \frac{\nu_0}{1 - v/u}. \tag{4.14}$$

Similarly, an observer moving towards a stationary source with velocity v will receive a frequency increased by the rate at which the observer covers wavelengths of distance. The observed frequency is therefore

$$\nu_{obs} = \nu_0 + \frac{v}{\lambda} \quad \text{or}$$

$$\nu_{obs} = \nu_0 \left(1 + \frac{v}{u} \right). \tag{4.15}$$

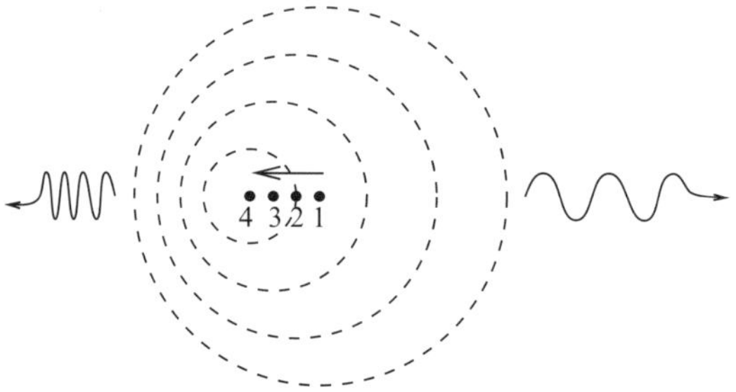

Fig. 4.10   The Doppler effect. A moving source emits a periodic wave, represented by the broken circles. The waves are bunched together in the direction of motion and spread out behind.

These two equations, Eq. (4.14) and (4.15), are nearly identical for small velocities, but they differ increasingly at large velocities. For light, however, unlike sound, there can be no difference between motion of the source and motion of the observer, which must give the same Doppler shift. The two equations are reconciled by a relativistic correction, which results from the different ways in which the source and the observer measure frequency. Time as measured by a moving clock is dilated as compared with time measured by the same clock when it is stationary. According to special relativity, an atomic oscillator at frequency ν_0, moving with velocity v relative to a stationary observer, will appear to the observer to be oscillating at a lower frequency ν_1, given by

$$\nu_1 = \nu_0 \left(1 - \frac{v^2}{c^2}\right)^{1/2}. \tag{4.16}$$

The observed frequency in Eq. (4.14) therefore becomes

$$\nu_{obs} = \nu_0 \left(\frac{1 + v/c}{1 - v/c}\right)^{1/2}. \tag{4.17}$$

It is easy to verify that the same result is obtained for the moving observer, since the observer's *clock* goes slow by the same relativistic factor. In terms of wavelength, since $\lambda = c/\nu$

$$\lambda_{obs} = \lambda_0 \left(\frac{1 - v/c}{1 + v/c}\right)^{1/2}. \tag{4.18}$$

The full relativistic formula is essential in the context of astronomical measurements of distant galaxies, which may be receding with velocities approaching the velocity of light. The observed wavelengths λ_{obs} of spectral lines are *redshifted* from their original wavelength λ_0 by a factor of up to 5, corresponding to a velocity $v = 0.96c$ (note that the non-relativistic formula would indicate a velocity $v = 0.8c$).

4.8 DOPPLER RADAR

In a Doppler radar (Fig. 4.11), one form of which is the radar used by police for measuring the speed of traffic, a transmitter T sends out a constant sinusoidal wave; say, with wavelength $\lambda = 3$ cm (frequency $\nu = 10$ GHz). The reflected wave is received at R and added to part of the transmitted wave; the relative phase of these two waves is determined by the distance TXR. If the target X moves with velocity v, this distance changes at a rate $2v$, and the beat frequency will be $2v/\lambda$. For wavelength of 3 cm and a velocity $v = 50$ km h$^{-1} \simeq 14$ m s^{-1},

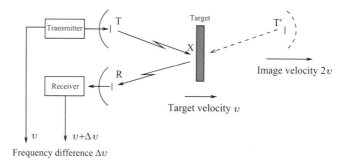

Fig. 4.11 A Doppler radar system. The reflected wave from a target receding with velocity v is at a frequency lower by $2v/\lambda$. This is measured as a beat against the transmitted frequency.

the beat frequency is 93 Hz; this may be used to give a direct reading of the velocity of the target.

Alternatively, we can consider the transmitted and reflected waves as adding to form a standing wave pattern in front of the moving reflecting target. The beat frequency is then the effect of the standing wave pattern moving past the radar receiver. Another way of looking at the same problem is also shown in Fig. 4.11, where the signal reaching the receiver may be considered to have originated in an image T′ of the transmitter, which moves with velocity $2v$ away from the radar. The beat is now between the frequency ν and the Doppler shifted frequency $\nu(1 - 2v/c)$, giving a beat frequency $2v/\lambda$ as before.

Doppler radar measurements give velocity, not distance. Distance, or range,[3] is measured by the time of flight of a reflected radar pulse, which travels at the group velocity.

4.9   ASTRONOMICAL ABERRATION

Astronomers are familiar with an effect in the propagation of light that is related to the Doppler shift but is purely geometrical and does not depend on wavelength or frequency. If a source of light appears to be in a certain direction, and if the observer then starts to move *transverse* to this direction, what change does he see? The change is easily assessed for slower wave motions, such as water waves seen from a moving boat, since we have only to compound the observer's motion with the wave motion. A vector sum of the two velocities gives the effective motion of the waves past the boat. The effect is observed as a periodic shift in the position of stars as the Earth follows its orbit round the Sun, known as *astronomical aberration* and illustrated in Fig. 4.12. As for the Doppler frequency shift, in the previous section, the calculation should include

---

[3] *Radar* is an acronym for radio detection and ranging.

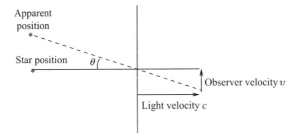

Fig. 4.12 Astronomical aberration. An observer moves transverse to a light wave at velocity v. The vector sum of their velocities gives an angular shift $\theta \approx v/c$.

a relativistic correction if the velocities are large, but the simple vector sum for velocity v transverse to light with velocity c gives an angular shift θ given by

$$\tan \theta = \frac{v}{c}. \tag{4.19}$$

The relativistic correction concerns the measurement of the angle θ. The analysis so far has been carried out from the point of view of a stationary observer. The moving observer uses a scale of length that is contracted by the relativistic factor $(1 - v^2/c^2)^{1/2}$ in the direction of motion. The observer therefore measures an angle θ given by

$$\tan \theta = \frac{v}{c} \left(1 - \frac{v^2}{c^2}\right)^{1/2}, \tag{4.20}$$

which reduces to the simple relation

$$\sin \theta = \frac{v}{c}. \tag{4.21}$$

For a star in the Earth's orbital plane the maximum peak-to-peak amplitude is 20.4 arcsec. Astronomical aberration was discovered by James Bradley in 1725, when he was attempting to measure stellar parallax, a perspective effect in which the position of a star varies according to the position of the Earth in its orbit rather than its velocity. The discovery of aberration was important in establishing both the finite velocity of light and the orbital motion of the Earth.

4.10   FOURIER SERIES

A simple harmonic oscillation with an infinite extent in time and space is an idealised concept; in practice, we deal with waves covering a range of frequencies and also travelling in various directions. To handle these cases we need to add a *spectrum* of waves distributed in frequency and in angle. Both frequency

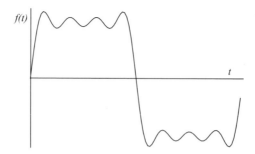

Fig. 4.13 A periodic square wave built up from a harmonic series of sine waves, including only the fundamental with the 3rd, 5th and 7th harmonics.

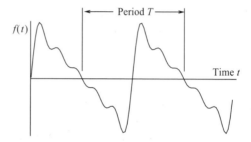

Fig. 4.14 Sawtooth wave built up from Fourier components, including only the fundamental with the 3rd, 5th and 7th harmonics.

spectra and angular spectra are conveniently expressed in Fourier terminology, which we explore first in the domain of waveform and frequency.

Figure 4.13 shows how a periodic square waveform may be built up from a *harmonic series*[4] of sine waves; only the odd harmonics are needed, with amplitudes $2/\pi n$. Only a small number of harmonics are needed to produce a recognizable square wave, although in theory a sharply defined square wave requires an infinite series. Note that all the components are in phase at the rising edge of the square wave. A different periodic wave, the sawtooth wave of Fig. 4.14, can be constructed from sine components with amplitudes $a_n = 1/\pi n$, where n is an integer. The more components that are added, the sharper is the sawtooth waveform.

The square wave, using sine harmonics, and the sawtooth, using cosine harmonics, are examples of a more general theorem due to Fourier. This states that any *periodic* function $f(t)$ that repeats with period T can be expressed in terms of a harmonic series of sine and cosine waves as

$$f(t) = \sum_{n=-\infty}^{+\infty} \left[\frac{a_n}{2} \cos\left(\frac{2\pi nt}{T}\right) + \frac{b_n}{2} \sin\left(\frac{2\pi nt}{T}\right) \right]. \tag{4.22}$$

---

[4] Until now we have used the term *harmonic* to mean a single angular frequency ω. A harmonic series is a sum of terms with frequencies $n\omega$, where n is any integer.

The *Fourier synthesis* of a periodic waveform consists of adding together a fundamental frequency component and harmonics of various amplitudes. Since $\cos\theta = \cos(-\theta)$, and $\sin\theta = -\sin(-\theta)$, the Fourier series can be written equivalently as

$$f(t) = \tfrac{1}{2}a_0 + \sum_{n=1}^{+\infty} a_n \cos\left(\frac{2\pi nt}{T}\right) + \sum_{n=1}^{+\infty} b_n \sin\left(\frac{2\pi nt}{T}\right). \qquad (4.23)$$

The derivation of the frequencies and amplitudes of the components of a periodic waveform is *Fourier analysis*. This is achieved for the harmonic series of Eq. (4.23) as follows. The coefficients a_n and b_n are obtained by multiplying Eq. (4.23) by $\cos(2\pi nt/T)$ and $\sin(2\pi nt/T)$ respectively and integrating over one period:

$$a_n = \frac{2}{T}\int_0^T f(t)\cos\left(\frac{2\pi nt}{T}\right)dt,$$

$$b_n = \frac{2}{T}\int_0^T f(t)\sin\left(\frac{2\pi nt}{T}\right)dt. \qquad (4.24)$$

Note that any constant term in $f(t)$ appears as $\tfrac{1}{2}a_0$, while b_0 is always zero.

In the exponential notation the Fourier series in Eq. (4.23) becomes

$$f(t) = \frac{a_0}{2} + \sum_{n=1}^{\infty}(a_n - ib_n)\exp\left(i\frac{2\pi nt}{T}\right). \qquad (4.25)$$

So far we have considered functions in time, and Fourier harmonic series involving time and frequency. But we can also have periodic functions in space, such as the diffraction gratings described in Chapter 10. Any periodic function in the space domain with period L may be represented by the Fourier series

$$f(x) = \tfrac{1}{2}a_0 + \sum_{n=1}^{+\infty} a_n \cos\left(\frac{2\pi nx}{L}\right) + \sum_{n=1}^{+\infty} b_n \sin\left(\frac{2\pi nx}{L}\right). \qquad (4.26)$$

The coefficients are then

$$a_0 = \frac{2}{L}\int_0^L f(x)dx,$$

$$a_n = \frac{2}{L}\int_0^L f(x)\cos\left(\frac{2\pi nx}{L}\right)dx,$$

$$b_n = \frac{2}{L}\int_0^L f(x)\sin\left(\frac{2\pi nx}{L}\right)dx. \qquad (4.27)$$

This spatial form of the Fourier series and Fourier transform occurs in many areas of optics and photonics.

4.11   MODULATED WAVES: FOURIER TRANSFORMS

Beats and standing waves are simple examples of *modulated waves*, whose amplitude varies periodically with time or space. Fourier theory also allows us to analyse non-periodic, i.e. *aperiodic*, modulation in the same way. A short burst of waves, such as a pulse of laser light, can be considered as the sum of waves with a continuous range of frequencies rather than a single frequency. In general, both periodic and aperiodic modulation of a cosine wave can be expressed in terms of a *modulating function g(t)*, so that the wave is $g(t)\cos(2\pi\nu_1 t)$. If the modulating function $g(t)$ is a sinusoid, the wave can be decomposed into two components with frequencies above and below ν_1; these would, for example, be the two frequencies producing a beat at the modulating frequency. If the wave is not fully modulated, so that the amplitude does not go to zero, there is also a component at ν_1; the spectrum of the modulated wave then consists of a *carrier* and two *sidebands*.

Fourier analysis is a general technique for relating the form of a variable to its spectrum; for example, it relates the spectrum of a sound to its actual waveform, and it gives a precise description of the wavelength (or frequency) components in a pulse of laser light. The relation, which was set out in Eq. (4.23) for the discrete harmonic components of a periodic wave, must now be extended to include a continuous spectrum and aperiodic modulation. The relation between a variable $f(t)$ and its frequency spectrum $F(\nu)$ is the integral

$$f(t) = \int_{-\infty}^{+\infty} F(\nu)\exp(2\pi i\nu t)\mathrm{d}\nu. \tag{4.28}$$

This integral expresses the fact that the waveform can be made up of an infinite set of frequency components, constituting a spectrum. There is an inverse relation, which shows how the spectrum can be derived from the waveform; this is

$$F(\nu) = \int_{-\infty}^{+\infty} f(t)\exp(-2\pi i\nu t)\mathrm{d}t. \tag{4.29}$$

These *Fourier transforms* apply equally well to non-periodic as to periodic variables. We apply them first to the modulated wave $g(t)\cos(2\pi\nu_1 t)$, whose spectrum is found from Eq. (4.29) as follows:

$$
\begin{aligned}
F(\nu) &= \int_{-\infty}^{+\infty} g(t)\cos(2\pi\nu_1 t)\exp(-2\pi i\nu t)\mathrm{d}t \\
&= \int_{-\infty}^{+\infty} g(t)\tfrac{1}{2}\{\exp[-2\pi i(\nu - \nu_1)t] + \exp[-2\pi i(\nu + \nu_1)t]\}\mathrm{d}t \\
&= \tfrac{1}{2}G(\nu - \nu_1) + \tfrac{1}{2}G(\nu + \nu_1),
\end{aligned}
\tag{4.30}
$$

where $G(\nu)$ is the Fourier transform of the modulating function $g(t)$.

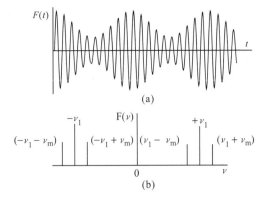

Fig. 4.15   Cosinusoidal modulated wave (a) and its spectrum (b).

The simplest example is the spectrum of a cosinusoidally modulated wave, as in Fig. 4.15(a). The full spectrum of the unmodulated wave (the carrier wave) has components at ν_1 and $-\nu_1$; the spectrum of the modulating cosine wave at any frequency ν_m similarly has components at $\pm\nu_m$. The resulting spectrum of the modulated wave is shown in Fig. 4.15(b); there are now sidebands separated from the original components by ν_m. This result is as expected from the consideration of beating between two cosine waves, which are now seen as the sidebands on either side of the carrier.

4.12   MODULATION BY A NON-PERIODIC FUNCTION

In Fig. 4.16 the carrier oscillation is limited in time by a time-limited modulation function $f(t)$, which is shown as either a *Gaussian* or the abrupt *top-hat* function. Following Eq. (4.29) the spectrum of either of these time-limited waves is found from the spectrum of the modulating function. A top-hat function with height h and width b, centred on the origin at $t = 0$ is written

$$f(t) = h\left(-\frac{1}{2}b < t < +\frac{1}{2}b\right),$$

$$f(t) = 0 \text{ elsewhere.}$$

(4.31)

The Fourier integral equation (4.29) then gives the spectrum

$$F(\nu) = \int_{-b/2}^{+b/2} h\exp(-2\pi i\nu t)\,dt$$

$$= \frac{h}{2\pi i\nu}\left\{\exp\left(+2\pi i\nu\frac{b}{2}\right) - \exp\left(-2\pi i\nu\frac{b}{2}\right)\right\}$$

$$= hb\frac{\sin\psi}{\psi} = hb\,\text{sinc}\,\psi, \quad \text{where} \quad \psi = \pi\nu b.$$

(4.32)

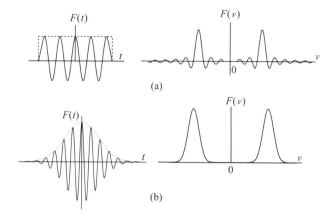

Fig. 4.16 A wave whose duration is limited by (a) a 'top-hat' function (b) a Gaussian function.

This Fourier transform of a top-hat function is a *sinc function*. Figure 4.16 shows the full spectrum of the time-limited wave, with positive and negative components each with the shape of a sinc function. The width of the sinc function is inversely proportional to the width of the top-hat.

The sinc function is frequently encountered in physics and in communication engineering. We shall see later that the spectrum of a waveform abruptly started and stopped has an intrinsic width inversely proportional to the length of the wavetrain; it also has sidebands that extend on either side of the main spectral component. A smooth modulation, avoiding the abrupt start and stop, has a wider main spectral component but lower side-lobes. The most important example of such a smooth modulating function is a *Gaussian*. The Gaussian function in Fig. 4.16 is written

$$f(t) = h \exp\left(-\frac{t^2}{\sigma^2}\right). \tag{4.33}$$

Evaluation by the same process, using the identity

$$\int_{-\infty}^{+\infty} \exp\left(-a^2 x^2\right) dx = \sqrt{\pi}/a \dots (a \neq 0) \tag{4.34}$$

gives the spectrum

$$F(\nu) = h\sigma\sqrt{\pi} \exp\left(-\pi^2 \nu^2 \sigma^2\right). \tag{4.35}$$

The transform of a Gaussian function is therefore another Gaussian, whose width $1/\pi\sigma$ is inversely proportional to the original width σ. Applying the general relation Eq. (4.30) gives the spectra shown in Fig. 4.16. We see that the Gaussian modulation of an oscillation gives Gaussian spectral lines, whose width is inversely proportional to the duration of the wavetrain. Similar analyses may be applied to a wave limited in space, showing that the spectrum of a wave group, such as the group of waves associated with a single particle or photon, depends on the length of the wave group. Extreme examples of short wave groups are found in pulsed light from lasers, where the pulse may be only a few wavelengths long, and consequently has a very wide spectral range (Chapter 17).

Gaussian modulating functions are also encountered in the lateral spread of concentrated light beams, and especially those from lasers. The lateral spread of the beam is related by a Fourier transform to the angular divergence of the beam, which is similarly described by a Gaussian function (see Chapter 15).

4.13   CONVOLUTION

We now state the convolution theorem that enables us to find the Fourier transform of a further class of functions: those that are obtainable by convolving together two functions, say $f(t)$ and $g(t)$. The convolution $C(t)$ of two functions $f(t)$ and $g(t)$ is defined by the equation

$$C(t) = \int_{-\infty}^{+\infty} f(\tau)g(t - \tau)\mathrm{d}\tau. \qquad (4.36)$$

This equation is often written symbolically as

$$C(t) = f(t) * g(t). \qquad (4.37)$$

The convolution equation is useful in Fourier analysis of any function, whether it is of time, distance or angle. It occurs naturally in the response of any optical instrument, such as a telescope or spectrometer that is intended to 'resolve' light either according to direction or wavelength; any such instrument has a limited resolving power that inevitably modifies and degrades the image or spectrum. For example, a photograph of a point source taken by an astronomical telescope appears to show the light originating from an extended source, which is the result of diffraction. Let a cross-section of this apparent source have brightness distribution $g(\theta)$. Then a photograph of an object that has an actual brightness distribution $f(\theta)$ has a blurred image $h(\theta)$ made by a combination of the two functions. The image at θ is made up of contributions from a range of angles covered by the blurring function. At a point α the true brightness is $f(\alpha)$, while the blurring function centred on the point θ is $g(\theta - \alpha)$. The resultant at θ is the integral over α:

$$h(\theta) = \int_{-\infty}^{+\infty} f(\alpha)g(\theta - \alpha)\mathrm{d}\alpha. \qquad (4.38)$$

This is a convolution of the source function $f(\alpha)$ with the instrumental response $g(\theta)$.

We now write the Fourier transforms of the functions $f(t)$, $g(t)$, $h(t)$ as $F(\nu)$, $G(\nu)$, $H(\nu)$; using the definitions of convolution and Fourier transform it can easily be shown that if $h(t) = f(t) * g(t)$, then

$$H(\nu) = F(\nu)G(\nu).$$ (4.39)

This is the convolution theorem, which may be stated as follows:

The Fourier transform of the convolution of two functions is the product of their individual transforms.

We will apply the convolution theorem in diffraction theory (Chapter 7), where the functions $f(t)$, etc. are replaced by amplitude distributions across a diffraction aperture; their transforms then represent the corresponding diffraction patterns.

4.14 DELTA AND GRATING FUNCTIONS

When a single pulse becomes infinitely narrow, its transform becomes infinitely wide. It is convenient to describe an infinitely narrow function as a *delta function*. We may, for example, describe the envelope of a very short pulse of light travelling along an optical fibre as a delta function, implying that it has an infinitely wide spectrum and can be used in a communication circuit with a very wide bandwidth. The broadening of the pulse as it travels, or in the detector circuits, can be regarded as a series of convolution processes, with a corresponding reduction in bandwidth and limitation of usefulness of the communication circuit.

The delta function $\delta(x - x_0)$ is an infinitely short function centred on x_0, and with unit area. For any reasonable function $f(x)$ its convolution with a delta function takes the form:

$$\int_a^b f(x)\delta(x - x_0)\mathrm{d}x = f(x_0) \quad \text{for} \quad a < x_0 < b$$

$$= 0 \text{ otherwise.}$$ (4.40)

A regular sequence of delta functions is known as the *grating function*. This may be regarded as a periodic function with an infinite series of harmonics: in fact, its Fourier transform is another infinite sequence of delta functions. The periodicities in the grating function and its transform are related inversely.

4.15   AUTOCORRELATION AND THE POWER SPECTRUM

The full description of a spectrum must contain the amplitude and the phase of all components. However, it is often only necessary to consider intensity (or power, or luminance, for example), which is proportional to the square of the amplitude and contains no phase information. This intensity distribution may also be expressed as a spectrum and is usually referred to as the *power spectrum*. The power is a real quantity; for a harmonic component $F(\nu)$, which may be a complex quantity, we obtain the power by multiplying by the complex conjugate[5] $F^*(\nu)$.

If the amplitude $A(t)$ of a time-varying quantity is convolved with itself, the result is the Fourier transform of its power spectrum. Although this follows from Section 4.13, the result is so valuable that we set out a proof as follows.

Convolving a function with itself, or self-convolution, is also known as an autocorrelation. We define the autocorrelation function Γ_{11} as

$$\Gamma_{11}(\tau) = \int A(t+\tau)A^*(t)\mathrm{d}t. \tag{4.41}$$

Now take the Fourier transform of this and manipulate

$$
\int \Gamma_{11}(\tau)\exp(2\pi i\tau\nu)\mathrm{d}\tau
$$

$$
= \int\left\{\int A(t+\tau)A^*(t)\mathrm{d}\tau\right\}\exp(2\pi i\tau\nu)\mathrm{d}\tau
$$

$$
= \left\{\int A^*(t)\exp(-2\pi i\tau\nu)\mathrm{d}\tau\right\} \times \left\{\int A(t+\tau)\exp\{2\pi i(t+\tau)\nu\}\mathrm{d}\tau\right\}
$$

$$
= F^*(\nu)F(\nu). \tag{4.42}
$$

This is the power spectrum.

4.16   WAVE GROUPS

Modulated waves and, in particular, wave groups such as those of Fig. 4.16(b), are of great importance in many branches of physics. In view of this we now consider modulated waves in a simple physical way, so as to illuminate the mathematical results of the Fourier approach. We start with two wave components only. The addition of two waves travelling in the $+z$ direction with equal amplitude a but different angular frequencies $\omega \pm \Delta\omega/2$ and wave numbers $k \pm \Delta k/2$ is expressed as

---

[5] The complex conjugate of $A + iB$ is $A - iB$, and the product is $A^2 + B^2$.

$$y = a \exp\left[i\left\{\left(\omega + \frac{\Delta\omega}{2}\right)t - \left(k + \frac{\Delta k}{2}\right)x\right\}\right] + a \exp\left[i\left\{\left(\omega - \frac{\Delta\omega}{2}\right)t - \left(k - \frac{\Delta k}{2}\right)x\right\}\right]$$

$$= a \exp\{i(\omega t - kx)\}\left[\exp\left\{i\left(\frac{\Delta\omega t}{2} - \frac{\Delta kx}{2}\right)\right\} + \exp\left\{-i\left(\frac{\Delta\omega t}{2} - \frac{\Delta kx}{2}\right)\right\}\right]$$

$$= 2ia \sin\left(\frac{\Delta\omega t - \Delta kx}{2}\right) \exp\{i(\omega t - kx)\}. \tag{4.43}$$

The exponential term is a wave at the centre frequency, and the sine term is a modulation of the wave in time at angular frequency $\Delta\omega/2$ and in space with wave number $\Delta k/2$. The real part of Eq. (4.43), shown in Fig. 4.16(b) is

$$y_{\text{real}} = -2a \sin\left(\frac{\Delta\omega t}{2}\right) \sin(\omega t - kz). \tag{4.44}$$

The wave at the centre frequency moves as before with a velocity

$$v = \frac{\omega}{k}. \tag{4.45}$$

This is known as the *phase velocity* of the group. Note that the argument of $\sin(\omega t - kz)$ is constant. The modulation moves with a different velocity, such that $\sin(\Delta\omega t - \Delta kz)$ is constant; this is known as the *group velocity* v_g, given by the ratio

$$v_g = \frac{\Delta\omega}{\Delta k}. \tag{4.46}$$

Any pair in a group of waves can be analysed in this way, so we may deduce that the whole group will move with the same velocity as the sinusoidal modulation pattern. In the limit, a group must be considered as an infinite series of waves all with angular frequencies and wave numbers near ω and k. The group velocity v_g is then the differential

$$\boxed{v_g = \mathrm{d}\omega/\mathrm{d}k.} \tag{4.47}$$

Note the distinction between group velocity and the phase velocity ω/k.

The limitation of the duration of a wave, or the limitation of its extent in space, requires the superposition of an infinite series. Two waves differing by $\Delta\nu$ in frequency reinforce over a time $\tau \approx 1/\Delta\nu$, which is the time between successive beat minima. A group lasting for time τ must consist of sinusoidal waves spread over a range $\Delta\nu \approx 1/\tau$, so that they are in phase during the time τ, and outside this time their relative phases become random. Similarly, a limitation of a group of waves to a spatial extent of length L implies that the group contains a range of wavelengths such that the component waves become out of step outside the group; if $L = n\lambda$, then the range of $\Delta\lambda$ is given by $\lambda/\Delta\lambda \approx n$.

4.17   AN ANGULAR SPREAD OF PLANE WAVES

The wave groups discussed in previous sections are limited in extent only along the direction of travel. A wave packet describing a particle, or a simple light beam, must also have a limited extent laterally. Can this also be regarded as a result of superposing plane waves? The longitudinal extent of a wave group is governed by the range of wavelengths of the plane waves constituting the group: we now show that the lateral extent is determined by a spread in wave *directions* rather than by a spread in *wavelength*.

Consider first the addition of two plane waves, with velocity c and wavelength λ, crossing at an angle 2θ, as in Fig. 4.17. Along the broken lines in this figure the two waves add in phase, making a wave progressing at velocity $c \sec \theta$. This resultant wave pattern shows a cosine variation of amplitude *across* the wavefront, i.e. perpendicular to the direction of propagation, with zero amplitude halfway between the maxima on the broken lines. For small θ, the maxima are separated by a distance λ/θ. If we now add more pairs of waves, with the same wavelength but crossing at different values of θ, we add to the resultant wave pattern further cosine components with different scales λ/θ. Following the same idea as in the longitudinal limitation of the wave group, we see that these different scales of lateral variation can add to produce a wave limited in space transverse to the direction of propagation.

A wavefront limited in this way to a lateral extent D requires a range of crossing waves with angles from zero to λ/D. The required distribution of wave amplitude with angle depends on the shape of the distribution of amplitude across the wavefront: following the example of the wave group we may expect the relation to be given again by a Fourier transform. This is explored in more detail in Chapter 7.

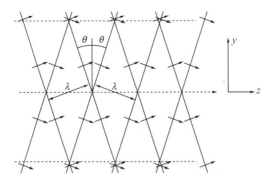

Fig. 4.17   Two plane waves crossing at angle 2θ. The waves add in phase along the broken lines, which are spaced by $\lambda/\sin\theta$, and are always in antiphase halfway between the broken lines.

4.18 FURTHER READING

R. N. Bracewell, *The Fourier Transform and its Applications*, 2nd edn, McGraw-Hill, 1986.

D. C. Champeney, *Fourier Transforms and their Physical Applications*, Academic Press, 1973.

J. W. Goodman, *Introduction to Fourier Optics*, McGraw-Hill, 1968.

E. G. Steward, *Fourier Optics: an Introduction*, Wiley, 1987.

NUMERICAL EXAMPLE 4

4.1 Draw a diagram showing phasors that represent the following oscillations: $2\sin\omega t$, $3\cos\omega t$, $4\cos(\omega t + \pi/4)$. Write down their x and y components and find the amplitude and phase of the phasor representing the sum of all three.

PROBLEMS 4

4.1 Demonstrate the equivalence of the following expressions for group velocity v_g, when phase velocity $v = c/n$:

$$v_g = \frac{c}{n + \omega\, dn/d\omega} = \frac{d\omega}{dk} = v - \lambda\frac{dv}{d\lambda}.$$

4.2 A plane wave propagates in a dispersive medium with phase velocity v given by

$$v = a + b\lambda,$$

where a and b are constants. Find the group velocity.

Show that any impulsive waveform will reproduce its shape at times separated by intervals of $\tau = 1/b$, and at distance intervals of a/b.

(*Hint*: consider any pair of component waves separated in wavelength by $\delta\lambda$ as in Section 4.16.)

4.3 Calculate the group velocity for the following types of waves, given the variation of phase velocity v with wavelength λ:

(i)   Surface water waves controlled by gravity: $v = a\lambda^{1/2}$

(ii)   Surface water waves controlled by surface tension: $v = a\lambda^{-1/2}$

(iii) Transverse waves on a rod: $v = a\lambda^{-1}$

(iv) Radio waves in an ionized gas: $v = (c^2 + b^2\lambda^2)^{1/2}$.

4.4 The refractive index for electromagnetic waves propagating in an ionized gas is given by

$$n^2 = 1 - \omega_p^2/\omega^2$$

where ω_p, the plasma frequency, is determined by the density of the gas. Show that the product of the group and phase velocities is c^2.

4.5 The negative half-cycles of a sinusoidal waveform $E = E_0 \cos \omega t$ are removed by a half-wave rectifier. Show that the resulting wave is represented by the Fourier series

$$E = E_0 \left(\frac{1}{\pi} + \frac{1}{2}\cos \omega t + \frac{2}{3\pi}\cos 2\omega t - \frac{2}{15\pi}\cos 4\omega t + \cdots \right).$$

Show that a full-wave rectifier, which inverts the negative half-cycles, has an output

$$E = E_0 \left(\frac{2}{\pi} + \frac{4}{3\pi}\cos 2\omega t + \text{even harmonics} \right).$$

4.6 A Gaussian function with height h and width σ is $f(t) = h \exp -(t^2/2\sigma^2)$. Show that its Fourier transform is

$$F(\nu) = (2\pi)^{1/2}\sigma h \exp(-2\pi^2\nu^2\sigma^2).$$

(You will require the integral $\int_{-\infty}^{\infty} \exp(-x^2)dx = \pi^{1/2}$.)

4.7 Show that the Fourier transform $F(\nu)$ of an isosceles triangular function, height h, base width b, is

$$F(\nu) = \frac{hb}{2}\frac{\sin^2 \psi}{\psi^2},$$

where

$$\psi = \frac{\pi b}{2}\nu.$$

4.8 A decaying wavetrain is represented by

$$f(t) = a\exp\left(-\frac{t}{\tau}\right)\exp i\omega_0 t.$$

Show that the Fourier transform of $f(t)$ is

$$F(\omega) = \frac{2\pi a}{1/\tau + i(\omega - \omega_0)},$$

and hence that the energy spectrum for ω close to ω_0 is given by

$$|F(\omega)|^2 = \frac{4\pi^2 a^2}{(1/\tau)^2 + (\omega - \omega_0)^2}.$$

4.9 What is the spectrum of the amplitude modulated wave

$$f(t) = (A + B\cos qt)\cos pt,$$

where p is the carrier frequency and q the modulating frequency?

4.10 Find the spectrum of the frequency-modulated wave

$$f(t) = A\cos(ptB\cos\omega t)$$

when B is small.

(*Hint*: expand the waveform as a power series in $B\cos\omega t$, and neglect B^2.)
What distinguishes this spectrum from the amplitude-modulated spectrum?

4.11 Show that the energy in the sum of two oscillations is equal to the sum of their individual energies, provided that they differ in frequency and a suitable time average is taken.

4.12 The relativistic Doppler effect. A signal received from an oscillator with frequency ν moving in a space vehicle with velocity V directly away from an observer has an apparent frequency ν' where

$$\nu' = \nu\left(\frac{1 - V/c}{1 + V/c}\right)^{1/2}.$$

Compare $\nu - \nu'$ with the value from the non-relativistic formula $\nu' = \nu(1 - V/c)$ for an oscillator at 6 GHz moving at $6\,\mathrm{km\,s^{-1}}$ in the line of sight.

Find also the transverse Doppler frequency shift $\nu - \nu_t$ for a velocity of $6\,\mathrm{km\,s^{-1}}$ in the line of sight where

$$\nu_t = \nu(1 - V^2/c^2)^{1/2}.$$

4.13 In the solar spectrum the same Fraunhofer line at 600 nm appears at wavelengths differing by 0.004 nm at the pole and at the edge of the disc near the equator. Find the velocity at the equator, and deduce the rotation period, given that the Sun's distance is 500 light-seconds and that it subtends an angle of 30′ at the Earth.

4.14 The Crab Pulsar emits a precisely periodic pulse train whose frequency is close to 30 Hz. It lies in a direction close to the ecliptic plane, in which the Earth orbits round the Sun. Calculate the peak-to-peak variation in the observed pulse frequency due to the Earth's annual motion, given that the Sun's distance from the Earth is 500 light-seconds.

4.15 The mean radius of the orbit of the Earth is 1.5×10^{11} m. Find the amplitudes of astronomical parallax and aberration for a star at a distance of 10 light-years situated in the plane of the orbit. What is the phase relation between these two periodic motions?

4.16 A ray of light falls at angle of incidence i on a mirror surface moving normally to its surface with a velocity v small compared with c. Use Huygens' construction to show that the angle of reflection r differs from i by approximately $(2v/c)i$ for small i.

5

Electromagnetic waves

The ether, this child of sorrow of classical mechanics . . .
Max Planck, quoted by Jean-Pierre Luminet in *Black Holes*.

The wave theory of light, which was applied so successfully in the nineteenth century to the phenomena of propagation, interference and diffraction, was naturally thought of in the same way as water waves and sound waves, which were obviously waves in a medium. Maxwell showed that light was an electromagnetic wave. But what was the medium through which light propagated? The ether, as it was called, had no observable properties. Attempts to detect it by measuring the motion of the Earth through it all failed, and it became clear that the description of electromagnetic waves did not depend in any way on the existence of the ether. Maxwell's equations, which are the basis of our understanding of electromagnetic waves, are relations between electric and magnetic fields, and not between these fields and some all-pervading medium.

In this chapter we first show how electromagnetic waves may be derived from the fundamental laws of electricity and magnetism, as formulated in Maxwell's equations. We then consider the flow of energy in an electromagnetic wave, and what happens when an electromagnetic wave meets a boundary, where it may be partly reflected and partly transmitted, depending on the materials at the boundary, the angle of incidence of the wave and its polarization. The quantum nature of light, by which the electromagnetic wave is also to be viewed as the propagation of elementary particles with zero mass, photons, provides an alternative approach to the interaction between light and matter; the transport of energy by the photons averages to that of the classical electromagnetic wave, and the momentum associated with a photon leads to a radiation pressure at an interface between media. These are examples of the dual nature of light; only the quantum picture, however, can account for the spectrum of black-body radiation, which we consider at the end of this chapter.

5.1   MAXWELL'S EQUATIONS

We start with the set of four equations, known as *Maxwell's equations*, which encapsulate the basic laws of classical electrodynamics.[1] They relate electric and magnetic fields to two different kinds of sources: first, by charges and currents, and secondly through induction, in which a changing magnetic field induces an electric field and a changing electric field induces a magnetic field. Both the variables in an electromagnetic wave, the electric and magnetic fields $\mathbf{E}$ and $\mathbf{B}$, are vector quantities, and we use vector notations throughout. We confine our analysis to isotropic and homogeneous materials, and mainly to non-conducting materials with linear properties. The full Maxwell's equations in vector form[2] are:

$$\text{div } \mathbf{D} = \rho, \qquad \text{div } \mathbf{B} = 0, \tag{5.2}$$

$$\text{curl } \mathbf{E} = -\frac{\partial \mathbf{B}}{\partial t}, \quad \text{curl } \mathbf{H} = \mathbf{J} + \frac{\partial \mathbf{D}}{\partial t}. \tag{5.3}$$

Here, ρ is the free charge density and $\mathbf{J}$ the free current density. The vector fields $\mathbf{D}$ and $\mathbf{H}$ are needed for general cases of dielectric polarization and magnetisation. In this book we mainly deal with linear isotropic materials where $\mathbf{D} = \epsilon\epsilon_0\mathbf{E}$, $\mathbf{B} = \mu\mu_0\mathbf{H}$; the constants ϵ_0 and μ_0 are the *permittivity* and *permeability* of free space, and ϵ and μ are the dielectric constant and the magnetic permeability of the medium. In most of optics ρ and $\mathbf{J}$ are zero and μ is unity; the four Maxwell equations then become

$$\text{div } \mathbf{E} = \frac{\rho}{\epsilon_0}, \qquad \text{div } \mathbf{B} = 0, \tag{5.4}$$

$$\text{curl } \mathbf{E} = -\frac{\partial \mathbf{B}}{\partial t}, \quad \text{curl } \mathbf{B} = \epsilon\epsilon_0\mu\mu_0\frac{\partial \mathbf{E}}{\partial t} + \mathbf{J}. \tag{5.5}$$

In a non-conducting material ($\mathbf{J} = 0$) with no free charge ($\rho = 0$)

$$\text{div } \mathbf{E} = 0, \qquad \text{div } \mathbf{B} = 0, \tag{5.6}$$

$$\text{curl } \mathbf{E} = -\frac{\partial \mathbf{B}}{\partial t}, \quad \text{curl } \mathbf{B} = \epsilon\epsilon_0\mu\mu_0\frac{\partial \mathbf{E}}{\partial t}. \tag{5.7}$$

Equations (5.7) are Faraday's law of electromagnetic induction and the complementary law of magnetoelectric induction introduced by Maxwell.

---

[1] See, for example I. S. Grant and W. R. Phillips, *Electromagnetism*, 2nd edn, Wiley, 1990.
[2] In Cartesian coordinates the divergence and curl of a vector $\mathbf{F}$ are

$$\text{div } \mathbf{F} = \nabla . \mathbf{F} = \frac{\partial F_x}{\partial x} + \frac{\partial F_y}{\partial y} + \frac{\partial F_z}{\partial z},$$

$$\text{curl } \mathbf{F} = \nabla \wedge \mathbf{F}$$
$$= \hat{\mathbf{l}}\left(\frac{\partial F_z}{\partial y} - \frac{\partial F_y}{\partial z}\right) + \hat{\mathbf{m}}\left(\frac{\partial F_x}{\partial z} - \frac{\partial F_z}{\partial x}\right) + \hat{\mathbf{n}}\left(\frac{\partial F_y}{\partial x} - \frac{\partial F_x}{\partial y}\right). \tag{5.1}$$

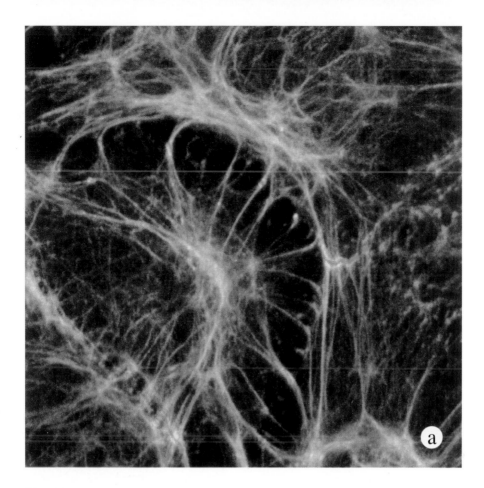

Plate 3.1 (a) Microphotography, using a confocal scanning microscope to scan and separate images at different depths. This three-dimensional network of cell-to-cell connections in cultured canine epithelia was stained with fluorescent dyes and scanned at 24 different focal depths. (b) Shown separately below are the individual sections, which can be superimposed to show the network from any aspect (S. Bagley, Paterson Institute, Manchester)

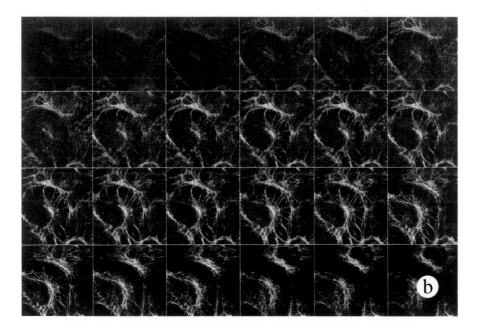

Plate 3.2 Cutaway section of a modern compact camera (the Minolta TC-1). The lens assembly is the G-Rokkor 3.5; the elements are all glass and three are multilayer coated; two of the inner lenses have aspheric surfaces. The camera is packed with instrumentation for automatic focusing and exposure control, optional flash and mechanisms for automatic film winding. (Minolta Co. Ltd)

Plate 11.1 LIGO, a gravitational wave detector in Louisiana, USA. This very large-scale optical interferometer (the arms are 4 km long) is based on the Michelson principle, but using multiple beam reflections as in the Fabry–Perot interferometer (Bernard Schutz)

M82
Merlin and VLA

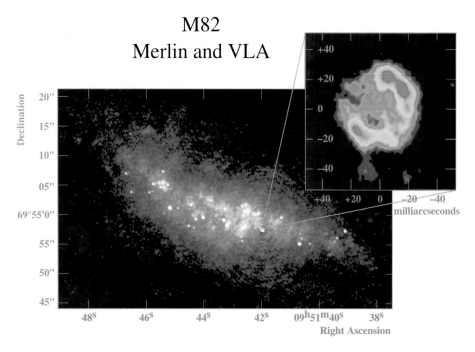

Plate 14.1 Aperture synthesis map of the radio emission at 20 cm wavelength from the galaxy M82. Radio interferometers with spacings from ten to several thousand kilometres provided the Fourier components, which were combined to produce this map. The prominent bright radio sources in this galaxy are supernova remnants; the ring-shaped remnant (shown inset) is mapped with a resolution of 1 milliarcsecond (T. Muxlow, Jodrell Bank Observatory)

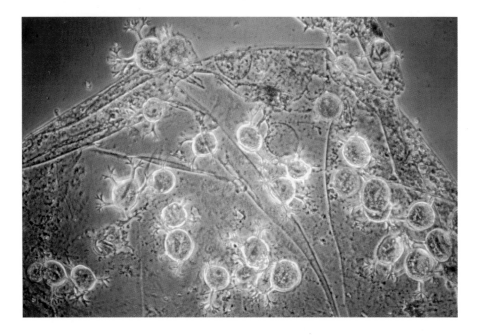

Plate 14.2 Dark-field and phase-contrast microscopy. The objects in this photo are almost completely transparent, but are made visible by refractive index differences that modify the phase of the light waves travelling through them. The circular objects (diameter 0.3mm) are protozoa on a gill plate of the fresh water shrimp *Gammamus pulex* (T. Allen, Paterson Institute, Manchester)

The properties of electromagnetic waves involve the interaction between the two fields expressed in the two laws of induction. We now eliminate one of the fields by combining the two equations (5.7). Taking the curl of both sides of the third Maxwell equation:

$$\text{curl curl } \mathbf{E} = -\text{curl} \frac{\partial \mathbf{B}}{\partial t}. \tag{5.8}$$

Since

$$-\text{curl} \frac{\partial \mathbf{B}}{\partial t} \equiv -\frac{\partial}{\partial t}(\text{curl } \mathbf{B}), \tag{5.9}$$

we can use the fourth equation to give:

$$\text{curl curl } \mathbf{E} = -\epsilon\epsilon_0\mu\mu_0 \frac{\partial^2 \mathbf{E}}{\partial t^2}. \tag{5.10}$$

Using the operational identity

$$\text{curl curl} \equiv \text{grad div} - \nabla^2,$$

and noting that div $\mathbf{E} = 0$ from the first Maxwell equation, we obtain

$$\boxed{\nabla^2 \mathbf{E} = \epsilon\epsilon_0\mu\mu_0 \frac{\partial^2 \mathbf{E}}{\partial t^2}.} \tag{5.11}$$

A similar derivation for $\mathbf{B}$ yields

$$\boxed{\nabla^2 \mathbf{B} = \epsilon\epsilon_0\mu\mu_0 \frac{\partial^2 \mathbf{B}}{\partial t^2}.} \tag{5.12}$$

These are the wave equations for an unattenuated electromagnetic field at any frequency and travelling in any direction. Comparison with the general wave equation (see Chapter 1)

$$\nabla^2 \psi = \frac{1}{v^2} \frac{\partial^2 \psi}{\partial t^2}, \tag{5.13}$$

gives the wave propagation velocity

$$\boxed{v = \left(\frac{1}{\epsilon\epsilon_0\mu\mu_0}\right)^{1/2}.} \tag{5.14}$$

All electromagnetic waves in free space ($\epsilon = \mu = 1$) travel with the same speed, which is a fundamental constant usually given the symbol c. For a medium with dielectric constant ϵ and permeability μ the wave velocity is

$$v = \frac{1}{\sqrt{\epsilon\epsilon_0\mu\mu_0}} = \frac{c}{\sqrt{\epsilon\mu}}.$$

(5.15)

Of the factors ϵ and μ that depend on the medium, the dielectric constant is usually the more important, since it is unusual to encounter light waves in media where μ differs appreciably from unity. In a dielectric, the ratio of the velocity in free space and the velocity in the medium is defined as the *refractive index* n of the medium. Hence, for a dielectric

$$n = \frac{c}{v} = \sqrt{\epsilon}.$$

(5.16)

The velocity for free space was evaluated by Maxwell using laboratory electrical measurements for ϵ_0 and μ_0. He obtained the velocity $3 \times 10^8 \, \text{m s}^{-1}$, in remarkable agreement with the measured speed of light. This led Maxwell to conclude that light was an electromagnetic disturbance that propagated according to the laws of electromagnetism.

5.2   TRANSVERSE WAVES

We stated in Chapter 1 that light is an electromagnetic wave with fields **E** and **B** oscillating *transversely* to the direction of propagation. We now show that the transverse nature of light follows directly from electromagnetic theory.

The wave equation (5.11) can represent waves of any frequency and any form, and it may be expressed in any system of coordinates. In Cartesian coordinates the vector field **E** has components

$$\mathbf{E} = \hat{\mathbf{l}}\,E_x + \hat{\mathbf{m}}\,E_y + \hat{\mathbf{n}}\,E_z,$$

(5.17)

where $\hat{\mathbf{l}}$, $\hat{\mathbf{m}}$, $\hat{\mathbf{n}}$ are unit vectors in the x, y, z directions. These three components can be independent solutions of Eq. (5.11). For a plane wave $\hat{\mathbf{l}}E_x$ travelling in the z-direction in free space

$$\frac{\partial^2 E_x}{\partial z^2} = \frac{1}{c^2}\frac{\partial^2 E_x}{\partial t^2}.$$

(5.18)

This has a general solution

$$E_x = E_0 f(z - ct),$$

(5.19)

representing a wave of any form travelling in the z-direction. Note that in free space all electromagnetic waves, with whatever waveform, travel with the same velocity. In the modern system of units (since 1983) the velocity of light in free space is a defined quantity set at 299 792 458 m s^{-1} exactly.[3]

Can there be a solution representing a longitudinally polarized wave? Consider a plane wave travelling in the z-direction, with

$$\mathbf{E} = \hat{\mathbf{n}}\, E_0 \cos(\omega t - kz)$$

As it is a plane wave there is no variation of the field $\mathbf{E}$ in the x- and y-directions; in the expansion of div $\mathbf{E} = 0$

$$\frac{\partial E_x}{\partial x} + \frac{\partial E_y}{\partial y} + \frac{\partial E_z}{\partial z} = 0, \tag{5.20}$$

the first two terms are zero, so that E_z must be independent of z and no such progressive wave can exist.

The $\mathbf{B}$ field is at right angles to the $\mathbf{E}$ field. This follows from the third Maxwell equation curl $\mathbf{E} = -\partial \mathbf{B}/\partial t$, where the only component of $\partial \mathbf{B}/\partial t$ is B_y. Both $\mathbf{B}$ and $\mathbf{E}$ are transverse to the direction of propagation, constituting a so-called TEM wave, in contrast to the TE and TM waves encountered, for example, in fibre optics (Chapter 18). (These are not plane waves, and there may be components along the direction of propagation.) The ratio B_y/E_x may in general be found from partial differentiation of Eq. (5.19). The relation between $\mathbf{E}$ and $\mathbf{B}$ is encapsulated in vector notation:

$$\mathbf{B} = \frac{1}{c}\hat{\mathbf{k}} \wedge \mathbf{E} \quad \text{(in free space).} \tag{5.21}$$

Here, $\hat{\mathbf{k}}$ is the *propagation vector*, which is a unit vector in the direction of propagation of the wave. Thus, for the above example, $E_x = cB_y$. In a dielectric, or any isotropic non-conducting medium, the relation becomes

$$\boxed{\mathbf{B} = \frac{1}{v}\hat{\mathbf{k}} \wedge \mathbf{E},} \tag{5.22}$$

where, as before

$$\boxed{v = \frac{c}{\sqrt{\epsilon\mu}} = \frac{c}{n}} \tag{5.23}$$

and $n \geq 1$. Since the wave speed is less than the speed in free space, for a fixed frequency the wavelength in a medium is less than that in free space. The wave

---

[3] See Chapter 11. The speed of light $c = (\epsilon_0 \mu_0)^{-\frac{1}{2}}$ is a fundamental constant with the defined value of 299 792 458 m s^{-1}. In Système International (SI) units the permeability of free space is given the value $\mu_0 = 4\pi \times 10^{-7}$ H m^{-1} (henry/metre), and it follows that the permittivity of free space is $\epsilon_0 = 8.854187 \times 10^{-12}$ F m^{-1} (farad/metre).

speed in the medium, and the refractive index, may vary with frequency; this is called *dispersion*. The splitting of colours in light refracted by a prism is due to dispersion by the glass in the prism.

5.3   ENERGY FLOW

The total energy per unit volume U contained in a system of electric and magnetic fields in an isotropic medium is[4]

$$U = \tfrac{1}{2}(\mathbf{D} \cdot \mathbf{E} + \mathbf{B} \cdot \mathbf{H}). \tag{5.24}$$

Thus, the energy density in a combination of electric and magnetic fields with magnitudes E and B may be written as $\epsilon\epsilon_0 E^2/2 + B^2/2\mu\mu_0$. In a harmonic wave we must take the average over a whole cycle. The energy is proportional to the square of the fields, so that for any wave component such as $\mathbf{E} = \hat{\mathbf{l}}E_0 \sin(kz - \omega t)$, the average square of the field is $\tfrac{1}{2}E_0^2$, where E_0 is the field amplitude. The mean energy density U is therefore

$$U = \tfrac{1}{4}(\epsilon\epsilon_0 E_0^2 + B_0^2/\mu\mu_0). \tag{5.25}$$

Since $B_0 = E_0/c$ and $c = (\epsilon_0\mu_0)^{-1/2}$, the two components are equal and the energy density may be written as

$$\boxed{U = \tfrac{1}{2}\epsilon\epsilon_0 E_0^2.} \tag{5.26}$$

The average energy crossing unit area per unit time in the z direction is the product $N = cU$:

$$N = \tfrac{1}{2}c\epsilon\epsilon_0 E_0^2 = \tfrac{1}{2}E_0^2\sqrt{\frac{\epsilon\epsilon_0}{\mu\mu_0}}. \tag{5.27}$$

In free space, $N = \tfrac{1}{2}E_0^2/Z_0$, where $Z_0 = (\mu_0/\epsilon_0)^{1/2} = \mu_0 c$ has the dimensions of resistance; it is known as the *impedance of free space*. In SI units the permeability of free space μ_0 is given the value

$$\mu_0 = 4\pi \times 10^{-7} \text{ henry/metre}$$

giving a value of 376.73 ohms (often quoted as 377 ohms) for Z_0. Using a root mean square (rms) of the field, $E_{rms} = (\bar{E}^2)^{1/2}$ in volts per metre, the energy flow is $E_{rms}^2/377$ watts m^{-2} (see Problem 5.3).

The energy flow is a vector known as the *Poynting vector* $\mathbf{N}$. For a plane wave in the direction of the unit wave vector $\hat{\mathbf{k}}$, and in a medium with relative permittivity ϵ and relative permeability μ, the Poynting vector is

---

[4] See, for example, I. S. Grant and W. R. Phillips, *Electromagnetism*, 2nd edn Wiley, 1990, p. 383.

$$\mathbf{N} = \tfrac{1}{2}\hat{\mathbf{k}} E_0^2 \sqrt{\frac{\epsilon\epsilon_0}{\mu\mu_0}}. \tag{5.28}$$

The most convenient formulation uses the auxiliary field $\mathbf{H}$, when the Poynting vector becomes

$$\boxed{\mathbf{N} = \mathbf{E} \wedge \mathbf{H}.} \tag{5.29}$$

5.4 REFLECTION AND TRANSMISSION: FRESNEL EQUATIONS

Snell's law, discussed in Chapter 1, relating the angles of incidence and refraction as a ray enters or leaves a refracting medium, tells only part of the story. A wave encountering a boundary between media with different refractive indices n_1, n_2 will not only be refracted, but will also be partly reflected. The ratios of the amplitudes of the reflected and transmitted waves to that of the incident wave are known as the amplitude *reflection* and *transmission* coefficients, r and t. For an electromagnetic wave, the *Fresnel* equations express the way in which these coefficients depend on the angles of incidence (θ_1) and refraction (θ_2), and on the polarization of the wave.

In Fig. 5.1 the reflected and refracted rays are shown for two cases of plane polarization, when the E-vector is (a) in the plane of incidence and (b) perpendicular to it. At the boundary, where the three rays meet, there must be a match between the components of the electric and magnetic fields on either side of the interface.

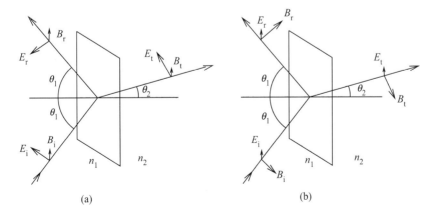

(a)   (b)

Fig. 5.1 Reflected and refracted rays at a boundary. The directions of the vector fields are shown for (a) E-vector in the plane of incidence; (b) B-vector in the plane of incidence.

The boundary conditions to be met[5] are

(i) the component of the **E** field parallel to the boundary, and
(ii) the components of the **H** and **B** fields parallel to the boundary

must be the same on both sides of the boundary. The subscripts $\parallel$ and $\perp$ used later in this section refer to whether the orientation of the **E** field is parallel or perpendicular to the plane of incidence, respectively.

Figure 5.1 defines the positive directions of the E- and B-vectors. Some of the directions shown are arbitrary and some are not. For example, E_r in Fig. 5.1(a) could have been shown pointing in the opposite direction. However, the E- and B-vectors are related by Eq. (5.21) to the direction of the wave, so any choice in the arbitrary direction of E_r defines a non-arbitrary direction for B_r. When writing the boundary conditions the directions of the fields in Fig. 5.1 define the positive reference directions. A negative value for one of the fields implies a 180° phase shift relative to the in-phase directions of Fig. 5.1.

For Fig. 5.1(a) the first boundary condition gives

$$E_i \cos \theta_1 - E_r \cos \theta_1 = E_t \cos \theta_2. \tag{5.30}$$

From Fig. 5.1(a) the magnetic fields are parallel to the interface, giving a second boundary condition

$$B_i + B_r = B_t. \tag{5.31}$$

Since the values of E and B are related by

$$E = \frac{c}{n} B, \tag{5.32}$$

where n is the refractive index of the medium, Eq. (5.31) gives

$$n_1 E_i + n_1 E_r = n_2 E_t. \tag{5.33}$$

Combining Eqs. (5.30) and (5.33) gives the reflection and transmission coefficients $r_\parallel$ and $t_\parallel$, which are ratios of *amplitudes*:

$$r_\parallel = \left(\frac{E_r}{E_i}\right)_\parallel = \frac{n_2 \cos \theta_1 - n_1 \cos \theta_2}{n_2 \cos \theta_1 + n_1 \cos \theta_2}, \tag{5.34}$$

$$t_\parallel = \left(\frac{E_t}{E_i}\right)_\parallel = \frac{2 n_1 \cos \theta_1}{n_2 \cos \theta_1 + n_1 \cos \theta_2}. \tag{5.35}$$

[5] See, for example, I. S. Grant and W. R. Phillips, *Electromagnetism*, 2nd Edn, Wiley, 1990, p. 392 *et seq.* In this section we are assuming non-magnetic dielectrics so that $\mu_1 = \mu_2 = 1$ and $\mathbf{B} = \mu_0 \mathbf{H}$. This means that a boundary condition for **H** is also obeyed for **B**.

A similar analysis for the E-vector perpendicular to the plane of incidence (Fig. 5.1 (b)) gives

$$r_\perp = \left(\frac{E_r}{E_i}\right)_\perp = \frac{n_1 \cos\theta_1 - n_2 \cos\theta_2}{n_1 \cos\theta_1 + n_2 \cos\theta_2} \qquad (5.36)$$

$$t_\perp = \left(\frac{E_t}{E_i}\right)_\perp = \frac{2n_1 \cos\theta_1}{n_1 \cos\theta_1 + n_2 \cos\theta_2}. \qquad (5.37)$$

Using Snell's law, $n_1 \sin\theta_1 = n_2 \sin\theta_2$, the amplitude reflection and transmission coefficients can be expressed in terms of angles only:

$$r_\parallel = -\frac{\tan(\theta_1 - \theta_2)}{\tan(\theta_1 + \theta_2)}, \quad t_\parallel = \frac{2\sin\theta_2 \cos\theta_1}{\sin(\theta_1 + \theta_2)\cos(\theta_1 - \theta_2)}, \qquad (5.38)$$

$$r_\perp = -\frac{\sin(\theta_2 - \theta_1)}{\sin(\theta_2 + \theta_1)}, \quad t_\perp = \frac{2\sin\theta_2 \cos\theta_1}{\sin(\theta_2 + \theta_1)}. \qquad (5.39)$$

Figure 5.2(a) is a typical plot of these reflection (r) and transmission (t) coefficients, for an air/glass boundary with $n = 1.5$ (where $n_1 \approx 1$).

Note that these coefficients are for amplitudes.[6] The rate of flow of energy across a surface, known as the *irradiance* (see Appendix), is proportional to the square of the amplitude, so that the *reflectance R* is r^2, as shown in Fig. 5.2(b). The *transmittance T* is found from

$$T = n\left(\frac{\cos\theta_2}{\cos\theta_1}\right)t^2, \qquad (5.40)$$

where the extra factor of $n = n_2/n_1$ accounts for power flow within a medium, and the geometric factor is due to the lateral compression of the wavefront. It is a useful exercise to check that $R + T = 1$ for both polarizations.

It will be seen from Eq. (5.38) that $r_\parallel$ goes through zero when $\theta_1 + \theta_2 = \pi/2$ (since $\tan(\pi/2) = \infty$), and that it changes sign. At this point, the angle of incidence is known as the *Brewster angle*, shown in Fig. 5.2(a). The change of sign indicates a phase reversal. Light reflected at the Brewster angle becomes completely linearly polarized, with the electric vector normal to the plane of incidence. It is this behaviour that makes polaroid glasses useful in reducing the glare of light reflected off a wet road, and in allowing fishermen to see into a

---

[6] Figure 5.2(a) shows that at normal incidence $r_\parallel$ and $r_\perp$ have different signs but equal magnitudes. The reason for this apparent discrepancy lies in the choices of the directions of the E-vectors in Fig. 5.1. As the text preceding Eq. (5.30) tries to make clear, the directions of E_r were arbitrary and the alternative choice will change the sign. The directions of E_r and E_i for normal incidence are in opposite directions in Fig. 5.1(a), but in the same direction in Fig. 5.1(b).

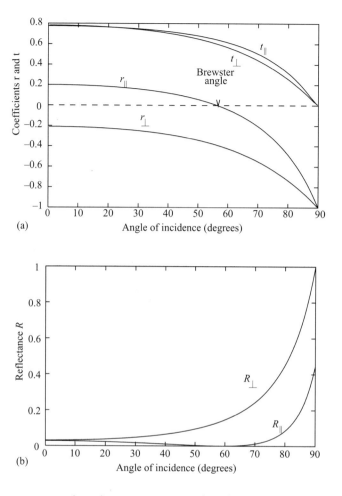

Fig. 5.2 (a) Reflection $(r_\parallel, r_\perp)$ and transmission $(t_\parallel, t_\perp)$ coefficients for light incident on an air/glass boundary with refractive index $n = 1.50$; (b) reflectance coefficients $R_\parallel, R_\perp$

lake despite the reflection of the bright sky in its surface. Similarly, a glass plate at the Brewster angle is completely transparent for light with the electric vector parallel to the plane of incidence; this is used in the windows of gas lasers to avoid reflection losses.

For normal incidence the reflection and transmission coefficients are independent of polarization, becoming simply

$$r = r_\parallel = r_\perp = \frac{n_2 - n_1}{n_1 + n_2},$$

$$t = t_\parallel = t_\perp = \frac{2n_1}{n_1 + n_2}. \tag{5.41}$$

The reflectance R and transmittance T are then

$$R = r^2 = \left(\frac{n_2 - n_1}{n_1 + n_2}\right)^2,$$

$$T = \frac{n_2}{n_1} t^2 = \frac{4 n_1 n_2}{(n_1 + n_2)^2}.$$

(5.42)

The reflectance loss of 4% at normal incidence for a typical air/glass surface with $n_2 = 1.5$ becomes a serious problem in the multi-component lenses of optical instruments such as cameras and telescopes. The losses can, however, be halved by coating the surface with a transparent layer with a lower refractive index $(n_1 n_2)^{1/2}$, as may be verified by substitution in Eq. (5.42). Further improvement can be achieved in very thin coated layers through the effects of thin film interference between reflections from the front and back of the coating (see Chapter 9). The reflectance can be reduced to zero for a chosen wavelength if the layer is made a quarter wavelength thick.

5.5   TOTAL INTERNAL REFLECTION: EVANESCENT WAVES

In Chapter 1 we saw that a ray meeting a boundary between media with higher and lower refractive indices at a large angle of incidence may be totally reflected; this is referred to as total internal reflection. In this case, there are two important extensions required to the Fresnel theory. The geometric ray approach merely shows total reflection, and makes no distinction between reflection at a dielectric and at a metallic surface. The boundary conditions are, however, quite different, since for the metallic conductor the tangential electric field is zero, while there is no such restriction on the tangential field at a dielectric surface. There are two consequences: first, there is an extension of the field across the boundary, and secondly, there is a phase shift in the reflected wave.

The wave field outside the dielectric boundary is an *evanescent wave*, whose amplitude falls exponentially with distance from the boundary. This field contains energy and transports it parallel to the boundary but not normal to it. The presence of this evanescent field is important in fibre optics, where light is confined to a thin glass fibre by total internal reflection. The energy flow is not confined to the core of the fibre, but extends to a cladding of lower refractive index glass into which, according to geometric optics, it cannot penetrate. No energy is lost by the evanescent wave unless there is absorption in the medium in which it is travelling. The cladding must therefore be thick enough to accommodate the evanescent wave, and it must also, like the core, be made of low loss material.

The analysis of reflection coefficients now involves matching the boundary of the evanescent wave to the incident and reflected waves. The reflection coefficients[7] then contain an imaginary component. Writing $n = n_2/n_1$

$$r_\parallel = \frac{n^2 \cos \theta_i - i(\sin^2 \theta_i - n^2)^{1/2}}{n^2 \cos \theta_i + i(\sin^2 \theta_i - n^2)^{1/2}}, \tag{5.43}$$

$$r_\perp = \frac{\cos \theta_i - i(\sin^2 \theta_i - n^2)^{1/2}}{\cos \theta_i + i(\sin^2 \theta_i - n^2)^{1/2}}. \tag{5.44}$$

The phase change on reflection $\phi(\theta)$ is found from these reflection coefficients:

$$\tan \frac{\phi_\parallel}{2} = -\frac{(\sin^2 \theta_i - n^2)^{1/2}}{n^2 \cos \theta_i} \tag{5.45}$$

$$\tan \frac{\phi_\perp}{2} = -\frac{(\sin^2 \theta_i - n^2)^{1/2}}{\cos \theta_i}. \tag{5.46}$$

Figure 5.3 shows the phase change for a glass/air interface where $n = 1.5$. Note that the difference $\phi_\parallel - \phi_\perp$ reaches 45°, so that the polarization of a linearly polarized ray can be changed substantially on reflection (see Chapter 6). This phase change on reflection can be used to produce circularly polarized light from plane polarized light.

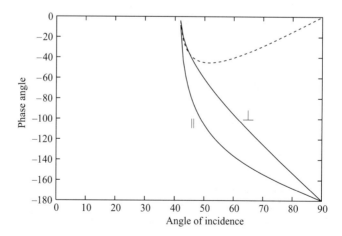

Fig. 5.3 The phase change at total internal reflection in a glass/air interface when $n_2 = n = 1.5$, for parallel and perpendicular polarizations. The broken line shows the difference between them.

---

[7] see M. Born and E. Wolf, *Principles of Optics*, 6th edn, Pergamon Press, Oxford, 1980, p. 48.

5.6   PHOTON MOMENTUM AND RADIATION PRESSURE

The reality of assigning a discrete momentum to a photon was demonstrated by Compton in 1922. He investigated the scattering of monochromatic X-rays by the electrons in a block of paraffin. An X-ray photon in collision with an electron will change direction, as in Fig. 5.4, and transfer part of its energy and momentum to the recoiling electron.

The X-ray photon leaves the scatterer with energy reduced by an amount depending on the angle of scatter. Taking account of the conservation both of momentum and of energy, the increase in wavelength $\lambda' - \lambda$ of the photon at the collision can be found from the dynamics of the collision.[8]

$$\lambda' - \lambda = \frac{h}{m_0 c}(1 - \cos\phi). \tag{5.47}$$

Subsequent experiments detected the individual recoil electrons in the Compton effect, but the measurement of the wavelength shift was in itself sufficient to establish the reality of this corpuscular behaviour of a photon, i.e. that photons behave like billiard balls.

When electromagnetic radiation meets a boundary between two media it exerts a pressure known as *radiation pressure*. This pressure is related to the flow of momentum in the radiation, and it is therefore most easily understood by considering the radiation in terms of photons. The momentum p carried by a photon is $p = h/\lambda$; this applies even when the photon is travelling in a dielectric medium, at a velocity v that is less than c. The flux of photons, i.e. the number n crossing unit area per unit time, is obtained from the Poynting vector divided by the photon energy:

$$n = \frac{N}{h\nu}. \tag{5.48}$$

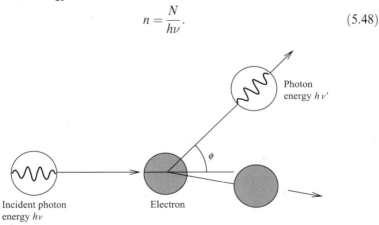

Incident photon
energy $h\nu$

Electron

Photon
energy $h\nu'$

Fig. 5.4   The Compton effect. An X-ray photon with energy $h\nu$ (wavelength λ) collides with an electron, loses energy and momentum, and emerges deviated through angle ϕ and with reduced energy $h\nu'$ (wavelength λ').

---

[8] See, for example, F. H. Read, *Electromagnetic Radiation*, Wiley, 1980, p. 230.

If all the photons are absorbed at the surface, then the radiation pressure P is given by Newton's second law as the rate of absorption of momentum:

$$P = n\frac{h}{\lambda} = \frac{N}{v} = \epsilon\epsilon_0 \bar{E}^2, \tag{5.49}$$

where $\bar{E}^2 = \frac{1}{2}E_0^2$ is the mean square field in the radiation. For total reflection the momentum transfer is doubled, and correspondingly the pressure is doubled because the direction of the photon is reversed: $P = 2\epsilon\epsilon_0\bar{E}^2 = \epsilon\epsilon_0 E_0^2$.

Radiation pressure may easily be demonstrated in the laboratory using intense laser light (Chapter 17). It is very important in astrophysics; the Sun and other stars are kept from gravitational collapse by the pressure of radiation from the interior. Even at the distance of the Earth the pressure of solar radiation may be important for artificial satellites, and may even be important in accelerating low-mass structures by solar sails. The pressure on a solar panel absorbing the whole incident solar energy ($1.4\,\mathrm{kW\,m^{-2}}$) is $4.7 \times 10^{-6}\,\mathrm{N\,m^{-2}}$; on a completely reflecting solar sail this value is doubled. Note that the force on one square metre of sail equals the gravitational force on a few milligrams weight (on Earth).

5.7   BLACK-BODY RADIATION

The quantized nature of radiation has a profound effect on the spectrum of thermal radiation, and we end this chapter by considering the spectrum of electromagnetic radiation from a black body.

A black body is one that completely absorbs any radiation incident upon it. A small hole in the surface of an isothermal enclosure, like a peephole into an oven, behaves in this way. The radiation within the enclosure reaches an equilibrium in which emission balances absorption, and the small sample of radiation that emerges from the hole is known as black-body radiation. We need to relate the spectrum and the intensity (the irradiance) of this radiation to the temperature of the enclosure.

Consider first the classical pre-1900 view of radiation and absorption, in which each small range of frequencies is continuously emitted and absorbed by an oscillator consisting of an electron in a resonant system. Each oscillator has an average energy kT, and, according to classical electromagnetic theory, it radiates energy at a rate proportional to kT and to ν^2. It must absorb at the same rate if the radiation is in equilibrium with its surroundings. The calculation of the equilibrium intensity involves the relation of the absorption cross-section of the oscillator to its rate of radiation, but the essential point is that the equilibrium intensity of the radiation is also proportional to $kT\nu^2$. The exact relation is the Rayleigh–Jeans formula[9]

---

[9] See, for example, F. Mandl, *Statistical Physics*, 2nd edn, Wiley, 1988.

$$\rho(\nu)d\nu = \frac{8\pi kT}{c^3}\nu^2 d\nu, \tag{5.50}$$

where $\rho(\nu)d\nu$ is defined as the energy per unit volume in a frequency range $d\nu$.

The problem with this classical calculation is the factor ν^2, which gives an intensity increasing indefinitely with frequency, which is obviously wrong. The radiation from an electric fire, for example, is concentrated in the red and infrared, and not in the ultraviolet. The solution was found by Planck (see Chapter 1), who abandoned the assumption that all oscillators would have an average energy of kT, and introduced an apparently arbitrary assumption that the energy of any oscillator at frequency ν could only exist in discrete units of $h\nu$, where Planck's constant $h = 6.626 \times 10^{-34}$ J s. This *quantization* gives the oscillator an average energy not of kT but kT multiplied by the factor

$$\frac{h\nu}{kT}\left(\exp\frac{h\nu}{kT} - 1\right)^{-1}.$$

The energy density of the black-body radiation spectrum in the frequency range $d\nu$ then becomes

$$\rho(\nu)d\nu = \frac{8\pi h\nu^3}{c^3}\frac{d\nu}{(e^{h\nu/kT} - 1)}. \tag{5.51}$$

This is the Planck radiation formula. The effect of the Planck term is seen in the solid line of Fig. 5.5, where the unmodified Rayleigh–Jeans curve, shown as a broken line, indicates a spectrum increasing indefinitely at high frequencies. Note that the Rayleigh–Jeans formula may be sufficiently nearly correct to be used at low frequencies when $h\nu/kT$ is small.

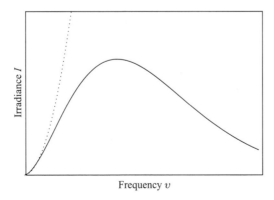

Fig. 5.5 The black-body radiation curve. The dotted curve shows the dependence expected without quantization.

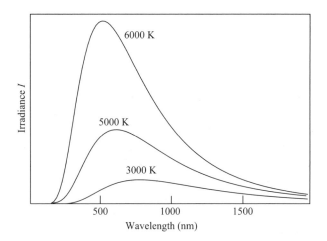

Fig. 5.6   Black-body radiation; intensity plotted against wavelength for different temperatures.

The black-body spectrum as a function of wavelength is

$$\rho(\lambda) = \frac{8\pi hc}{\lambda^5} \frac{1}{(e^{hc/\lambda kT} - 1)}. \tag{5.52}$$

This is plotted in Fig. 5.6 for a range of temperatures. The peak in each curve at λ_{max} is near the wavelength at which $h\nu = kT$, so that the product $\lambda_{max}T$ is a constant. This gives Wien's law:

$$\boxed{\lambda_{max}T = 2.897 \times 10^{-3}\text{m K,}} \tag{5.53}$$

where λ_{max} is in metres. It is interesting to note that Wien's law was formulated before quantization was introduced by Planck. It turns out that thermodynamic arguments alone can establish both Wien's law and another fundamental radiation law due to Stefan. This concerns the total energy integrated over a black-body spectrum, and is unaffected by quantization. *Stefan's law*, found experimentally in 1879 and derived from thermodynamics by Boltzmann in 1884, states that the total power radiated by a black body over all wavelengths is proportional to the fourth power of the temperature, giving

$$\boxed{W(T) = \sigma T^4,} \tag{5.54}$$

where W is the total power radiated per unit area, and $\sigma = 5.67 \times 10^{-8}$ W m^{-2} K^{-4} is the *Stefan–Boltzmann* constant.

The most perfect black-body radiation curve ever observed is that of the cosmic microwave background radiation, which is a relic of the concentrated

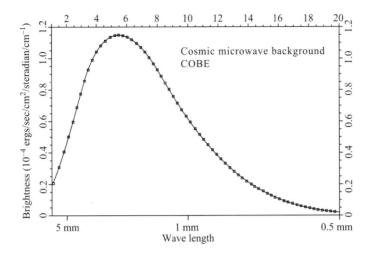

Fig. 5.7   The black-body spectrum of the microwave cosmic background radiation. The observational data from the COBE satellite fit precisely on the theoretical curve for a temperature of 2.73 K. (*Source*: J. C. Mather *et al.*, *Astrophysics Journal*, **420**, 439,1994.)

thermal radiation that filled the early Universe soon after the Big Bang. As the Universe expands, reducing the energy concentration in this radiation, the radiation cools but its spectrum remains that of a black body. At the present state of expansion the temperature of this radiation is 2.73 K, giving a spectrum peaking near 1 mm wavelength. The spectrum was measured from above the Earth's atmosphere, with a spectrometer on the COBE satellite (Fig. 5.7).

Wein's law gives a useful guide to the spectral range at which any hot body radiates most efficiently, even if it is not a perfect black body. The spectrum of solar radiation is a good approximation to that of a black body at 6000 K; the peak at 500 μm (1 μm $\equiv 10^{-6}$ m) comes within the visible spectrum, coinciding with the range of wavelengths that can penetrate the Earth's atmosphere and to which our eyes are sensitive. X-rays originate in hotter places, with temperatures of order 10^6 K; in astronomy most such sources are ionized gas clouds, such as the outer part of the solar atmosphere.

5.8   FURTHER READING

H. A. Haus, *Waves and Fields in Optoelectronics*. Prentice-Hall, 1984.
D. H. Staelin, A. W. Worgenthaler and J. A. Kong, *Electromagnetic Waves*. Prentice-Hall, 1994.

NUMERICAL EXAMPLES 5

5.1 A slab of GaAs crystal, used in a laser, has a refractive index $n = 3.6$. What fraction of the energy of radiation generated in the slab and incident normally on the top face is reflected? What is the transmittance for radiation from outside entering the slab at normal incidence?

5.2 Two glass slabs, with refractive indices 1.5 and 1.3, are glued together with a thick layer of transparent material with refractive index 1.4. Show that the light lost by reflection is approximately halved compared with a direct contact between the slabs.

5.3 A light wave in glass with refractive index 1.5 has a transverse electric field amplitude of 10 V/m. What is the associated magnetic field and the energy density?

5.4 At what wavelengths are the maximum outputs of radiation from black bodies at temperatures 3K, 20° C, 5800 K?

5.5 The average irradiance of solar radiation at the Earth is 1.4 kW m^{-2}. Most is absorbed; calculate the total force on the whole of the Earth. The mean radius of the Earth is 6.4×10^6 m.

PROBLEMS 5

5.1 What fraction of light is reflected at the surface of a lens with refractive index 1.5? Show how this may be improved by a suitable surface coating.

5.2 Compare the solar radiation pressure on the Earth (see Example 5.5 above) with the gravitational attraction of the Sun, and find the radius of a sphere with density the same as the Earth for these forces to balance. (*Hint*: the gravitational force can be found from the period of the Earth's orbit and its distance 1.5×10^{11} m from the Sun).

5.3 Following section 5.3, estimate the electric field due to normal illumination from a desk lamp.

5.4 Under what circumstances can two electromagnetic waves add so that the intensity of the sum is always equal to the sum of their two separate intensities?

5.5 Two plane waves exactly in phase combine to form a wave with double amplitude, i.e. with quadruple power. Where does the extra energy come from?

6

Polarization of light

I will found my enjoyments on the affections of the heart, the visions of the imagination, and the spectacle of nature.

Etienne Louis Malus, born Paris, 23 July 1775.
The discoverer of the polarization of light.

Linearly polarized light is a surprisingly common phenomenon in everyday circumstances. It can be detected by the use of the 'polaroid' material in glasses used, especially by motorists, to reduce glare in bright sunlight. Polaroid transmits light that is plane polarized in one direction only, and absorbs light polarized perpendicular to this direction. Light reflected from any smooth surface, such as a wet road or a polished table top, is partially linearly polarized; this is easily demonstrated by rotating the polaroid glass, which gives a change in brightness according to the change in angle between the plane of polarization and the transmission axis of the polaroid. The maximum effect is found for reflection at a particular angle of incidence, the 'Brewster angle' (section 5.4). The light of the blue sky, which is sunlight scattered through an angle, is also noticeably polarized. Insects such as honey bees can detect the polarization of the sky, and use its direction in relation to the sun for navigation.

Circular polarization is less easily observed, but it is important in several phenomena concerned with the propagation of electromagnetic waves in anisotropic media; for example, in the propagation of light in crystals such as quartz, and in some liquids such as sugar solution.

In this chapter we show how any state of polarization in a wave can be expressed in terms of elementary components, either plane or circularly polarized, and how the state of polarization may be changed by transmission through optically active materials.

6.1   POLARIZATION OF TRANSVERSE WAVES

In the last chapter we showed that light is a transverse electromagnetic wave. The *polarization* of the wave is the description of the behaviour of the vector **E** in the plane x, y, perpendicular to the direction of propagation z.

The plane of polarization is defined as the plane containing the ray, i.e. the z-axis, and the electric field vector.[1] The plane of polarization need not be constant at any point on the ray, but if the vector **E** does remain in a fixed direction, the wave is said to be *linearly* or *plane* polarized. If the direction of **E** changes randomly with time, then the wave is said to be *randomly* polarized, or unpolarized. The vector **E** can also rotate uniformly in the plane x, y at the wave frequency, as observed at a fixed point on the ray; the polarization is then *circular*, either right- or left-handed depending on the direction of rotation. A combination of plane and circular polarizations produces elliptical polarization. A *partially polarized* light wave can have a combination of polarized and unpolarized components.

In this chapter we consider the polarized components of a light wave, and show the relation between them. It is convenient to consider the components E_x and E_y on the orthogonal axes x and y as two independent plane-polarized waves with individual amplitudes and phases. The vector addition of these two components can produce any state of polarization of the actual electric field, depending on the relative phase of the two oscillations. If the two oscillations are in phase, then the successive vector sums are as in Fig. 6.1(a). The resultant is a vector at a constant angle to the x-axis. Two plane-polarized waves have combined to produce another wave that is also plane polarized.

If the two oscillations are in quadrature, so that their values of ϕ in equations of the form $E = a\cos(\omega t - kz + \phi)$ differ by $\pi/2$, the successive vector additions follow Fig. 6.1(b). Here, a is the amplitude on the x axis, and b the

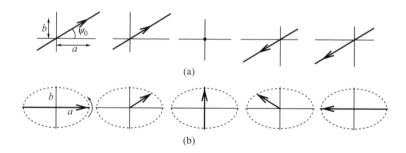

(a)

(b)

Fig. 6.1   Vector addition of two oscillating electric fields at successive moments through one half-cycle. The two fields have equal amplitudes and are mutually perpendicular: in (a) they are in phase and in (b) they are in quadrature. The resultant oscillation is linearly polarized in (a) and elliptically polarized in (b).

---

[1] It is to be emphasized that it is the plane of vibration of the **E** vector that is taken to define the plane of polarization; there is ambiguity and confusion over this in some of the older literature.

amplitude on the y axis. The resultant now rotates in the plane x, y, following a circle if the two amplitudes a, b are the same, or an ellipse if $a \neq b$. If the x oscillation is phase-advanced on the y oscillation, then the rotation is anti-clockwise; if it is retarded, then the rotation is clockwise. The two plane-polarized waves have combined to produce a wave that is elliptically polarized.

We now show how any state of polarization in a ray of light can be described in terms of elementary components of plane-polarized light. It is necessary to distinguish first between the polarized and unpolarized components of the ray. Note that when we consider states of polarization we are adding field components such as E_x and E_y, while for unpolarized light these components have a randomly changing phase relation and only the mean square of the amplitude is significant; it is the mean square of the amplitude that is proportional to the irradiance of the ray.

We now analyse the polarized component of a wave propagating along the z-axis in terms of linear components with any phase difference ϕ:

$$E_x = a\cos(\omega t - kz), \tag{6.1}$$
$$E_y = b\cos(\omega t - kz + \phi). \tag{6.2}$$

If $\phi = 0$, E_x and E_y combine vectorially to give a resultant field $\mathbf{E}$ with magnitude $(a^2 + b^2)^{1/2}$, and at an angle ψ_0 to the x-axis given by

$$\tan \psi_0 = \frac{b}{a}. \tag{6.3}$$

If $\phi = \pi/2$ and $a = b$, the wave is circularly polarized, and $\mathbf{E}$ describes a uniform circular motion in space; the handedness reverses when $\phi = -\pi/2$. In optics[2] the hand is defined looking back towards the source of the ray, when the wave vector in any one plane rotates clockwise for a right-handed circular polarization.

More generally, adding orthogonal vector oscillations when ϕ is not zero or $\pi/2$ produces an elliptical polarization whose major axis does not lie along x or y. We add the real parts of the oscillations, with components E_x and E_y

$$\frac{E_x}{a} = \cos \omega t, \quad \frac{E_y}{b} = \cos(\omega t + \phi) \tag{6.4}$$
$$= \cos \omega t \cos \phi - \sin \omega t \sin \phi. \tag{6.5}$$

Eliminating the time factor:

$$\frac{E_y}{b} = \frac{E_x}{a}\cos \phi - \left\{1 - \left(\frac{E_x}{a}\right)^2\right\}^{1/2} \sin \phi, \tag{6.6}$$

$$\frac{E_y^2}{b^2} + \frac{E_x^2}{a^2} - \frac{2E_x E_y}{ab}\cos \phi - \sin^2 \phi = 0. \tag{6.7}$$

---

[2] Unfortunately, this is the opposite of the convention for radio waves, where the handedness is defined looking *along* the direction of propagation.

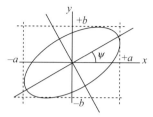

Fig. 6.2 Elliptically polarized oscillation, combining linearly polarized oscillations on the x and y axes, with amplitudes a and b, and with an arbitrary phase difference.

This equation describes an ellipse, as in Fig. 6.2. At any time, the resultant field vector $\mathbf{E}$ reaches to a point on the ellipse, moving around as time progresses. The ellipse is contained in a rectangle with sides $2a$ and $2b$. The position of the major axis of the ellipse, at an angle ψ to the x axis, is found as follows.

The amplitude $E_x^2 + E_y^2$ is maximum on the major axis. Therefore, at this point

$$E_x \, dE_x + E_y \, dE_y = 0. \tag{6.8}$$

Also from Eq. (6.7), for a fixed value of ϕ:

$$\left(\frac{E_x}{a^2} - \frac{\cos \phi}{ab} E_y \right) dE_x + \left(\frac{E_y}{b^2} - \frac{\cos \phi}{ab} E_x \right) dE_y = 0. \tag{6.9}$$

On the major axis we have $\tan \psi = E_y/E_x$, so that combining Eqs. (6.8) and (6.9)

$$\frac{1}{a^2} - \frac{\cos \phi}{ab} \tan \psi = \frac{1}{b^2} - \frac{\cos \phi}{ab} \cot \psi. \tag{6.10}$$

Since $\tan \psi - \cot \psi = -2 \cot 2\psi$, we find that

$$\tan 2\psi = \frac{2ab \cos \phi}{a^2 - b^2}. \tag{6.11}$$

The ratio of the maximum and minimum axes of the ellipse may be found by rotating the coordinate axes through the angle ψ. The axial ratio may in this way be shown to be given by R in

$$\frac{R}{1 + R^2} = \pm \frac{ab}{a^2 + b^2} \sin \phi. \tag{6.12}$$

The sign depends on the sense, or hand of rotation.

6.2 ANALYSIS OF ELLIPTICALLY POLARIZED WAVES

By a suitable choice of amplitudes, a, b, and relative phase ϕ, the vibration ellipse of Fig. 6.2 may be set with any axial ratio and with the major axis at any position angle. It follows that any elliptical oscillation or elliptically polarized wave may be analysed mathematically into two linearly polarized components at right angles, using axes at any angle to the major axis of the ellipse. The relative phase depends on this position angle; it is only important to note that the two components are in quadrature if they are aligned with the major and minor axes of the ellipse. The analysis may be done experimentally by using a device that will change the relative phases of two orthogonal components of a ray by a known amount. A particularly useful device is the 'quarter-wave plate', a transparent slice of anisotropic crystalline material in which the wave velocity differs between two perpendicular directions by such an amount that one component takes a quarter-period longer to propagate than the other (see section 6.4). This, in combination with a polaroid analyser, can be used for analysis of elliptically polarized light by turning the axes of the quarter-wave plate until a position is found where the emergent light is fully plane polarized.

A combination of an analyser and a quarter-wave plate can also be used to determine the state of polarization of an arbitrarily polarized wave. For this purpose it is convenient to think of the most general state of polarization as a combination of elliptical and random polarization. The procedure is as follows:

1. Using the analyser discussed above the amount of plane-polarized light can be determined by rotating this analyser. The remaining light when the analyser is set to admit a minimum intensity may be circularly or randomly polarized.

2. The quarter-wave plate is inserted before the analyser, and the orientations of both are changed independently to produce a minimum intensity. Elliptically polarized light will give zero intensity in these circumstances, since the quarter-wave plate when it is properly oriented will turn it into plane-polarized light that will be rejected by the analyser. Any remaining light at minimum intensity must have a random polarization.

6.3 POLARIZERS

Light from most sources is unpolarized. It can be converted into fully polarized light by the removal of one component, usually either plane or circularly polarized.

A simple example is the wire grid polarizer used originally for radio waves, but which can be demonstrated to work for infrared light at about $1\,\mu\mathrm{m}$ wavelength. This is simply a parallel grid of thin conducting wires whose diameter and spacing are small compared with the wavelength. In such a grid only the electric field component perpendicular to the wires can exist; for the

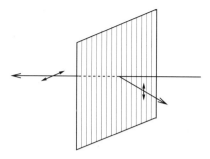

Fig. 6.3 A wire grid polarizer. The wires are spaced at less than a wavelength apart; light polarized parallel to the wires is reflected.

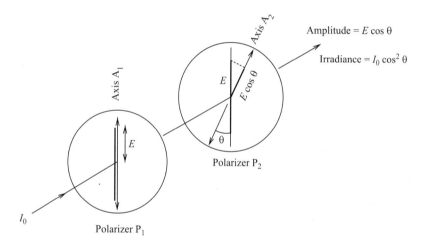

Fig. 6.4 Malus' law. Light from a polarizer with transmission axis vertical falls on a linear analyser with axis at angle θ. The irradiance is reduced by $\cos^2 \theta$.

component polarized parallel to the wires the grid acts as a reflecting plane. Only the plane of polarization perpendicular to the grid is transmitted (Fig. 6.3). Such devices are called *polarizers* when they are used to create polarized light, or *analysers* when they are used to explore the state of polarization, as in the previous section. The action of a linear analyser on a linearly polarized wave is shown in Fig. 6.4. Light from a linear polarizer with transmission axis A_1 is incident normally on a linear analyser with axis A_2 at angle θ to the plane of polarization defined by A_1. The amplitude of the transmitted wave is reduced by the factor $\cos \theta$, giving *Malus' law* for the irradiance I

$$I(\theta) = I_0 \cos^2 \theta. \tag{6.13}$$

The metallic wire grid has an analogue in the aligned molecular structure of polaroid sheet. This is a stretched film of polyvinyl alcohol containing iodine;

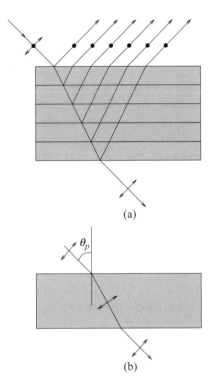

(a)

(b)

Fig. 6.5 Reflection and transmission at the Brewster angle θ_p: (a) reflection at a pile of plates; (b) transmission at a Brewster window.

the iodine is in aligned polymeric strings that absorb light polarized parallel to the direction of alignment. A general class of crystals, including the well-known material tourmaline, has the same property of selectively absorbing one plane of polarization. These materials are often referred to as *dichroics*.[3]

Light can also be made linearly polarized by reflection at the Brewster angle from a dielectric surface (Chapter 5). The reflection coefficient is, however, often inconveniently low; this can be overcome by the *pile-of-plates* polarizer shown in Fig. 6.5(a), where reflections from multiple layers of glass or other dielectric add to give almost complete reflection. The high transmission coefficient for the other polarization at the Brewster angle is used to make perfect non-reflecting windows, as shown in Fig. 6.5(b). Such *Brewster windows* are used in gas lasers (Chapter 15), where light from the laser cavity makes repeated passes through windows placed in front of mirrors at either end of the cavity. The emerging laser light is usually fully linearly polarized in a plane determined by the Brewster windows.

---

[3] The term *dichroic* originated in mineralogy, where it referred to the different *colours* of two polarized rays emerging from birefringent crystals. The colours arise from selective absorption; if the absorption over a large wavelength range is much larger in one polarization than the other, then the material is dichroic in the sense used in optics.

6.4   BIREFRINGENT POLARIZERS

In an *anisotropic* medium, and, in particular, many transparent crystals, the phase velocity of light varies with crystal orientation. The refractive index is then not a single number, but a quantity that varies with direction; it may be represented by a surface such as an ellipsoid. A further complication is that the refractive index in any one direction may be a function of the state of polarization of the light wave, so that a ray entering a crystal with random polarization will be split into two components that will be refracted differently. The medium is then said to be *doubly refracting* or *birefringent*. The two components may be plane polarized, as for propagation in crystals such as calcite, or they may be circularly polarized, as, under some circumstances, for the propagation of light in quartz. The refractive index of a crystal depends generally on the direction of polarization in relation to the crystal structure. Not all crystals behave in this way; those with a highly symmetric form, such as sodium chloride, are not birefringent.

The crystal structure of calcite ($CaCO_3$) has a single axis of symmetry, which coincides with an *optic axis*; this is also an axis of symmetry for the refractive index surface.[4] A point source of unpolarized light within a calcite crystal will generate two wavefronts, as shown in Fig. 6.6. One is spherical, and is known as the *ordinary wave*, or o-*wave*; the other is ellipsoidal, and is known as the *extraordinary wave*, or e-*wave*. (Note that these are wavefronts; the refractive index surface for the extraordinary ray is an oblate, not prolate, ellipsoid. The difference in refractive indices $n_e - n_o$ is negative for calcite, which is classified as *negative uniaxial*.) The polarizations of these two wavefronts are related to the crystal structure; in the o-wave the electric vector is everywhere normal to the optic axis, and in the e-wave it is perpendicular to the optic axis.

An unpolarized ray incident on a face of a calcite crystal will, in general, be refracted into two rays, propagating in different directions within the crystal,

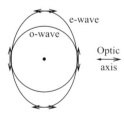

Fig. 6.6   Ordinary and extraordinary wavefronts radiating from a point source in a uniaxial crystal. The electric field of the e-wave is shown by the double-headed arrows; the polarization of the o-wave is out of the plane of the diagram. For propagation perpendicular to the optic axis the refractive index depends on the orientation of the E-vector in relation to the axis; along the axis both waves travel at the same velocity.

---

[4] Calcite is a *uniaxial* crystal; other crystals have more complex symmetries and their birefringence is correspondingly more complex.

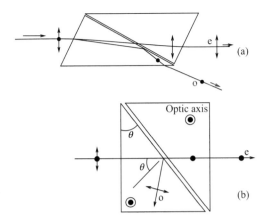

Fig. 6.7   Nicol and Glan–Foucault prisms: (a) selective reflection in the Nicol prism is obtained by using a transparent cement between the parts of the calcite crystal with refractive index 1.55, intermediate to the index for the e-ray (1.49) and the index of the o-ray (1.66); (b) in the Glan–Foucault polarizer the two prisms are spaced by an air gap. The calcite prisms require an angle $\theta = 38°$.

and with orthogonal plane polarizations. This separation is used in various forms of *birefringent polarizer*. In the *Nicol prism*, made of calcite, (Fig. 6.7) the two rays are separated at a layer of transparent cement within the calcite, arranged so that one of these rays is removed by total internal reflection. The single emergent ray is accurately linearly polarized. Figure 6.7 also shows the more commonly used *Glan–Foucault prism* in which there is no deviation at the first face, and the transmitted ray is undeviated overall. The space between the two prisms is usually airfilled (the *Glan-air polarizer*); polarization selection then requires simply that the prism angle θ is related to the two refractive indices n_o and n_e by

$$n_e < 1/\sin\theta < n_o \tag{6.14}$$

for the ordinary and extraordinary rays to be separated as shown. An increased field of view is obtained by cementing the prisms together (the *Glan–Thompson polarizer*), but the air-spaced version can handle larger irradiances, as is often required in high-powered laser systems.

In the Wollaston prism (Fig. 6.8) the optic axes of the two components are orthogonal as shown. The two polarized rays are both transmitted, but they are separated by a sufficient angle for them to be treated individually; for example, they may go to separate photoelectric detectors. Such devices are used in optical telescopes for measuring the plane-polarized component of starlight. The advantage over the Nicol and Glan–Foucault prisms is symmetry: both components are transmitted through similar paths in the crystal, and any absorption is the same for both.

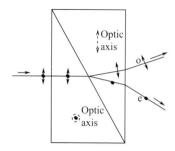

Fig. 6.8 Wollaston prism. The two parts are made of quartz, with the optic axis in the two directions orthogonal to the ray. The e-and o-rays are separated at the interface.

6.5   QUARTER- AND HALF-WAVE PLATES

The polarization state of light can also be analysed using components sensitive to circular and other states of polarization; many of these components depend on phase changes during propagation in anisotropic media rather than on selective refraction or absorption.

Consider the propagation of plane-polarized light incident normally on a parallel-sided thin slab of crystal such as calcite, cut so that the optic axis is in the plane of the slab. The component of the wave with an electric vector parallel to the optic axis travels faster than the perpendicular component; these are the e-and o-waves (for extraordinary and ordinary) introduced in the previous section. These two components are in phase as they enter the slab, but the e-wave travels faster and a phase difference δ grows as they travel. If the two refractive indices are n_e and n_o, the phase difference after a distance d is:

$$\delta = \frac{2\pi}{\lambda}(n_e - n_o)d \quad \text{rad.} \tag{6.15}$$

The amplitudes of the two components are $A\cos\theta$ and $A\sin\theta$, where θ is the angle between the incident plane of polarization and the optic axis and A is the amplitude in the incident ray (Fig. 6.9). Combining these two again with phase difference δ produces a different state of polarization (Fig. 6.10); for $\delta = \pi/2$ this is an ellipse with a principal axis along the optic axis, while for $\delta = \pi$ the polarization is again plane but rotated by angle 2θ. In the particular case where $\theta = 45°$ the ellipse becomes a circle, and circularly polarized light is produced; the opposite hand of circular polarization is obtained when $\theta = 135°$. For $\delta = \pi$ and $\theta = 45°$, the plane of polarization is rotated by 90°. The successive changes in polarization are shown in Fig. 6.11.

Crystal slabs giving $\delta = \pi/2$ and π are known as *retarders*, either quarter-wave or half-wave plates, respectively. We have already shown how a quarter-wave plate can be used in the analysis of polarized light. These components have important uses in manipulating polarization in optical systems.

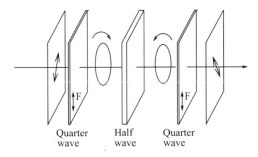

Fig. 6.9   A plane-polarized wave at 45° passes through a quarter-wave plate and becomes circular; the hand is reversed by a half-wave plate; and the orthogonal plane is produced by a second quarter-wave plate. The fast axis is indicated by F.

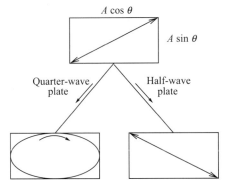

Fig. 6.10   A plane-polarized wave at angle θ to the fast direction of a quarter-wave and a half-wave plate, converted into an elliptical and a plane-polarized wave, respectively.

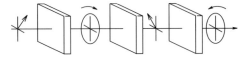

Fig. 6.11   Changes of polarization through a series of quarter-wave plates.

6.6   OPTICAL ACTIVITY

In many anisotropic media the refractive index is different for the two hands of *circular* polarization. This form of birefringence, known as *chirality*, has an important effect on plane-polarized light: along any line of sight the plane of polarization is rotated through an angle proportional to the birefringence in that direction. The same effect occurs in the propagation of linearly polarized radio waves through the ionized hydrogen of interstellar space, where the birefringence is due to the interstellar magnetic field.

The common characteristic of these particular media is that light of different hands of circular polarization travels with different velocities, so that if the two hands are propagated together, their phase relation changes progressively along the line of sight. The addition of two circularly polarized oscillations or waves, of equal amplitudes but with opposite hands, results in a plane-polarized wave whose plane depends on the relative phase of the two circular oscillations. The difference in propagation velocity therefore results in a rotation of the plane of polarization, as observed, for example, in the propagation of plane-polarized light in sugar solution.

The double refraction, or 'birefringence', of calcite depends on its anisotropic crystal structure. In some crystals, notably crystalline quartz (but not fused quartz), and in the molecules of many organic substances such as sugar, the molecular structure is a spiral. The refractive index for circularly polarized light then depends on the relation between the hand of polarization and the hand of the spiral structure. The phenomenon is useful both in the manipulation of polarization and in elucidating the molecular structure of so-called *optically active* materials.

A plane-polarized ray traversing an optically active crystal must then be thought of as the combination of two circularly polarized rays, which travel at different speeds. Their relative phases, which determine the position angle of the linear polarization, change along the ray path, and the plane rotates.

The rate of rotation of the plane of polarization in quartz, for light propagated along the optic axis, is $21°$ per millimetre. In liquids the rotation is normally less, but for the so-called 'liquid crystals' (LC), which are liquids in which molecules are partially oriented, as in a crystal lattice, the rotation may be very much larger. Cholesteric liquid crystals, in which the molecules have a helical structure, have rotations up to $40\,000°$ per mm. These are used in the digital displays of wrist-watches and other instruments.

6.7   INDUCED BIREFRINGENCE

Some materials can be made birefringent by an external electric or magnetic field. The effects can be understood at the atomic or molecular level, as in the permanently birefringent materials.

In the *Kerr effect* the birefringence is induced in a liquid by a transverse electric field. As in a uniaxial crystal, quarter-wave and half-wave plates can be created in a parallel-sided cell, although a cell several centimetres long may be needed in practice. The difference in refractive indices is related to the field $\mathbf{E}$ and the vacuum wavelength λ by

$$n_e - n_o = \lambda K E^2, \tag{6.16}$$

where K is the *Kerr constant* for the liquid. The *Kerr cell* is used to modulate a ray of plane-polarized light. A cell about $10\,cm$ long containing nitrobenzene

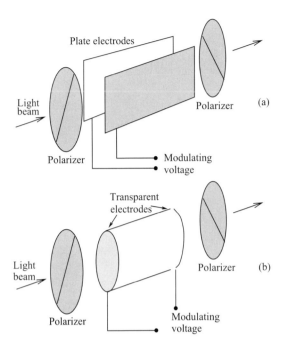

Fig. 6.12   (a) A Kerr cell; (b) a Pockels cell. In each a light beam is modulated by an electric field that induces birefringence, rotating the plane of polarization.

becomes a half-wave plate when a transverse field of around $10\,\text{kV cm}^{-1}$ is applied. If the incident polarization is at $45°$ to the field, then the emergent beam is rotated by $90°$ and it can be rejected by an analyser set in the original plane, as in Fig. 6.12(a). The Kerr cell can therefore be used as an electrically operated light switch.

The *Pockels effect* is a birefringence induced in a crystal by a longitudinal electric field. The classes of crystal that show this effect are also piezoelectric. Among the many exotic crystals developed specially for a large Pockels effect are barium titanate and potassium dideuterium phosphate (known as KD*P). A Pockels cell, as shown in Fig. 6.12(b), acts in a similar way to the Kerr cell; it is, however, more compact and is widely used for electrical modulation and the switching of light beams in communications systems.

The *Faraday effect* (Fig. 6.13) is induced optical activity, in which a longitudinal magnetic field can induce a rotation of the plane of polarization in an isotropic material such as glass. The angle of rotation ψ is proportional to the magnetic field strength B and the path length l, so that

$$\psi = VBl, \qquad (6.17)$$

where V is the *Verdet constant* for the medium. A particularly simple explanation is available for the propagation of a radio wave through a cloud of free

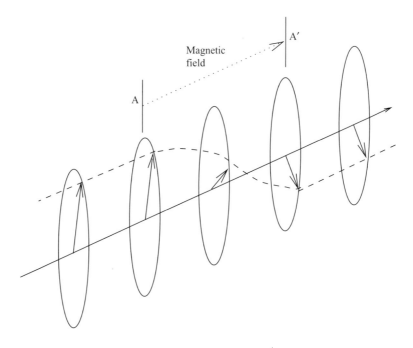

Fig. 6.13 The Faraday effect. Between the planes A, A', a longitudinal magnetic field separates the refractive index into different values for the two hands of circular polarization. The relative phases of the two circularly polarized components of the plane-polarized wave change, and the plane rotates.

electrons, such as in the ionosphere and in interstellar space, where the Faraday effect is easily demonstrated. The refractive index depends on the amplitude of the oscillation of the electrons in response to the electric field of the wave, which now includes a gyration around the steady magnetic field. The amplitude of the oscillation depends on the hand of the circular polarization as compared with the direction of natural gyration around the magnetic field.

A device that allows light to travel in one direction but not in the opposite direction, i.e. an *optical isolator*, can be made by placing a Faraday rotating medium between polarizers P_1 and P_2, which are set at 45° to each other. The longitudinal magnetic field in the Faraday rotator is arranged to give a rotation of 45°. Polarized light produced by the first polarizer, P_1, is rotated by 45° by the Faraday cell and is transmitted by the second polarizer, P_2. Light travelling from the opposite direction through P_2 receives a 45° rotation in the *same* direction, and is rejected by polarizer P_1. These devices find application in eliminating back reflections in optical fibre systems and in high-power laser amplifiers.

The familiar liquid crystal displays (LCD) of pocket calculators and wristwatches employ a different form of electro-optic modulator shown in Fig. 6.14. The active element is an LC, in which long organic molecules align naturally parallel to a liquid–glass interface, but can be realigned by an electric field.

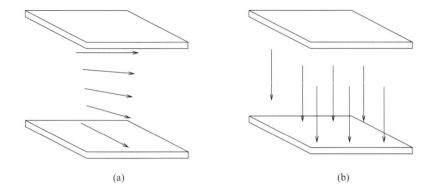

        (a)   (b)

Fig. 6.14 Molecular alignment in an LCD. The cell is typically about 5 μm thick: (a) with no field between the electrodes, the molecules align with the surface structures, which are arranged to give a twisted molecular structure; (b) an electric field aligns the molecules and removes the twist.

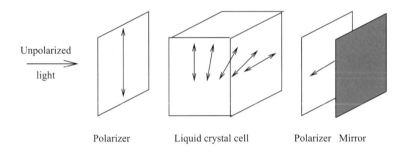

Polarizer   Liquid crystal cell   Polarizer Mirror

Fig. 6.15 An element of an LCD using reflected light.

The natural alignment is determined by conditions at the surface, so that a twisted structure can be produced in a thin cell, as in Fig. 6.14(a). The effect is now to rotate the plane of polarization, as in a quartz crystal, and in the arrangement of an LCD the reflected light from a mirror behind the cell is rejected by a polarizer. Application of an electric field rearranges the molecules as in Fig. 6.14(b), removing the spiral structure and allowing light to pass both ways through the cell.

    An illustration of the use of the LC cell as a reflective display device is shown in Fig. 6.15. The cell LC is placed between two polarizers, which are aligned to correspond to the directions of the molecular ordering on the two surfaces, and in front of a mirror. Incident light polarized by the first polarizer has its polarization direction rotated by the cell, passes through the second polarizer, is reflected by the mirror and again passes through both polarizers. With no electric field on the cell the image therefore appears bright. When a field is applied the direction of polarization of light is not rotated, light cannot travel in either direction through the cell and the image appears dark.

6.8   FORMAL DESCRIPTIONS OF POLARIZATION

The analysis of fully polarized light in terms of orthogonal components, either linear or circular, lends itself to a simple mathematical formulation in terms of *Jones vectors*. In this analysis the orthogonal linear components of the electric field that determine the state of polarization are written in column form

$$\begin{bmatrix} E_x \exp i\phi_x \\ E_y \exp i\phi_y \end{bmatrix},$$

where E_x, E_y are the scalar components on the x and y axes, with phases ϕ_x and ϕ_y. The Jones vector is a simplified and normalized form of this. For example, for $E_x = E_y$ and $\phi_x = \phi_y$, corresponding to a linear polarization at 45°, it is written

$$\begin{bmatrix} 1 \\ 1 \end{bmatrix}.$$

A phase difference appears as an exponential; for example, a circular polarization in which the y-component leads by 90° is written

$$\begin{bmatrix} 1 \\ \exp(-\pi/2) \end{bmatrix}.$$

or simply

$$\begin{bmatrix} 1 \\ -i \end{bmatrix}.$$

(Note that the minus sign arises from the direction of propagation in the wave form $f(kz - \omega t)$).

The Jones vector of the sum of two *coherent* light beams is the sum of their individual Jones vectors. The advantage of this formulation is that devices such as polarizers and wave plates can be specified by simple 2×2 matrices, the *Jones matrices*, and their operation on a polarized wave is found by matrix multiplication. As a reminder, the product

$$\begin{bmatrix} E_x \\ E_y \end{bmatrix} \begin{bmatrix} a_{11} & a_{12} \\ a_{21} & a_{22} \end{bmatrix}$$

gives the modified field components

$$E'_x = a_{11}E_x + a_{12}E_y, \tag{6.18}$$

$$E'_y = a_{21}E_x + a_{22}E_y. \tag{6.19}$$

The action of a series of components can be found from matrix multiplication, giving a single 2×2 matrix to represent the whole system. Matrices for the some of the polarizers and retarders dealt with in this chapter are tabulated below.

Jones matrices

Linear polarizers:

$$\text{horizontal } \begin{bmatrix} 1 & 0 \\ 0 & 0 \end{bmatrix}; \quad \text{vertical } \begin{bmatrix} 0 & 0 \\ 0 & 1 \end{bmatrix}; \quad 45° \ \frac{1}{2}\begin{bmatrix} 1 & 1 \\ 1 & 1 \end{bmatrix}.$$

Circular polarizer:

$$\text{left-hand } \frac{1}{2}\begin{bmatrix} 1 & i \\ -i & 1 \end{bmatrix}; \quad \text{right-hand } \frac{1}{2}\begin{bmatrix} 1 & -i \\ i & 1 \end{bmatrix}.$$

Polarization plane rotator:

$$\text{Rotation angle } \beta \ \begin{bmatrix} \cos\beta & -\sin\beta \\ \sin\beta & \cos\beta \end{bmatrix}.$$

Phase retarders: F is the fast axis of a quarter-wave plate (QWP).

$$\text{QWP, F vertical } \exp(i\pi/4)\begin{bmatrix} 1 & 0 \\ 0 & -i \end{bmatrix},$$

$$\text{QWP, F horizontal } \exp(i\pi/4)\begin{bmatrix} 1 & 0 \\ 0 & i \end{bmatrix}.$$

The Jones vectors apply only to fully polarized light. For partial polarization the appropriate analysis uses the *Stokes parameters*, which are functions of irradiance rather than fields. If the irradiance is measured through four different analysers: (i) passing all states; (ii) and (iii) linear analysers with axes at angles 0° and 45°; (iv) a circular analyser opaque to linear states, measuring respectively I_0, I_1, I_2, I_3, then the Stokes parameters are

$$S_0 = 2I_0, \tag{6.20}$$
$$S_1 = 2I_1 - 2I_0, \tag{6.21}$$
$$S_2 = 2I_2 - 2I_0, \tag{6.22}$$
$$S_3 = 2I_3 - 2I_0. \tag{6.23}$$

The Stokes parameters for two *incoherent* light beams are the sum of their individual Stokes parameters.

6.9   FURTHER READING

J. M. Bennett, *Handbook of Optics*: Polarization, Part 1, Chapter 5, Polarizers, Part 2, Chapter 3. McGraw-Hill, 1995.

E. Collett, *Polarized Light: Fundamentals and Applications*. Marcel Dekker, 1992.
A. Garrard and J. M. Burch, *Introduction to Matrix Methods in Optics*. Wiley, 1975.
D. S. Kliger, J. W. Lewis and C. E. Randall, *Polarized Light in Optics and Spectroscopy*. Academic Press, 1990.
W. A. Shurcliff and S. S. Ballard, *Polarized Light*. Van Nostrand, 1964.

NUMERICAL EXAMPLE 6

6.1 Calculate the thickness of a calcite quarter-wave plate for sodium D light ($\lambda = 589$ nm, given the refractive indices $n_o = 1.658$ and $n_e = 1.486$ for the two linearly polarized modes.

PROBLEMS 6

6.1 A pair of crossed polarizers, with axes at angles $\theta = 0°$ and $90°$, is placed in a beam of unpolarized light with flux density I_0, so that light emerges from the first with $I_1 = \frac{1}{2}I_0$ and from the second with $I_2 = 0$. A third polarizer is placed between the two at angle $\theta = 45°$. What then is I_2?
If the third polarizer rotates at angular frequency ω show that

$$I_2 = \frac{I_0}{16}(1 - \cos 4\omega t). \tag{6.24}$$

6.2 A plane-polarized wave propagates along the optical axis of quartz as two circularly polarized waves, so that the difference in refractive indices $n_L - n_R$ introduces a phase difference δ between the two. Show that the plane of polarization is rotated by angle $\delta/2$.
Calculate the thickness of quartz plate that will rotate the plane by $90°$ at wavelength 760 nm, given $|n_L - n_R| = 6 \times 10^{-5}$.

6.3 A printed page appears double if a doubly refracting crystalline plate is placed upon it. Why is it that a distant scene does not appear double when viewed through the same plate?

6.4 Why does a thin plate of doubly refracting crystal generally appear faintly coloured when it is placed between two polarizers?

6.5 Show that an elliptically polarized wave can be regarded as a combination of circularly and linearly polarized waves.

6.6 An elliptically polarized beam of light is passed through a quarter-wave plate and then through a sheet of polaroid. The quarter-wave plate is rotated to two positions where the polaroid shows the light to be plane polarized, and it is found that the plane of polarization is then at angles of $24°$ and $80°$ to the vertical. Describe the original elliptical polarization.

7

Interference and Fraunhofer diffraction

A man alike eminent in almost every department of human learning... [who] first established the undulatory theory of light, and first penetrated the obscurity which had veiled for ages the hieroglyphics of Egypt.
Tablet in Westminster Abbey commemorating Thomas Young (1773–1829).

We have seen in Chapter 1, where the idea of Huygens' secondary waves was introduced, that the future position of a wavefront may be derived from a past position by considering every point of the wavefront to be a source of secondary waves. If the wavefront effectively propagates along rays, then the geometric optics approach of Chapters 2 and 3 may be the most appropriate description of the progress of a wavefront, taking no account of the physical nature of the wave. We now turn to the phenomena of interference, and diffraction, where light is treated as a periodic wave, and ray optics provides a totally inadequate description.

Interference effects occur when two or more wavefronts are superposed, giving a resultant wave amplitude that depends on their relative phases. *Diffraction* is the spreading of waves from a wavefront limited in extent, occurring either when part of the wavefront is removed by an obstacle, or when all but a part of the wavefront is removed by an *aperture* or *stop*. The general theory that describes diffraction at large distances is due to Fraunhofer,[1] and is referred to as *Fraunhofer diffraction*.

In this chapter we start with simple examples of interference and diffraction that can be understood either intuitively or by using the phasor ideas and constructions of Chapter 2. We will find, however, that this simple approach is transcended by a general theory that uses Fourier analysis to analyse diffraction at any aperture.

---

[1] Fraunhofer (1787–1826), optician in Munich, known mainly in his lifetime for his skill in making telescope lenses and for solar spectroscopy. The dark absorption lines in the solar spectrum were named 'Fraunhofer lines'.

7.1   INTERFERENCE

Figure 7.1 shows two monochromatic plane waves, with amplitude A and wavelength λ, propagating at angles $\pm\theta$ to the z-axis. In the figure at a particular moment of time the solid and dotted lines correspond to positive and negative maxima. The two waves combine to give resultant positive and negative maxima of $+2A$ and $-2A$ where two solid and two dotted lines intersect. Where a solid line intersects with a dotted line the resultant is zero. Along the line OY in the y direction the resultant varies from $2A$ to zero to $-2A$ to zero to $+2A$ and so on. The intensity of the resultant, the square of the amplitude, varies as $4A^2$, 0, $4A^2$, 0, $4A^2$, etc. The pattern of intensity forms *interference fringes* spaced uniformly across the propagation direction z. We now find the shape and spacing of these fringes.

We have already noted that in any harmonic wave with a plane wavefront the phase changes linearly with distance along the direction of the wave, changing by 2π in distance λ; the phase is constant across the wavefront. Along the direction OY a distance y has a component $y \sin \theta$ in the direction of the wave, so that the phase change relative to $y = 0$ is

$$\phi = \pm 2\pi \frac{y \sin \theta}{\lambda}, \tag{7.1}$$

where the plus and minus signs correspond to the two waves.

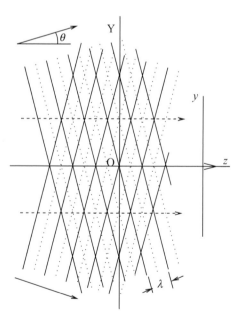

Fig. 7.1   Interference between two plane waves. The waves are crossing at angles $\pm\theta$ to the z-axis.

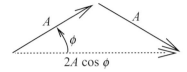

Fig. 7.2 Phasor diagram for crossing waves.

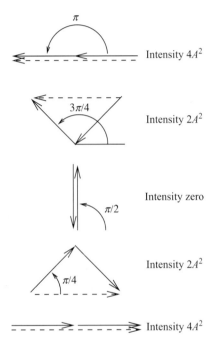

Fig. 7.3 Phasor diagrams across the interference fringes of Fig. 7.1, at intervals of $\pi/4$ in the phase ϕ.

The phasor diagram of Fig. 7.2 shows the phasors for the two crossing waves at a general point y on the line OY, with a phase difference 2ϕ given by Eq. (7.1). Phasor diagrams for $\phi = 0$, $\pi/4$, $\pi/2$, $3\pi/4$ and π are shown in Fig. 7.3, with the corresponding intensities. The intensity is given by the square of the resultant amplitude:

$$I = \left(A_{\text{resultant}}\right)^2 = \left(2A \cos \phi\right)^2 \tag{7.2}$$
$$= 4I_0 \cos^2 \phi, \tag{7.3}$$

where I_0 is the intensity of each plane wave alone. The variation along the y-axis of the intensity, which for light is the luminance, is the cosine curve shown in Fig. 7.4. This is known as a pattern of *cos squared* fringes.

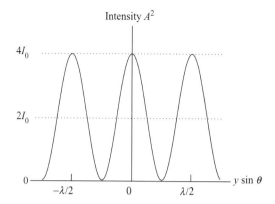

Fig. 7.4   The pattern of intensity along the y-axis for the crossing waves. Note that each wave alone would give an intensity I_0. The average intensity is $2I_0$, and the peak is $4I_0$.

7.2 YOUNG'S EXPERIMENT

A simple example of interference between two crossing waves is Young's experiment, which provided the first demonstration of the wave nature of light. Two closely spaced narrow slits, A and B in Fig. 7.5, transmit two elements of a light wave from a single source. The two sets of waves spread by diffraction, then overlap and interfere. If they then illuminate a screen, there will be light and dark bands across the illuminated patch; these are the interference fringes. Figure 7.5 shows that the geometry of the fringes becomes simpler at an increasing

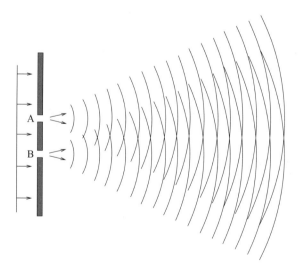

Fig. 7.5   Young's experiment. Light from the two pinholes or slits A, B spreads by diffraction, and the two sets of waves overlap and develop an interference pattern.

distance from the slits, where the two sets of waves from A and B behave like two sets of nearly plane waves crossing at a small angle, like the plane waves of Fig. 7.1. We will consider the effect for monochromatic light.

The light incident on the slits is a plane wave, so that each slit is the source of identical expanding waves. Consider the sum of the two waves at a point P, which is sufficiently far from A and B for the amplitudes of the two waves to be taken as equal. There is a phase difference between the two waves depending on the small difference l between the two light paths (Fig. 7.6(a)), so that the waves add as in Fig. 7.6(b). The path difference is $l = d \sin \theta$, giving a phase difference $\phi = 2\pi l / \lambda$, and the intensity I at P varies with the phase difference as in Eq. (7.3), giving

$$I = 4I_0 \cos^2 \frac{\pi l}{\lambda}, \qquad (7.4)$$

where I_0 is the intensity of each wave at P. The phase reference may be taken at O, half-way between the slits. The waves are then advanced and retarded on the reference by $\pi l / \lambda$, as in the phasor diagram of Fig. 7.6(b). Thus, where constructive interference occurs the intensity is four times that due to one slit, or *twice* the intensity due to two slits if interference did not happen. The conditions for constructive and destructive interference are as follows:

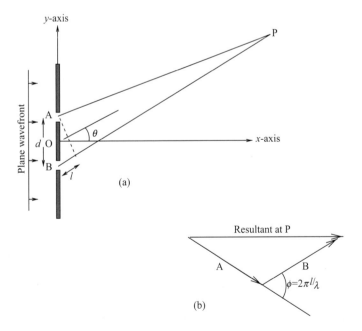

(a)

Resultant at P

(b)

Fig. 7.6 Young's experiment: (a) two wavelets spread out from the pair of slits, and interfere. Constructive or destructive interference occurs at P according to whether the path difference l is $N\lambda$ or $(N + \frac{1}{2})\lambda$; (b) phasor diagram for the sum at P, at a large distance from the slits. The phase reference (giving a horizontal phasor) is at O and the phase difference between the two waves is $\phi = 2\pi l / \lambda$.

$$\text{constructive } l = d \sin \theta = \pm N\lambda,$$
$$\text{destructive } l = d \sin \theta = \pm(N + \tfrac{1}{2})\lambda, \tag{7.5}$$

where N is an integer. N is called the *order* of the interference; it is the number of whole wavelengths difference in the paths to points where constructive interference takes place. The bright and dark 'fringes' that appear on a screen placed anywhere to the right of the double slit system can therefore each be labelled according to their order. The highest possible order is the integral part of d/λ. The fringes are spaced uniformly in $\sin \theta$ at intervals $N\lambda/d$, and at a distance z the linear spacing is $z\lambda/d$.

The condition that the distance is sufficient to allow the use of the simple relation $l = d \sin \theta$ is important; it is equivalent to the condition that the phase of any elementary wave from the screen is a linear function of distances x and y in the plane of the screen. This is the condition for *Fraunhofer diffraction*, of which two-slit interference is a special case. Under this condition, the whole of the screen, whatever the pattern of apertures, can be considered as a single diffracting object.

The amplitude of the Fraunhofer diffraction pattern of a pair of slit sources is the function

$$A(\theta) = A(0) \cos\left(\frac{\pi d}{\lambda} \sin \theta\right). \tag{7.6}$$

$A(0)$ is the amplitude of the diffracted wave when $\theta = 0$; negative values in Eq. (7.6) represent a reversal of phase. The intensity is the square of $A(\theta)$, giving $\cos^2$ fringes

$$I(\theta) = I(0) \cos^2\left(\frac{\pi d}{\lambda} \sin \theta\right). \tag{7.7}$$

Notice that the average intensity across several fringes is $2I_0$ (see Fig. 7.4), because the peaks of the upper half of the $\cos^2$ fringes just fill the troughs of the lower half. This is an example of an important general principle of all interference and diffraction effects: the energy is *redistributed* in space by these effects, but remains in total the same. This rather obvious remark enables some not so obvious predictions to be made; for example, if light is diffracted *into* the geometrical shadow of an object, the intensity must begin to fall *outside* the geometrical shadow to compensate. We examine this in the next chapter; it is an example of *Fresnel diffraction*.

A Young's double slit interference can easily be made with two parallel scratches on an overexposed photographic film. Fringes will be seen if a distant street lamp is viewed through the double slit; for slits 0.5 mm apart the angular spacing λ/d will be about $4'$.

7.3 DIFFRACTION AT A SINGLE SLIT

The simplest case of diffraction is for a single slit, width w, illuminated by a plane wavefront with uniform amplitude. The slit is perpendicular to the paper in Fig. 7.7(a), so we take as our basic element a strip of width dy, small compared with the wavelength. The diffraction pattern is observed at a large distance from the slit. Then each strip contributes an equal amplitude proportional to dy to the total at P, but the phase of each contribution depends on y as

$$\phi(y) = \frac{2\pi y \sin\theta}{\lambda}. \qquad (7.8)$$

The contributions of the elementary strips can be added in the phasor diagram, Fig. 7.7(b). The central strip, at O, is taken as the phase reference origin, appearing as a horizontal phasor in the diagram. Contributions from below O have phases retarded on this reference; these have been placed on the left-hand side of the phasor diagram. Contributions from above O appear on the right-hand side, and the diagram then becomes an arc of a circle. The resultant is a chord, representing the amplitude $A(\theta)$ of the resultant wave in direction θ. When $\theta = 0$ the phasor diagram is a straight line, representing the maximum amplitude $A(0)$. Hence, from Fig. 7.7 we see that

$$\frac{A(\theta)}{A(0)} = \frac{\sin\psi}{\psi}, \qquad (7.9)$$

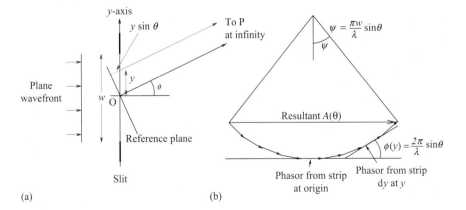

Fig. 7.7   Diffraction at a slit: (a) each elementary strip dy contributes an equal phasor with a phase varying linearly with y; (b) the phasor diagram is then part of a circle, allowing the resultant to be easily calculated.

where $\psi = \pi w \sin \theta / \lambda$ is the phase of the component from one edge of the slit. The function $\sin \psi / \psi$ is named sincψ. The intensity correspondingly varies as

$$\frac{I(\theta)}{I(0)} = \left(\frac{\sin \psi}{\psi}\right)^2. \tag{7.10}$$

Using exponential functions, we can express the sum of the elementary contributions as the integral

$$A(\theta) = \int_{-w/2}^{+w/2} a \exp(i\phi) dy, \tag{7.11}$$

which integrates directly to give the result Eq. (7.9).

We now see the vital connection between diffraction and the Fourier transform. In Chapter 4 we set out the Fourier transform in terms of time and frequency in Eq. (4.28), which we now repeat:

$$f(t) = \int_{-\infty}^{+\infty} F(\nu) \exp(2\pi i\nu t) d\nu. \tag{7.12}$$

In terms of spatial variables x, u, where u is a periodicity, the Fourier transform is

$$f(x) = \int_{-\infty}^{+\infty} F(u) \exp(2\pi ixu) du. \tag{7.13}$$

This is identical to Eq. (7.11) if $x = \sin \theta / \lambda$ and $u = y$. The integral need not extend to infinity, as there is no contribution from beyond the edges of the slit at $y = +w/2$ and $-w/2$. The integral (7.11) is, in fact, the Fourier transform of the 'top-hat' function, which was evaluated in section 4.12 and found to be the sinc function $(\sin \psi)/\psi$, as also found in Eq. (7.9). This equality between the Fraunhofer diffraction pattern and the Fourier transform of the aperture function is universal.

The sinc function is important in many branches of physics. By considering the behaviour of the phasor diagram in Fig. 7.8 as ψ changes we can easily see its main properties. At $\theta = 0$, $\psi = 0$ and the phasor is a straight line; sinc 0 is indeed unity. As ψ moves away from zero the phasor diagram begins to curve into the arc of a circle; the resultant, the chord, shortens but remains parallel to the contribution from the centre of the slit. Thus, the amplitude decreases but the phase remains the same. When $\psi = \pi$, so that $\sin \theta = \lambda/w$, the phasor diagram is a closed circle and the amplitude is zero. As ψ increases beyond π there is again a resultant, but it is in the opposite direction; the amplitude is now negative, or we may say that the phase has changed by π in going through the zero. At approximately $\psi = 3\pi/2$ the resultant reaches its extreme negative

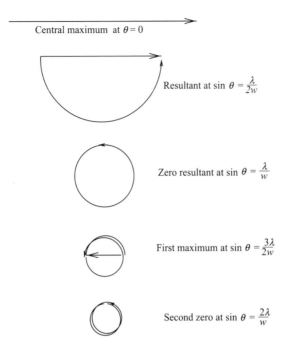

Central maximum at $\theta = 0$

Resultant at $\sin \theta = \frac{\lambda}{2w}$

Zero resultant at $\sin \theta = \frac{\lambda}{w}$

First maximum at $\sin \theta = \frac{3\lambda}{2w}$

Second zero at $\sin \theta = \frac{2\lambda}{w}$

Fig. 7.8   As the direction θ of diffraction moves away from zero, the phasor, straight at $\theta = 0$, curls up as shown. Each zero corresponds to the phasor being wrapped around an integral number of times.

value and then begins to shorten as ψ increases. At $\psi = 2\pi$, so that $\sin \theta = 2(\lambda/w)$, the resultant is again zero.

It is easy to see that this oscillatory behaviour continues, giving zeros at exactly $\sin \theta = \pm\lambda/w, \pm 2(\lambda/w), \pm 3(\lambda/w)$, etc. corresponding to $\psi = \pm\pi, \pm 2\pi, \pm 3\pi$, etc. The values of ψ to give the maximum and minimum values, where $dA/d\psi = 0$, may be found by differentiation:

$$\frac{dA}{d\psi} = \frac{d}{d\psi}\left(\frac{\sin \psi}{\psi}\right) = \frac{\psi \cos \psi - \sin \psi}{\psi^2} \tag{7.14}$$

so that

$$\frac{dA}{d\psi} = 0, \tag{7.15}$$

where $\psi = \tan \psi$.

This intrinsic equation is best solved either graphically or numerically. If n is an integer the extremes for large n are at

$$\psi = \pm\left(n + \frac{1}{2}\right)\pi. \tag{7.16}$$

For small values of n the maxima and minima are somewhat closer in than the values given by Eq. (7.16). For example, the first extremes come at $\psi = 1.43\pi$

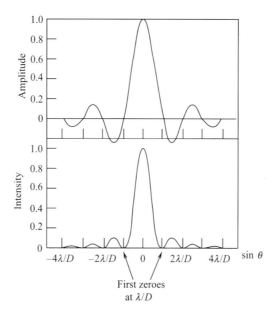

Fig. 7.9   The amplitude function $\sin(\pi w \sin\theta/\lambda)/(\pi w \sin\theta/\lambda)$ and its square, the intensity function, for diffraction at a slit of width w.

rather than at $\pm 1.5\pi$. In terms of the phasor diagram these extremes correspond to the same length of phasor being formed into a circle by being wrapped around approximately $1\frac{1}{2}$, $2\frac{1}{2}$, times, etc. Similarly, the zeros are given exactly by the same length of phasor being wrapped around 1, 2, 3, etc. times, and the central maximum by its being straight. Remember that the change in $\sin\theta$ between the zeros on each side of the central maximum is twice that between subsequent zeros. In amplitude the first subsidiary lobe is negative and about 22% of the central lobe.

The eye or a photographic plate is directly sensitive to the intensity, and thus proportional to the amplitude squared. The amplitude and intensity are plotted in Fig. 7.9. In intensity the first subsidiary maximum is only about 5% of the main maximum, and has fallen to 0.5% by the fourth.

7.4   THE GENERAL APERTURE

Having looked at a simple but important case of Fraunhofer diffraction, we now go on to make an important generalization. We generalize in two ways:

(i) by considering diffraction in two dimensions;

(ii) by allowing the complex amplitude in the aperture to be non-uniform; that is to say, it can have an arbitrary distribution of amplitude and phase.

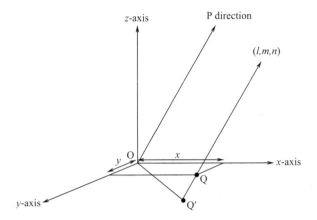

Fig. 7.10 A general aperture in the *xy*-plane. The contribution from Q in the P-direction specified by the direction cosines (l, m, n) is advanced in phase compared with those from the origin by $(2\pi/\lambda)QQ' = (2\pi/\lambda)(lx + my)$.

Let the aperture be any shape in the *xy*-plane. Then the direction P of interest may be specified[2] by the direction cosines (l, m, n). We choose a reference plane through the origin O perpendicular to the direction P given by (l, m, n). As before, Q is a general point in the aperture and Q′ the point in which the line in the direction (l, m, n) to P cuts the reference plane. The distance from Q to the distant point P is shorter than that from O by QQ′. From Fig. 7.10,

$$QQ' = lx + my. \tag{7.17}$$

Hence, on account of path difference the phase of light from Q at P will be advanced by $(2\pi/\lambda)(lx + my)$.

So much for the extension to two dimensions. Now let both the amplitude and phase of the wavefront in the aperture be functions of (x, y). Mathematically, we can express this by letting the amplitude be a complex function of position $F(x, y)$. An element d*x*d*y* at (x, y) will then make a contribution to the amplitude at P of $F(x, y)$ d*x*d*y*, but rotated in phase by $(2\pi/\lambda)(lx + my)$. Expressing the sum of all such components over the aperture by a double integral,

$$A(l, m) = C' \int\int F(x, y) \exp\left[-\frac{2\pi i}{\lambda}(lx + my)\right] dx\, dy. \tag{7.18}$$

Be clear what this integral represents. Each element of the aperture d*x* d*y* contributes a phasor of length $F(x, y)$ d*x*d*y* and phase given by the initial phase of $F(x, y)$ advanced by the phase due to the path difference QQ′ (the

---

[2] The direction cosines are the cosines of the angles between the direction they define and the coordinate axes.

minus sign in Eq. (7.18) is due to a phase *advance*). The complex integral is just a way of arriving at the resultant in the phasor diagram. Equation (7.18) allows the calculation of the complex amplitude at a distant point P in terms of the complex amplitude $F(x, y)$ in the aperture. The dimensions must be the same on each side, and this is taken care of by the constant C' with dimensions $[\text{length}]^{-2}$. For present purposes we are concerned with the form of the diffraction pattern rather than its absolute value.

Equation (7.18) has the form of a Fourier transform in two dimensions, using the pairs of transform variables x/λ and l, y/λ and m. The coordinates x/λ and y/λ in the aperture are lengths measured in wavelengths, while l and m are sines of angles measured from the normal to the aperture. Thus, we have the important general result:

The Fraunhofer diffraction pattern in amplitude of an aperture is the Fourier transform of the complex amplitude distribution across the aperture.

7.5   RECTANGULAR AND CIRCULAR APERTURES

We can now apply the general expression of Eq. (7.18) to some particularly important examples of diffracting apertures, using Fourier transforms directly. For the first three of the following examples we need only use the one-dimensional form of Eq. (7.18).

Uniformly illuminated single slit

As in section 7.3, a uniformly illuminated slit of width w is represented by an amplitude distribution

$$F(x) = F_0, \quad \text{for } |y| < w/2$$
$$= 0, \quad \text{for } |y| > w/2. \tag{7.19}$$

The Fourier transform, i.e. the amplitude function $A(\theta)$, is

$$A(\theta) = A(0) \operatorname{sinc}\left(\frac{\pi w \sin \theta}{\lambda}\right). \tag{7.20}$$

The intensity distribution is the diffraction pattern shown in Fig. 7.9.

Two infinitesimally narrow slits

The Young's double slit of section 7.2 is represented by the amplitude distribution

$$F(x) = F_0(\delta(x - d/2) + \delta(x + d/2)), \tag{7.21}$$

where the narrow slits are represented by two delta functions (section 4.14) located at $(x - d/2)$ and $(x + d/2)$. The Fourier transform, i.e. the double slit interference pattern, is

$$A(\theta) = A(0) \cos\left(\frac{\pi}{d} \sin\theta\lambda\right). \tag{7.22}$$

Two slits with finite width

For slits with width w, separated by d, the results of the previous sections combine to give the amplitude distribution

$$A(\theta) = A(0) \operatorname{sinc}\left(\frac{\pi w \sin\theta}{\lambda}\right) \cos\left(\frac{\pi}{d} \sin\theta\lambda\right). \tag{7.23}$$

Note that this result is an example of the convolution theorem (section 4.13); the double slit pattern is the convolution of a top-hat function with the narrow double slit, and the resultant Fourier transform is the product of the two individual transforms.

Uniformly illuminated rectangular aperture

Here, $F(x, y) = F_0$, within the aperture sides extending from $-a/2$ to $+a/2$ and $-b/2$ to $+b/2$. The aperture sides are aligned along the x and y axes. The diffraction pattern in terms of direction cosines l and m is

$$A(l, m) = C'F_0 \int_{-a/2}^{+a/2} \int_{-b/2}^{+b/2} \exp\left[-\frac{2\pi i}{\lambda}(lx + my)\right] dx\, dy. \tag{7.24}$$

The double integrals in Eq. (7.24) are separable, so

$$A(l, m) = C'F_0 \int_{-a/2}^{+a/2} \exp\left(-\frac{2\pi ilx}{\lambda}\right) dx \int_{-b/2}^{+b/2} \exp\left(-\frac{2\pi imy}{\lambda}\right) dy. \tag{7.25}$$

The two integrals both give sinc functions:

$$\int_{-a/2}^{+a/2} \exp\left(-\frac{2\pi ilx}{\lambda}\right) dx = -\frac{\lambda}{2\pi il}\left[\exp\left(-\frac{2\pi ilx}{\lambda}\right)\right]_{-a/2}^{+a/2} \tag{7.26}$$

$$= \frac{\sin(\pi la/\lambda)}{\pi la/\lambda} \tag{7.27}$$

with a similar expression for the y integral. The amplitude $A(0, 0)$ at the centre of the pattern where l and m are zero is $C'F_0ab$, so the amplitude $A(l,m)$ is given in terms of that at $(0, 0)$ by

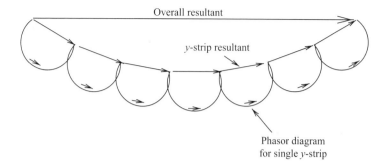

Fig. 7.11   Phasor diagram for a rectangular aperture for diffraction in a direction off both l and m axes. The diagram corresponds to a point on the central maximum close to the first minimum in the m-direction, but fairly far up the central maximum in the l-direction.

$$A(l,m) = A(0,0) \frac{\sin(\pi la/\lambda)}{\pi la/\lambda} \frac{\sin(\pi mb/\lambda)}{\pi mb/\lambda}. \qquad (7.28)$$

The intensity is given by Eq. (7.28) squared. Thus, the expression for the Fraunhofer diffraction pattern of a uniformly illuminated rectangular aperture is proportional to the product of the expressions for the separate diffraction patterns of two crossed slits. Along the l and m axes the subsidiary maxima have the same values as those of each single slit, since one of the product terms of Eq. (7.28) is unity on each axis. Faint subsidiary maxima exist in the four quadrants. For these, both product terms are of the order of a few percent, and the brightest of them, at approximately $(\pm 1.5\lambda/a, 1.5\lambda/b)$, is $(0.047)^2$ or 2.2×10^{-3} of the intensity at the centre.

Physically, we can see why this is so by considering the phasor diagram shown in Fig. 7.11. Along the l-axis each strip parallel to the y-axis is at the same phase all over. Phasors from each of these strips combine to give the diffraction pattern as in the single slit; the same applies to the m-axis. If we consider a general point (l, m), however, then the phasor from a strip parallel to the y-axis is already bent into the arc of a circle because of the phase along the strip. The total phasor diagram is thus constructed of phasors already bent, and hence comes out very small.

Uniformly illuminated circular aperture

How can we now apply the general theorem (Eq. (7.18)) to a circular aperture? First, we note that the diffraction pattern must be circularly symmetrical. The result is simply written as the Fourier transform

$$A(l,m) = C'F_0 \int\int \exp\left[\frac{2\pi i}{\lambda}(lx + my)\right] dx\, dy. \qquad (7.29)$$

Evaluating this integral is messy (because the limits link x and y). It is simpler to change directly to polar coordinates (r, θ) in the aperture and (w, ϕ) in the diffraction pattern. Now $r \cos \theta = x$; $r \sin \theta = y$; and an elementary area is $r dr d\theta$. In the diffraction pattern coordinates, $w \cos \phi = l$; $w \sin \phi = m$, so that $w = \sqrt{(l^2 + m^2)}$ = sine of the angle that (l, m) makes with the axis $(l, m = 0)$. The amplitude for a circular aperture is now given by

$$A(w, \phi) = C' F_0 \int_0^a \int_0^{2\pi} r \, dr \, d\theta \exp\left[-\frac{2\pi i}{\lambda} rw \cos(\theta - \phi)\right]. \qquad (7.30)$$

The integral in Eq. (7.30) is only soluble analytically in terms of Bessel functions. However, most integrals can only be solved in terms of some sort of tabulated functions; it is merely that Bessel's are less familiar than the sines needed to perform the integrals of Eq. (7.27), for example. This being done, we have

$$A(w, \phi) = A(0, 0) \frac{2J_1(2\pi aw/\lambda)}{2\pi aw/\lambda}. \qquad (7.31)$$

The square of the right-hand side of Eq. (7.31) gives the intensity pattern for Fraunhofer diffraction at a circular aperture.

$$I(w, \phi) = I(0, 0) \left[\frac{2J_1(2\pi aw/\lambda)}{2\pi aw/\lambda}\right]^2. \qquad (7.32)$$

This famous and important result was first derived by George Airy in 1835 at about the time he became Astronomer Royal. It is of especial importance to astronomers since it is the pattern produced in the focal plane of an ideal telescope with a circular lens (or mirror) by a plane wavefront from a distant star. The circular edge of the objective lens of the telescope limits the aperture, and it is the angular width of the diffraction patterns due to two nearby stars that determines whether or not they can be distinguished.

Airy's pattern both in amplitude and intensity is plotted in Fig. 7.12. At first sight it looks like the similar plots for the slit in Fig. 7.9, but there are several differences. Most important, it is a *ring* system, so that the plots are radial sections of a pattern possessing circular symmetry. The first zero is at $1.22 \lambda/D$ (where $D = 2a$ is the diameter of the aperture), compared with λ/w for a slit of width w. The zeros are not equally spaced but tend to a separation of λ/D for large values of w. The first subsidiary maximum of intensity is lower: 1.75% compared with 4.72% for the slit.

The Airy diffraction pattern is often quoted in relation to the angular resolving power of telescopes and similar optical instruments. If, for example, a double star is to be seen as two clearly distinguishable images, each image must be smaller than the separation between them. The resolving power of the

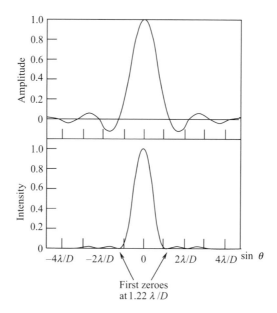

Fig. 7.12   The amplitude and intensity functions $2\{_1(\pi d \sin\theta/\lambda)\}/(\pi d \sin\theta/\lambda)$ and its square for diffraction at a circular aperture of diameter D. The substitutions have been made in the expression in Eq. (7.31) to allow direct comparison with Fig. 7.9 for a single slit.

telescope is therefore ideally $1.22\lambda/D$, where D is the aperture; for example, for $D = 2.4\,\mathrm{m}$ (the Hubble Space Telescope) the angular resolution in visible light is around 0.05 arcsec. (This is unattainable for a comparable terrestrial telescope, because of random refraction effects in the atmosphere.)

7.6   THE FIELD AT THE EDGE OF AN APERTURE

In the diffraction theory of this chapter, and indeed in most of the later diffraction theory, we have assumed that the wave can be described by a scalar variable, and made no mention of polarization. It is not usually necessary to calculate diffraction separately for each component of the polarization of the vector wave, but we can easily see one situation where this is necessary. It concerns the assumption, made in Section 7.5, that the diffraction of a plane wave at a slit may be calculated as if the amplitude of the wave were uniform over the whole slit.

Suppose the diffracting slit is made of perfectly conducting metal sheet. Then the electric field must be zero in the sheet; immediately outside the sheet the component parallel to the slit edge must also be zero. Only at distances greater than about one wavelength from the edge can the field reach its full value. The wavefront is therefore narrower for polarization parallel to the edge than it is for polarization perpendicular to the edge, and the width of the diffraction patterns will correspondingly be somewhat different. This effect is only important if the scale of the diffracting object or slit is not large compared with one wavelength.

Evidently there can be considerable complications introduced by the behaviour of the wavefront close to a diffracting object. The full solution of such problems involves a detailed description of the wavefront, which must accord with the boundary conditions at the edge of the object. When the wave is prescribed, the diffraction pattern can be calculated either by the simple theory of this chapter, or in more difficult cases by a full wave theory due to Kirchhoff. This theory is discussed briefly in the next chapter. Fortunately it is often possible to proceed without the full rigour of the Kirchhoff theory.

7.7 FURTHER READING

M. Francon, *Optical Image Formation and Processing*. Academic Press, 1979.
K. J. Gasvik, *Optical Metrology*. Wiley, 1995.
S. G. Lipson, H. Lipson and D. S. Tannhauser, *Optical Physics*, 3rd edn. Cambridge University Press, 1993.
W. H. Steel, *Interferometry*. Cambridge University Press, 1983.

NUMERICAL EXAMPLES 7

7.1 A Young's slit experiment has two very narrow slits separated by 1.1 mm. At what angles are the first and second order fringes for red and blue light (700 nm and 450 nm, respectively)?

7.2 If the slits in the previous example are each of width 0.01 mm, how many red fringes might one see easily?

7.3 A simple demonstration of diffraction and interference can be made by scratching lines through the emulsion of an undeveloped photographic plate, and looking through the lines at a distant bright light with the plate held close to the eye. Find the angular breadth of the pattern given by a sodium lamp ($\lambda = 589$ nm) with a slit width of 0.1 mm. What will be the effect for two such slits 1 mm apart?

7.4 What is the limiting angular resolution of the astronomical telescope with objective diameter $D_1 = 40$ cm described in Numerical Example 3.1?

PROBLEMS 7

7.1 A single slit width D is made into a double slit by obscuring its centre with a progressively wider opaque strip, leaving two slits with width a. Draw phasor diagrams for the single slit diffraction pattern at the edge of the main maximum and at the first zero, and show how these are changed as the opaque strip is widened.

Sketch the diffraction and interference patterns for the single slit and double slit on the same scales of intensity and angle.

7.2 Two pinholes 0.1 mm in diameter and 0.5 mm apart are illuminated from behind by a parallel beam of monochromatic light, wavelength 500 nm. A convex lens of diameter 1 cm and focal length 1 m is placed 110 cm from the holes. Describe the pattern formed on a screen placed (i) 1 m; (ii) 11 m from the lens.

7.3 Show that the slit interference pattern in Fig. 7.10 can be observed using a source of light that is extended along a line parallel to the slit. What is observed when the line source is not parallel to the slit?

7.4 Estimate the smallest possible angular beam width of (i) a paraboloid radio telescope, 80 m in diameter, used at a wavelength of 20 cm; (ii) a laser operating at a wavelength of 600 nm, with an aperture of 1 cm.

7.5 An aperture in the form of an equilateral triangle diffracts a plane monochromatic wave. The side of the triangle is 20 wavelengths long. Find the directions of the zeros of the diffraction pattern closest to the normal.

7.6 From the asymptotic expansion for large z,

$$J_1(z) = \frac{\sin z - \cos z}{(\pi z)^{1/2}},$$

show that the angular distance between diffraction minima far from the axis of a circular aperture, diameter d, is approximately λ/d.

7.7 The altitude of aircraft approaching land is controlled by a 'glide path' in which a radio transmitter of wavelength 90 cm forms interference fringes. The fringes are formed from a transmitter at height h above a conducting ground plane. Find the height h for a maximum signal to be received along a path at $3°$ elevation from the airfield.

7.8 An image of a narrow slit, illuminated from behind by light of wavelength 500 nm, is formed on a screen by a convex lens of focal length 100 cm. The slit is 200 cm from the lens.

A second slit, parallel to the first, now limits the beam of light to a width of 0.5 mm. This slit is placed successively (i) 100 cm from the screen; (ii) in contact with the lens; (iii) 100 cm from the first slit. What is the width between the first zeros of the diffraction pattern in each case?

7.9 Calculate approximate values for the theoretical angular resolution of (i) a 100-m radio telescope working at $\lambda = 5$ cm; (ii) the unaided human eye, aperture 4 mm, at $\lambda = 500$ nm; (iii) an 8-m diameter optical telescope at $\lambda = 1\,\mu$m; (iv) an optical interferometer with 100-m baseline at $\lambda = 500$ nm; (v) a radio interferometer, working at $\lambda = 10$ cm, with baseline 6000 km.

8

Fresnel diffraction

Augustin Jean Fresnel (1788–1827), ... *unable to read until the age of eight,* ... *the first to construct multiple lenses for lighthouses* ... *was enabled in the most conclusive manner to account for the phenomena of interference in accordance with the undulatory theory.*
Encyclopaedia Britannica.

Diffraction as treated in Chapter 7 is concerned with the *angular* spread of light leaving an aperture of arbitrary shape and size; if the light then falls on a screen at a large distance, the pattern of illumination is described adequately by this angular distribution. But if the screen is close to the aperture, so that one might expect to see a sharp shadow at the edges, we see instead diffraction fringes, whose analysis involves a theory introduced by Fresnel. A famous prediction of this theory was that the shadow of a circular object should have a central bright spot; the demonstration that this indeed exists was, and still is, a powerful argument in support of the wave theory of light.

In this chapter we set out the formal distinction between Fraunhofer and Fresnel diffraction, and show how the diffraction fringes at the edge of a geometric shadow may be analysed using Fresnel integrals rather than the Fourier integrals appropriate to Fraunhofer diffraction.

8.1 FRAUNHOFER AND FRESNEL DIFFRACTION

In any diffraction problem we find the amplitude and phase of the light wave at a point by adding all contributions by every possible path from a source to that point. In Chapter 7 Fraunhofer diffraction was characterized by a linear variation of the phase of contributions from elements across an aperture. At a point close to an aperture or an obstacle the phase of these contributions will no longer vary linearly with distance across the aperture, and quadratic terms must be introduced. This is typical of Fresnel diffraction; the results are no longer given by Fourier transforms as in Fraunhofer diffraction.

The general problem is illustrated in Fig. 8.1. Each element of the wavefront at the aperture is considered as the source of a secondary Huygens' wavelet; the resultant amplitude and phase at any point P is determined by summing these wavelets, as in a phasor diagram. The phase of each wavelet is behind that of the wave at Q by an amount depending on the distance PQ and the wavelength. This summation of Huygens' wavelets taking account of their phase is the Huygens–Fresnel diffraction theory; it was Fresnel who contributed the essential idea of the interference of the Huygens' secondary wavelets. Figure 8.1(a) shows how the distance PQ, and accordingly the phases of the wavelets, varies according to the position of the source Q; (b) shows the linear variation typical of Fraunhofer diffraction; and (c) shows the quadratic variation typical of Fresnel diffraction.

The transition from Fresnel to Fraunhofer diffraction is illustrated for slit diffraction in Fig. 8.2. To determine the relative phases of contributions across

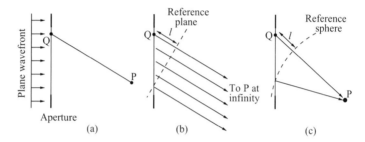

Fig. 8.1 (a) The amplitude and phase at P may be considered as the sum of Huygens' wavelets from points such as Q in the aperture. In Fraunhofer diffraction the phase varies linearly across the aperture, as in (b); in Fresnel diffraction it varies quadratically, as in (c).

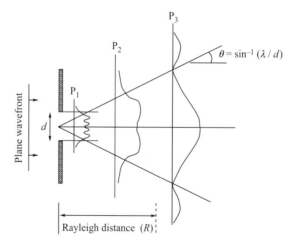

Fig. 8.2 Transition from Fresnel to Fraunhofer diffraction. A portion of a plane wave W passes through a slit, width d. Intensity distributions across the wave are shown for planes P_1 (close to the slit), P_2 (just inside the Rayleigh distance), and P_3 (beyond the Rayleigh distance).

the aperture for a point on any plane P_3 considerably beyond the distance R, the linear approximation of Fig. 8.1(b) is sufficient. But for a point on the plane P_1 inside the distance R, equal distances lie on a spherical surface rather than a plane, as in Fig. 8.1(c). For a point such as P_3 the difference between a sphere and a plane becomes unimportant, since if the maximum deviation is less than about $\lambda/8$ it has little effect on the phasors that add to give the resultant at P_3. The distance R, dividing the two regimes, is known as the *Rayleigh distance*; for an aperture width d it is given by

$$R = \frac{d^2}{\lambda}. \tag{8.1}$$

We return to this definition in section 8.4 below.

In section 8.6 we consider two further factors that may complicate some applications of Fresnel diffraction theory:

(i)   The distances from P of elements of the aperture may vary sufficiently to have a significant effect on wave amplitude as well as phase.

(ii) The line to P from different parts of the aperture may make considerably different angles with the normal to the surface of the aperture; this *inclination factor* also may have a significant effect on amplitude.

Fortunately there are two cases of particular interest that can be solved without detailed analysis of these factors, and that are well illustrated by graphical means as well as by simplified integrals. These are diffraction at a straight edge and at circular holes or obstacles.

8.2   SHADOW EDGES – FRESNEL DIFFRACTION AT A STRAIGHT EDGE

One of the most interesting predictions of the wave theory of light is that there should be some light within a geometric shadow, and interference fringes just outside it. The effect, seen in the photograph Fig. 8.3 and in the irradiance plot of Fig. 8.4, is that at the geometrical edge of a shadow the intensity is already reduced to a quarter of the undisturbed intensity, falling monotonically to zero within the shadow. Outside the shadow the intensity increases to more than its undisturbed value and oscillates with increasing frequency as it approaches a uniform value. These are the 'fringes': a name that has been extended to many other types of interference phenomena.

Consider the diffraction of a plane wave incident normally on a straight-edged obstacle. We will evaluate the contributions of wavelets from strips parallel to the edge of the obstacle to the wave at a point P at distance s from the obstacle and on the edge of the geometric shadow. We take as phase reference the phase of a wave from the closest point in the plane of the incident wave. The extra path length from a strip distant h from this point gives a phase delay of

$$\phi(h) = \frac{2\pi}{\lambda}\left[(s^2 + h^2)^{1/2} - s\right] \approx \frac{\pi h^2}{\lambda s}. \tag{8.2}$$

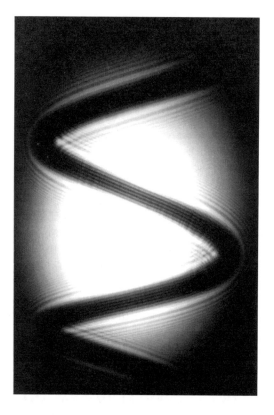

Fig. 8.3 Fresnel diffraction fringes at the shadow edges of a spiral spring. (Paul Treadwell, University of Manchester.)

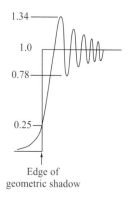

Fig. 8.4 Fresnel diffraction; irradiance distribution for a straight edge.

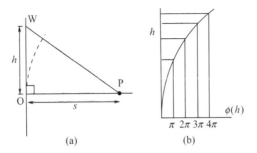

(a)   (b)

Fig. 8.5 (a) Fresnel diffraction. Portions of the wavefront W contribute to the wave at P according to their amplitude (proportional to dh and phase relative to the contribution from O (proportional to h^2); (b) the decline in area of the half-period zones.

The approximation, in which the phase ϕ increases as the square of h, is valid when $h \ll s$. If we divide the contributions in equal increments of phase, the corresponding increments of h decrease as h increases. The plot of $\phi(h)$ in Fig. 8.5 is marked off at intervals of π in phase, showing the decreasing width of zones across which the phase reverses. Zones marked off in this way at intervals of π in phase are known as 'half-period zones'.

We can now construct a phasor diagram made up of the contribution of infinitesimal strips to the resultant at P. The contribution of an infinitesimal strip of width dh at h has a phase $\phi(h)$ given by Eq. (8.2) and an amplitude proportional to dh. The phasors from each contribution may be added geometrically by adding their components along two axes x and y. Taking the x-axis as the phase reference these components are

$$dx = dh \cos \frac{\pi h^2}{\lambda s} \quad \text{and} \quad dy = dh \sin \frac{\pi h^2}{\lambda s}. \tag{8.3}$$

The x and y components of the phasor at P resulting from contributions from the origin up to any value of h are now given by the integrals of the expressions of Eq. (8.3). As h increases, the tip of the phasor traces a spiral, with the property that the angle $\phi(h)$ that its tangent makes with the x-axis is proportional to the square of the distance along it from the origin. (It is instructive to notice here that if $\phi(h)$ were simply proportional to the distance along it, then the spiral would become a circle through the origin with its diameter along the y-axis. This is why the corresponding Fraunhofer phasor diagrams are circular!)

It is usual when plotting this spiral to do so in terms of a dimensionless variable v, which is a distance along the spiral. It is related in the present case to the variable h by

$$v = h \left(\frac{2}{\lambda s} \right)^{1/2}. \tag{8.4}$$

Then the coordinates x, y of any point a distance v along the spiral are

$$x = \int_0^v \cos\frac{\pi v'^2}{2}\,dv', \quad y = \int_0^v \sin\frac{\pi v'^2}{2}\,dv'. \tag{8.5}$$

The integrals of Eqs. (8.5) are called *Fresnel integrals*, and the plot of their value as v varies is called *Cornu's spiral*, shown in Fig. 8.6. The oscillations of irradiance in the fringes correspond to the turns in the spiral, which for large v contracts to a point $Z(x = \frac{1}{2}, y = \frac{1}{2})$.

The edges of the half-period zones correspond to points on the spiral where its tangent is parallel to the x-axis. The Mth such position is given by

$$\phi(h) = M\pi = \frac{\pi}{2}v^2, \tag{8.6}$$

or

$$v = \sqrt{(2M)}. \tag{8.7}$$

The phasor representing the wave at P, on the edge of the geometric shadow, is the resultant of the whole spiral from the origin to Z. Now imagine removing the obstacle and opening the other half-plane. Then the half of the plane wavefront that was covered by the obstacle can clearly be treated similarly, and contributes another branch of the Cornu spiral in the third quadrant. The whole curve is shown in Fig. 8.6. The resultant Z'Z when there is no screen at all is clearly double the resultant OZ with the screen in place. This explains why the

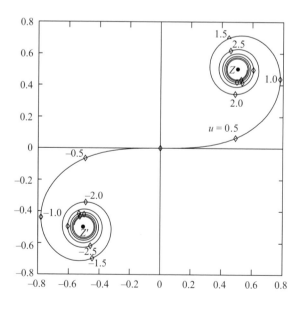

Fig. 8.6 Cornu's spiral.

irradiance at the position of the geometric shadow is a quarter of that of the undisturbed wave.

Now starting with the undisturbed wave, consider how the amplitude and intensity at P vary as a half-plane is moved from infinity across the plane wave. Starting at Z', a growing proportion of the spiral is deleted (Fig. 8.7). The resultant, instead of being Z'Z, is DZ. D moves around the spiral, so the amplitude begins to show oscillations above and below its undisturbed value. Each extreme represents the deletion of one half-turn of spiral, corresponding to a movement in by one half-period zone. If w is the coordinate of the edge of the plane, the rate of oscillation increases as w^2; as the edge moves in the spiral gets bigger and the oscillations become larger and less rapid. The last minimum and the last maximum are 0.88 and 1.16 of the undisturbed wave amplitude, giving irradiances of 0.78 and 1.34 of the unobstructed irradiance. From here on the resultant moves smoothly, arriving at the origin at just half the amplitude and a quarter the irradiance. As w becomes positive the resultant becomes a short vector joining Z to a point on the spiral around Z that rotates, shortening smoothly in length and reducing rapidly in size as P gets deeper into the geometric shadow.

It is quite easy to observe the first bright fringe around the edge of a shadow in white light, although of course the further ones get progressively out of step due to the large range of wavelengths. For example, the shadow cast by the back of a chair placed half-way across a room, illuminated with a car headlamp bulb at one side of the room shows the bright fringe quite convincingly around its shadow on the opposite wall. If one looks back from a position in the shadow area towards the obstacle, the edge of the obstacle appears bright. This is the light that is diffracted into the shadow; it appears to originate at the edge itself, and it is sometimes referred to as an 'edge wave'.

An interesting point to notice is that whilst the scale of Fresnel fringes is determined by the wavelength and the distance s from the edge to the plane on which the shadow is observed, the ratio of the oscillations to the undisturbed irradiance is always the same. So these effects still occur even at short X-ray wavelengths.

8.3 DIFFRACTION OF CYLINDRICAL WAVEFRONTS

The treatment in the last section was of the diffraction of a plane wavefront at an edge. It is very easy to extend this to the diffraction of cylindrical wavefronts such as the wavefront emerging from a slit. If the source of the wavefront is r from the diffracting screen as shown in Fig. 8.8, the extra phase in the path that passes a distance h from the centre line is given by

$$\phi(h) = \frac{2\pi}{\lambda} h^2 \left(\frac{1}{2s} + \frac{1}{2r} \right). \tag{8.8}$$

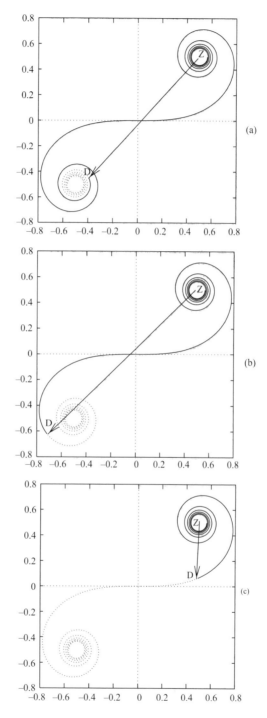

Fig. 8.7 Edge diffraction. Phasor diagrams for successive positions of a shadow edge: (a) a diffraction minimum; (b) the first diffraction maximum; (c) inside the geometric shadow.

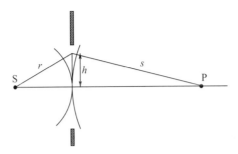

Fig. 8.8 Geometry for the diffraction of cylindrical wavefronts.

This is similar to Eq. (8.2) with the addition of the $h^2/2r$ term to take account of the curvature of the wavefront before diffraction. The same argument can be followed through with this slightly more complicated expression. It turns out that the Cornu spiral can be applied as before if instead of the change of variable in Eq. (8.4) the substitution

$$v^2 = \frac{2h^2}{\lambda} \frac{(r+s)}{rs} \qquad (8.9)$$

is made. The same spiral can then be used with just this change of scale factor. For example, if $r = s$ the diffraction pattern would be scaled by $\frac{1}{\sqrt{2}}$ compared with the plane wave ($r = \infty$) case. For simplicity we shall go on to discuss the Fresnel diffraction of plane waves by slits, but all the results are easily adapted to cylindrical waves by the change of scale given by Eq. (8.9).

8.4 FRESNEL DIFFRACTION BY SLITS AND STRIP OBSTACLES

The Cornu spiral will now be seen as the phasor diagram obtained by adding the contributions at some point of infinitesimal strips across the whole of a plane or cylindrical wavefront. If an obstacle or a slit deletes some of the spiral, the remainder allows us easily to obtain the amplitude and phase. From this point of view it is natural to work in terms of v as variable, always remembering that v is related to h, the actual dimensional coordinate perpendicular to the strips, by Eq. (8.4) for a plane wave or Eq. (8.9) for a cylindrical wave.

In the case of slits it is again conventional to think of the slit being moved past P as was done in the case of the half-plane; the contribution of the uncovered portion of the wavefront is then represented by a segment of the Cornu spiral with a fixed length v_s. Moving the slit relative to P, the point of observation, moves v_s along the spiral. To illustrate this, Fig. 8.9 shows successive positions

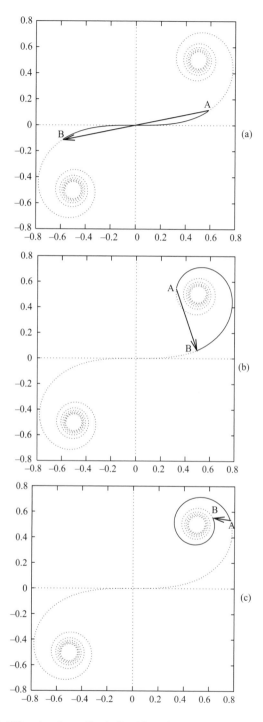

Fig. 8.9 Fresnel diffraction by a slit. A fixed length $v_s = 1.2$ of the Cornu spiral forms phasor diagrams (a) at the centre of the diffraction pattern; (b) and (c) at increasing distance off-centre.

of a segment with length $v_s = 1.2$ moved along the spiral in unit steps. This corresponds to a slit too narrow for the undisturbed brightness ever to be attained. As the free length of phasor moves out from the centre – here it is almost straight – to the spiral portions, its resultant decreases monotonically until the spiral is of small enough diameter for it to be wrapped once around. From then the resultant *increases* until it is wrapped around $1\frac{1}{2}$ times, after which it again decreases, repeating the process in a series of fringes getting smaller and smaller. This process is highly reminiscent of the *Fraunhofer* diffraction at a single slit, as indeed it should be! As a glance at Eq. (8.4) shows, to make v_s small at a given λ, we must make s, the distance from the slit to the observation point, large, which is just the condition for Fraunhofer diffraction.

We can now see the basis of the Rayleigh criterion for the minimum distance from the slit for Fraunhofer diffraction to apply. If the slit width covers a range $v_s = 2$ the phasor diagram is just beginning to be seriously bent in the centre: this is evident by inspection of the Cornu spiral. Equation (8.4) then may be used to give the distance s in terms of slit width h and wavelength λ. Then

$$2 = h\left(\frac{2}{\lambda s}\right)^{1/2} \quad \text{or} \quad s = \frac{1}{2}\frac{h^2}{\lambda}, \tag{8.10}$$

which is half the Rayleigh distance (Section 8.1) and Fresnel effects should still be appreciable. Similarly, if $v_s = 1$, the bending of the phasor at the centre of the spiral is negligible, its resultant being 99.4% of its unbent length, and a similar calculation shows we are at twice the Rayleigh distance.

As v_s increases the irradiance in the centre goes up, reaching the undisturbed intensity at $v_s \approx 1.4$, and rising to 1.8 times the undisturbed intensity at $v_s \approx 2.4$. This great increase in brightness can be thought of as the coherent superposition of the bright fringes near each edge diffracting separately. The general character of the diffraction from wider and wider slits will now be clear. As the fixed length of phasor slides along the Cornu spiral there are two conditions:

(i) *In the geometrically bright area*, the two ends of the phasor are in opposite parts of the spiral. The intensity is of the same order as the undisturbed intensity but because the resultant joins the ends of the phasor that are independently going around different spirals, complicated beating effects may be seen, as shown in Fig. 8.10.

(ii) *In the geometrical shadow*, the two ends of the phasor are on the same part of the spiral. The intensity is low but, because the ends of the phasor are on the same spiral, fringes are produced having maxima if the two ends are opposite, and minima if they are close.

Between these conditions is the rapid transition through the edge of the geometric shadow, where for a change of position of about one unit of v the edge of the phasor sweeps from one arm of the spiral to the other.

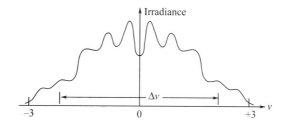

Fig. 8.10   A slit diffraction pattern in the Fresnel region.

The Cornu spiral can be used in a similar way to analyse the effect of strip obstacles. Here, a limited portion of the spiral is *removed*, so that if this is in the centre the two coils Z and Z′ move closer together. In the centre of the shadow of a strip obstacle there is always some light, although it rapidly becomes less as the strip is made wider. Similarly, as we see in the next section, the centre of the shadow of a perfectly circular object contains a narrow spot of light; but this spot has the *same* irradiance as the unobstructed light.

8.5   SPHERICAL WAVES AND CIRCULAR APERTURES: HALF-PERIOD ZONES

This section deals with Fresnel diffraction in axially symmetrical systems, as, for example, along the axis of a circular diffracting disc or hole. In the limit, a large enough circular hole offers no obstacle at all, so the case of free-space propagation is also covered.

In Fig. 8.11 the wave amplitude at a point P due to a point source P_0 is to be calculated by integrating all contributions originating from a spherical surface surrounding P_0. The limit of the integral will depend on the size of the diffracting aperture, whose circular edge lies on the sphere. A reference sphere radius b, centred on P, touches the surface on the axis; the phase of a contribution from an annulus at a distance h from the axis is then given to a first approximation by

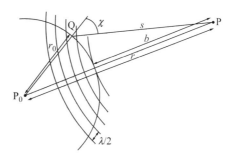

Fig. 8.11   Fresnel's half-period zone construction.

$$\phi = \frac{\pi h^2}{\lambda}\left(\frac{1}{r_0} + \frac{1}{b}\right). \tag{8.11}$$

The area of an annulus between h and $h + dh$ is $ds = 2\pi h dh$. Differentiating Eq. (8.11) gives

$$d\phi = \frac{2\pi}{\lambda}\left(\frac{1}{r_0} + \frac{1}{b}\right)h\,dh. \tag{8.12}$$

The element of area ds is therefore proportional to $d\phi$, so that the integral over the surface is conveniently carried out in terms of ϕ. The wave amplitude at P is then given by an integral of the form

$$A(P) = c\int_{\text{aperture}} \exp(-i\phi)d\phi, \tag{8.13}$$

where c is a constant depending on the amplitude of the source, and on r_0, b, λ.

For an aperture that is a circular hole with radius ρ the integral giving the wave amplitude on the axis is between 0 and

$$\phi_0 = \frac{2\pi}{\lambda}\rho^2\left(\frac{1}{r_0} + \frac{1}{b}\right) \tag{8.14}$$

giving

$$A(P) = ic\{\exp(-i\phi_0) - 1\}. \tag{8.15}$$

The wave amplitude at P is therefore proportional to $\sin(\phi_0/2)$, and the irradiance is proportional to $\sin^2(\phi_0/2)$. This means that if the radius of the hole is increased progressively from zero, then the irradiance at P increases until ϕ_0 reaches $\pi/2$, and decreases and increases cyclically thereafter. The successive annuli opened up between these turning points are known as the Fresnel half-period zones.

The contributions making up the integral (8.13) are shown in the phasor diagram, Fig. 8.12. This is shown as an open spiral, but it should be more nearly a circle; indeed, it should be exactly a circle given the approximation we have made in neglecting distance and inclination effects (see Numerical Example 8.3). Ultimately, these effects shrink the circle to a point at its centre, giving a resultant amplitude for free space that is half the amplitude obtained when a single zone is exposed.

A circular disc, acting as an obstacle, requires the integral (8.13) to be carried out from a limit ϕ_0 out to infinity. The contribution from large values of ϕ tends to zero and the integral becomes

$$A(P) = c\{-\exp(i\phi_0)\}. \tag{8.16}$$

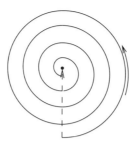

Fig. 8.12 The spiral phasor diagram for a spherical wavefront. (The spiral form is exaggerated in this diagram: the phasor diagram is nearly circular.)

Surprisingly, the modulus of $A(P)$ is independent of ϕ_0; the wave amplitude at a point on the axis behind any circular obstacle is the same as the unobstructed wave amplitude.

The prediction that there should be a bright spot at the centre of the shadow of a circular disc was first made by Poisson[1] on reading a dissertation by Fresnel on diffraction, submitted to the French Academy of Sciences in 1818. When the test was made, and the bright spot was found, the wave theory of light was firmly and finally established.

A circular aperture in which alternate half-period zones are blacked out, called a *zone plate*, is shown in Fig. 8.13. Alternate semicircles are now removed from the phasor diagram, and a large concentration of light appears at P. Figure 8.14 shows how the phasors from the half-period zones add in phase. The zone plate is acting like a lens; for any given wavelength the relation between object and image distance r_0 and b conforms to a simple lens formula. An improved zone plate can be made by reversing the phase of alternate zones instead of blacking them out; this is done by a change in thickness of a transparent plate. As shown in Fig. 8.14(c), the amplitude at the focal point is then doubled.

The zone plate used as a lens is particularly useful at X-ray wavelengths, where there is no transparent refracting material that can be used to make conventional lenses. It has also been used on a minute scale in electron optics to produce an electron lens only $0.7\,\mu m$ in diameter and with a focal length of $1\,mm$.[2] Each transparent zone in this lens consisted of an array of holes only a few nanometres in diameter, drilled through a thin inorganic film; there were 4000 holes altogether in the complete lens. This astonishing achievement has a

---

[1] Siméon Denis Poisson (1781–1840), celebrated French mathematician. His fame was predicted by his teacher M. Billy in a couplet due to Lafontaine:

Petit Poisson deviendra grand
Pourvu que Dieu lui prête vie.

[2] Y. Ito, A. L. Blelock and L. M. Brown, *Nature* **394**, 49, 1998.

practical application: the same lens pattern can be reproduced many times, allowing multiple beams of electrons or X-rays to be used in the fabrication of electronic circuits on silicon chips.

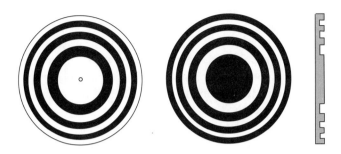

Fig. 8.13   Zone plates. Alternate half-period zones are either blacked out as shown above or reversed in phase (as seen in the cross-section on right).

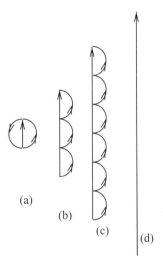

Fig. 8.14   Phasor diagrams for (a) a circular aperture containing an odd number of half-period zones; (b) a zone plate with three clear zones; (c) a zone plate of the same size as (b) but with phase-reversal instead of obscured zones; (d) a perfect lens.

8.6   FRESNEL–KIRCHHOFF DIFFRACTION THEORY

In both Fresnel and Fraunhofer theory we have assumed that a diffracted wave amplitude can be calculated from the sum of secondary waves originating at an aperture. We have assumed also that the wave can be represented by a scalar, so that polarization can be neglected; and we have assumed that the amplitude distribution across an aperture is that of the undisturbed wavefront. These latter assumptions may be improved in specific cases; for example, we know that at the edge of a slit in a metal sheet the electric field must be perpendicular to the conducting surface, so that the parallel component is zero near the edge of the slit. The diffraction pattern will therefore be narrower for a wave polarization perpendicular to the slit length than for the wave polarized parallel to the slit length. Such cases can be dealt with by the application of boundary conditions in determining the amplitude distribution across the aperture. Some fundamental questions still remain, however, which were clarified by Kirchhoff and added to the Fresnel theory.

One of the problems of the Huygens–Fresnel principle was to assign to each wavelet an *inclination factor*, which would give it unit amplitude in the forward direction and zero backwards. Fresnel assumed, incorrectly, that it was also zero at 90° to the forward direction. The inclination factor is obtained explicitly in Kirchhoff's analysis, which involves not only the amplitude and phase on a diffracting surface but also their differentials *along* the wave normal.

A harmonic wave from a point source in a homogeneous medium travels at the same speed in all directions, but with an amplitude decreasing inversely with distance. At a diffracting aperture, distance r_0 from the source, the wave amplitude of this spherical wave can be written as $(A_0/r_0) \exp(ikr_0)$. Figure 8.15 shows a small element of the aperture at Q with area da, which is the origin of a wave reaching a field point P at a further distance r, giving a contribution at P with the form

$$dA = A_0 \, da \frac{1}{r_0} \exp(ikr_0) \frac{1}{r} \exp(ikr). \qquad (8.17)$$

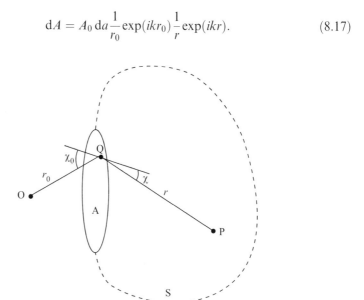

Fig. 8.15   Fresnel–Kirchhoff theory. An aperture forms part of the surface S enclosing P.

The Fresnel–Kirchhoff analysis adds two further factors, an inclination factor and a change in phase, giving the diffracted wave amplitude $A(P)$ at P as the integral over the diffracting surface S

$$A(P) = -\frac{ik}{4\pi}\int_S A_0 \frac{\exp ik(r + r_0)}{rr_0}(\cos\chi_0 + \cos\chi)da. \qquad (8.18)$$

Inside the integral sign the exponential term determines the phase of each component from an area da, while the amplitude is proportional to $1/r$, the distance of P from the area da. The factor $(\cos\chi_0 + \cos\chi)$ is the inclination factor, where χ_0 is the angle to the normal of the incident wave at the diffracting surface, and χ is the angle to the normal at P. Outside the integral the factor $-ik/4\pi$ normalizes the amplitude and phase of $A(P)$; the factor $-i = \exp(-i\pi/2)$ accounts for a 90° phase shift of the diffracted wave relative to the incident wave.

In most of the diffraction problems encountered in this chapter the surface S may be made to coincide with a wavefront, so that the incidence angle χ_0 is zero. The inclination factor $(\cos\chi_0 + \cos\chi)$ then becomes $(1 + \cos\chi)$. The propagation of Huygens' wavelets forwards but not backwards is now clear, since the inclination factor becomes zero for $\chi = 180°$. The correct factor for $\chi = 90°$ is not zero, but half the forward amplitude; we should point out, however, that diffraction through such a large angle is very dependent on the boundary conditions at the edge of the aperture.

This integral may look formidably complicated, and, indeed, it can be so for an arbitrary shape of diffracting aperture or obstacle. As we have seen, however, the evaluation of Eq. (8.18) can be greatly simplified in many practical situations.

8.7   BABINET'S PRINCIPLE

A consequence of the Kirchhoff theory, due to Babinet, concerns complementary diffracting screens. Consider a surface S_1 with some open and some opaque areas, and a complementary surface S_2 in which all the apertures are made opaque, and all the opaque regions are made open. With neither screen in place the complex amplitude at a point beyond the screen can be regarded as $A_1 + A_2$, the two diffracted amplitudes from S_1 and S_2. If P is outside the unobstructed light beam, so that $A_1 + A_2 = 0$, then it follows that $A_1 = -A_2$. If either screen diffracts light so that it reaches P, then the complementary screen also diffracts to give exactly the same irradiance at P.

Babinet's principle applies to any situation where light is diffracted by an obstacle or aperture into an otherwise dark region. For example, if a small obstruction is placed in a large parallel light beam, the light diffracted out of the beam is the same as that which would be diffracted out of the beam by an aperture of the same shape and size. Astronomical photographs often show this effect as a cross-like diffraction pattern extending from images of bright stars:

this is due to a support structure for a secondary mirror, forming an obstructing cross in the telescope aperture. The diffraction pattern is the same as would be obtained from crossed slits of the same dimensions in an otherwise totally obscured telescope aperture.

8.8   FURTHER READING

O. S. Heavens and R. W. Ditchburn, *Insight into Optics*. Wiley, 1991.

NUMERICAL EXAMPLES 8

8.1 The beam shape of a 15-m diameter millimetre wave telescope is to be measured from beyond the Rayleigh distance. Calculate this distance for a wavelength of 0.5 mm.

8.2 The Cornu spiral (Fig. 8.6) represents a phasor diagram giving the amplitude and phase of contributions at a point P from strips of a plane wave at a distance s from the nearest component. When $s = 10\,000\lambda$, how large is h (in wavelengths) for the phase of the contribution to be 5π behind that of the component at $h = 0$? Where on the Cornu spiral is this contribution, and what is the value of v?

8.3 Equation (8.18) gives the inclination factor of Fresnel–Kirchhoff theory. What is the percentage change in this factor for the contributions in Example 8.1 from $h = 0$ to $h = 224\lambda$?

PROBLEMS 8

8.1 Consider the possibility of observing optical Fresnel diffraction when a star is occluded by the Moon. Calculate for wavelength 600 nm (i) the width of the first half-wave zone at the Moon; (ii) the angular width of a star just filling this zone. Compare these with the size of irregularities on the Moon's surface, and the actual angular width of bright stars (Moon's distance $= 376 \times 10^6$ km).

8.2 A distant point source of light is viewed through a glass plate dusted with opaque particles. The light now appears to have a diffuse halo about 1° across. Use Babinet's principle to explain this and estimate the size of the particles.

8.3 Compare the intensities of light focused from a point source by a zone plate and by a lens of the same diameter and focal length. What change is made by reversing the phase of alternate zones rather than blacking them out? Where has the remaining energy gone?

8.4 An infinite screen is made of polaroid, and divided by a straight line into two areas in which the polaroid is oriented parallel to and perpendicular to the division. Describe the diffraction pattern due to the edge when unpolarized light is incident normally on the screen.

8.5 A shadow edge for demonstrating Fresnel diffraction is made by depositing a metallic film on glass. What will be the effect of using a film that transmits one-quarter of the light intensity?

9

Interference by division of amplitude

... diversely coloured with all the Colours of the Rainbow; and with the microscope I could perceive, that these Colours were arranged in rings that in-compassed the white speack or flaw, and were round or irregular, according to the shape of the spot which they terminated; and the position of Colours, in respect of one another, was the very same as in the Rainbow.
Hooke, 1665, on interference colours in a flake of mica.

Interference between two beams of light usually requires them to be derived from the same source. There are two ways of achieving this. The first is to divide a wavefront into separate sources, as in Young's double slit or by using a diffraction grating. The second, which is the subject of this chapter, is to divide the amplitude of the wavefront by partial reflection, obtaining identical wavefronts that can be brought together by different paths. The most familiar example of interference by this *division of wavefront* is the pattern of coloured fringes seen in soap bubbles and thin oil films.

In this chapter we analyse the geometry of interference by division of wavefront in the examples of thin films provided by Newton's rings and thin oil films. The two wavefronts often have different amplitudes; we define a *fringe visibility* that describes the resulting fringes. We show how fringes that are sharply defined rather than cosinusoidal are obtained from the multiple beams of parallel reflecting surfaces; these are the *Fabry–Perot fringes* that have many applications in spectrometers, length measurement and laser resonators.

9.1 NEWTON'S RINGS

As an introduction to this type of interference we shall consider Newton's rings. These may be observed with the optical system shown in Fig. 9.1. A long-focus lens is placed in contact with a flat glass plate, and illuminated at vertical incidence by the source reflected by the glass plate tilted at 45°. Observations

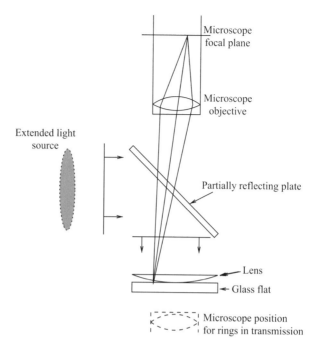

Fig. 9.1 Newton's rings. The rings may conveniently be observed in reflection with the system shown. The much lower visibility rings seen in transmission may be seen from below.

are made with the microscope. The wavefront travelling downwards is partially reflected at each boundary of the lens and plate. The two wavefronts that interfere to cause Newton's rings arise from partial reflection at the lower surface of the lens and the upper surface of the plate. As they are both derived from the same source, the two waves can interfere with each other. Their relative phases are determined by two factors:

(i) The reflection coefficients of the two boundaries are of opposite sign (as shown in Section 5.4)

(ii) There is an extra path traversed by the wave reflected by the flat surface.

The first factor provides a phase difference of π, so that the centre of the pattern is dark, with the two reflections in antiphase. (No reflection is to be expected anyway from the central area where the glass is in contact and effectively continuous.) Succeeding rings are light and dark as the extra path length is $(N - 1/2)\lambda$ or $N\lambda$, and the two reflections become in and out of phase. If the radius of curvature of the bottom face of the lens is R the condition for brightness at r from the axis is

$$\text{path difference} = 2h = (N - 1/2)\lambda, \tag{9.1}$$

where the value of h in terms of R and r can be determined approximately by an expansion:

$$h = R(1 - \cos\theta) = R\left(\frac{\theta^2}{2} + \dots\right) \approx \frac{r^2}{2R}. \tag{9.2}$$

This approximation, shown in Fig. 9.2, often appears in optics; it is important to be clear under what circumstances it is applicable. The expression

$$h = \frac{r^2}{2R} \tag{9.3}$$

is a *parabola* that approximates to a circle of radius R for small r. Usually this approximation will be good enough if the deviation between circle and parabola is small compared with a wavelength. If the expansion of Eq. (9.2) is taken to one more term, this condition becomes

$$R\frac{\theta^4}{4!} \ll \lambda, \tag{9.4}$$

so if $R = 100$ cm and $\lambda = 5 \times 10^{-5}$ cm, the condition is $\theta \ll 5.8 \times 10^{-2}$ rad or a few degrees. Now at $r = 1$ cm, $h = 0.005$ cm so $2h = 0.01$ cm. The order N of the ring is then $(10^{-2}/(5 \times 10^{-5}) = 200$. In this example the approximation is good for many tens of rings.

These ring-fringes are localized in the sense that to see them the microscope must be focused on the boundary between the lens and optical flat. When this condition is satisfied the light reflected from a point on the lens surface is brought ultimately to an image point by the same optical path, even though

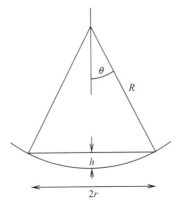

Fig. 9.2 The intersecting chord theorem. The sagittal distance h is related to the chord $2r$ and radius R by $r^2 = h(2R - h)$; for $h \ll 2R$ the sagittal distance is $h \approx r^2/2R$.

it may have arrived from an extended source. The light reflected from the corresponding point below on the flat is likewise brought to the same focal point, the only difference being that the path for all rays from the flat is longer by approximately $2h$ than that from the lens surface. Combining Eqs. (9.1) and (9.3) gives

$$r = \{R\lambda(N - 1/2)\}^{1/2} \tag{9.5}$$

as the condition for brightness. The ring system then has a dark centre surrounded by rings getting more and more crowded as r increases.

The visibility V of interference fringes, sometimes called fringe contrast, is defined in terms of the maximum and minimum irradiances as

$$V = \frac{I_{max} - I_{min}}{I_{max} + I_{min}}. \tag{9.6}$$

The visibility of Newton's rings is not intrinsically close to unity, as in the case of Young's fringes formed by two equal amplitude waves. It is, instead, determined by the relative amplitudes of the two waves reflected from the lower surface of the lens and the upper surface of the flat. In practice, these are nearly the same, and the ring system is well defined and of high visibility. Quite the opposite is true in the case of Newton's rings seen in transmission rather than reflection; the interference is now between a directly transmitted wave and a much smaller wave that has been twice reflected (see Chapter 5 for the coefficient of reflection). The transmitted fringes are complementary: bright where those reflected are dark, but the visibility is very low. A system for observing them is shown in Fig. 9.1. One way to see that they must arise is from a consideration of the conservation of energy. Most of the light incident on the system is transmitted, but due to the interference effects discussed above some areas (the bright rings) reflect some light back, whilst others (the dark rings) reflect hardly at all. Hence, as the energy not reflected back is transmitted, the complementary low-visibility system is seen in transmission. The well-known property of a photographic negative to look like a positive if seen in reflected light from the front is somewhat analogous to this. The high transmission (transparent) parts look dark compared with the low transmission (opaque) parts.

If, instead of a monochromatic source, a white light source is substituted, then a dark spot is still seen in the centre of the reflected pattern. The scale of the pattern is proportional to wavelength, and the overlapping ring systems in all the different wavelengths get increasingly out of step, giving a white field only a few orders away from the centre. Between the dark and white regions is a system of coloured rings in a sequence called Newton's colours, since the various colours are added or subtracted according to their wavelength and the distance between the surfaces.

9.2   INTERFERENCE EFFECTS WITH A PLANE-PARALLEL PLATE

The case of Newton's rings is only one example of interference effects observed between two beams derived by the division of amplitude by partial reflections from two surfaces. Consider first a monochromatic point source S illuminating a parallel-sided slab of transparent material of refraction index n, shown in Fig. 9.3. Then if the direct path is excluded there are two paths from S to P corresponding to reflections from the front and back of the slab, and P will be light or dark according to whether the optical paths (taking into account the phase change at the upper reflection) differ by an odd or even number of half wavelengths. Clearly, there will be circular symmetry about the line SN through the source normal to the slab. Geometrically, the system is like Young's double slit, the interference being between light from the two images S_1 and S_2 of S, in the front and back surfaces of the slab. The fringes are non-localized in space, and a photographic plate in any plane parallel to the slab (for example) would record circular fringes centred on SN. The order of the fringes is high if the thickness of the plate is large compared with the wavelength, so that the source must be highly monochromatic if fringes are to be observed. Similarly, if the source is not a point, but extended, the fringes from different parts of the source will overlap and the pattern may be lost.

A surprising change is made in the system by inserting a lens as shown in Fig. 9.4, and observing the distribution of brightness in its focal plane. The lens

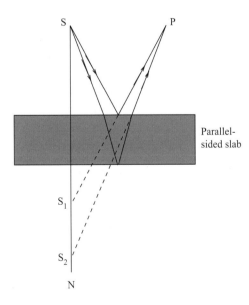

Fig. 9.3   A point source S has two images S_1 and S_2 in the front and back surfaces of a parallel-sided slab. The interference in the light from these gives non-localized fringes similar to Young's.

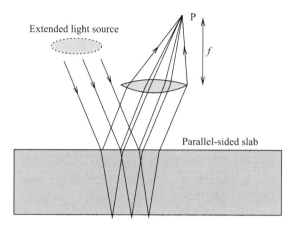

Fig. 9.4 The effect of introducing a lens that brings all *parallel* rays to a single point P is to allow fringes of equal *inclination* to be observed with an extended source.

brings together at a point P all the rays that leave the plate at a particular *angle*; the figure shows two of these for each of three points in the source. The source may now be extended, since the path difference for all pairs of rays reflected in the front and back faces of the slab is the same if they leave the plate at the same angle, regardless of which part of the extended source they come from. Each element of the extended source thus contributes twice to the light wave at P, once by reflection at the back, and once by reflection at the front. These contributions interfere either constructively or destructively, as determined by their different path lengths.

Since only the angle of incidence determines the brightness or darkness, these fringes are called *fringes of equal inclination*. It is easy to see from Fig. 9.5 that the conditions for bright and dark fringes are related to the angle of refraction r, inside the slab, by

$$2nh \cos r = \left(N + \tfrac{1}{2}\right)\lambda \text{ for bright fringes,} \tag{9.7}$$

$$2nh \cos r = N\lambda \quad \text{ for dark fringes.}$$

The visibility of the fringes is again not necessarily unity. Let A_f and A_b be the amplitudes of light arriving at P from the front and back of the slab. Then the irradiance will be proportional to

$$A^2 = A_f^2 + A_b^2 + 2A_f A_b \cos \phi, \tag{9.8}$$

where

$$\phi = \frac{4\pi nh \cos r}{\lambda} + \pi. \tag{9.9}$$

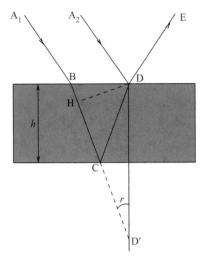

Fig. 9.5 The path difference for two rays A_1 BCDE and A_2 DE. The paths are equal up to the points D and H, where DH is perpendicular to BC. The path difference is therefore HCD, which by construction is seen to be equal to HD$'$. Since HCD is within the film, the extra path is $2nh \cos r$.

From Eq. (9.6) putting $\cos \phi = 1$ for maximum irradiance and $\cos \phi = -1$ for minimum irradiance,

$$V = \frac{2A_f A_b}{A_f^2 + A_b^2}. \qquad (9.10)$$

So with a thick slab of material and a monochromatic extended source, fringes may be observed in the focal plane of the lens in Fig. 9.4. The irradiance profile of the fringes follows a squared cosine function, as in Eq. (7.2) for Young's double slit; they are often referred to as $\cos^2$ fringes, but their visibility is less than unity. Notice that the present discussion has excluded multiple reflections, a good approximation unless special measures are taken to increase the reflectivity. This point is returned to in section 9.5.

9.3 THIN FILMS

The parallel-sided slab discussed in the previous section can produce fringes of a very high order; that is to say, the path difference between the interfering beams might be many thousands of wavelengths. Thin films in which the thickness is only a few wavelengths display interesting and somewhat different interference phenomena, which do not depend on the film having parallel sides. Interference fringes now appear in the surface of the film; their position is

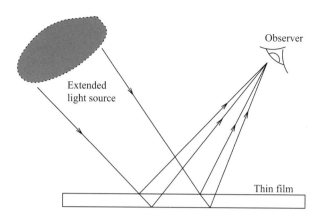

Fig. 9.6   Fringes are seen in a thin film by reflection of an extended source of light.

determined mainly by the thickness of the film and very little by the angle at which they are seen.

Consider first the familiar observation of light from the sky reflected in a thin oil film on water (Fig. 9.6). Each colour forms a set of interference fringes across the film, the positions of the fringes depending on wavelength and the thickness of the film. Each part of the film reflects two waves, one from the front and one from the back of the oil film; the phase difference of these two waves depends both on angle r and thickness h, but in practice mainly on h so that the interference effects appear to outline the contours of equal thickness in the film.

There are, in fact, two extreme cases for interference in thin films. If the thickness is completely uniform, then only a variation of the angle r can change the path difference; fringes will therefore be seen outlining directions where r is constant, as in the thick slab already discussed. The fringes will be very broad, since a large change in $\cos r$ corresponds to a small change in path difference when h is small. These are *fringes of constant inclination*. Alternatively, as with the oil film or a soap bubble, the thickness may vary rapidly from place to place, while $\cos r$ changes little; *fringes of constant thickness* are then seen. These are known as *Fizeau fringes*.

The fringes of constant thickness seem to be located within the film, since their position is determined by h and not by r; by contrast, the fringes of constant inclination are seen in fixed *directions*, and therefore appear to be at infinite distance. In practice, the fringes seen in oil films are also determined in small part by the angle r, so that they are located just behind the film, as can be verified by the observer moving his head from side to side, looking for parallactic motion of the fringes across the film.

Fringes of equal thickness have many practical applications, since they allow measurements of thickness to be made to a fraction of a wavelength.

9.4   MICHELSON'S SPECTRAL INTERFEROMETER

An instrument that has been very influential in the development of interference optics, both theoretical and practical, is the Michelson interferometer (not to be confused with his stellar interferometer, Chapter 11). Its simplest form is shown in Fig. 9.7, which may seem at first sight to bear little relationship to the optical systems discussed in connection with thin films in the last section. In fact they are closely related.

Light from the extended source S is split in amplitude by a half-silvered mirror D, and the two beams are then reflected by the mirrors M_1 and M_2. A further partial reflection in D sends the two beams out towards the observer at P. The observer can see the source S reflected simultaneously in the two mirrors. He can also see an image M_2' of the mirror M_2 close to the surface of M_1. The mirror M_2 is equipped with screws to allow its inclination to be adjusted, so that M_2' can be made parallel with M_1; M_1 itself is mounted on a screw-controlled carriage so that the perpendicular distance between M_1 and M_2' can be adjusted over a considerable range.

When M_1 and M_2' nearly coincide, the view of the source S is exactly the same as the view reflected in a thin film, and fringes will be seen in the surface of M_1. The equivalent thin film is the space between M_1 and M_2', so that the path difference between two rays reaching P from a point on S is determined by the

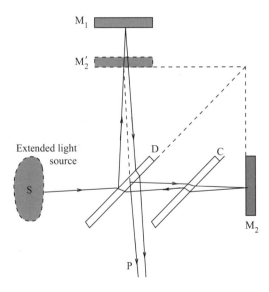

Fig. 9.7   Michelson's interferometer. Observations are made from P either directly or with a microscope or camera. An observer at P sees the extended light source S reflected in a 'thin film' consisting of the mirror M_1 and the image M_2' of the mirror M_2. Fringes of constant thickness are generally seen in this thin film, although if M_1 and M_2' are made accurately parallel the fringes take the form of a circular pattern of fringes of constant inclination.

thickness of this space. Exact equality of the paths is ensured for the situation when M_1 and M_2' coincide by the insertion of the glass plate C, which is the same thickness as D. Each path then traverses a glass plate twice, and the paths will remain equal for all wavelengths even if the refractive index of the glass is dispersive.

Suppose that the distance between M_1 and M_2' is small, but that M_1 is not parallel to M_2'. Since the mirrors are accurately plane the fringes will then be light and dark lines across the mirrors; these are fringes of constant thickness. The angle of M_2 can then be adjusted, and M_1 and M_2' can be made parallel to a small fraction of a wavelength; the linear fringes of equal thickness are then replaced by fringes of equal inclination. For a bright fringe of order n at angle θ, with a path difference d

$$2d \cos \theta = n\lambda. \tag{9.11}$$

Adjustment of the position of M_1 will now make the fringes expand outwards from the centre if the distance $M_1 M_2'$ is decreasing, and grow outwards from the centre if it is increasing. Finally, a position can be attained when the field is uniformly bright all over. This corresponds to the planes of M_1 and M_2' coinciding. The lightness or darkness of the field in this condition depends on the value of the phase change ϕ at the reflection in D. If $\phi = \pi$ and the division in amplitude by D has been accurately equal, then the field will be black. Suppose that M_2 is now tipped about its centre so that the linear fringes of equal inclination are seen. The centre black fringe then corresponds to the zero order. If the monochromatic source S is now replaced by a white light, then the central dark fringe will remain, and on either side of it a few fringes will be seen. These will display Newton's colours, merging into white light a few fringes away on each side when the overlapping of the different coloured fringes is complete. So substitution of a white light source is useful since it allows identification of the zeroth-order fringe. The visibility of fringes in mono- chromatic light is unity even for high-order fringes.

The basic properties then of the Michelson interferometer are the ability:

(i) to make both arms equal in optical length to within a fraction of a wavelength;
(ii) to measure changes of position as measured on a scale (the positions of M_1) in terms of wavelength by counting fringes;
(iii) to produce interference fringes of a known high *order* (number of wave- lengths difference in path lengths).

The last of these is the basis for the use of Michelson's interferometer for measuring the width and shape of spectral lines, as discussed in Chapter 14.

A closely related interferometer is the Mach–Zehnder interferometer described in Chapter 11. Here, the two beams are again separated by a mirror system, but in an arrangement that is convenient for inserting optical components into one of the beams to measure differences in optical paths by observing fringe displacements.

9.5   MULTIPLE BEAM INTERFERENCE

In the discussion of interference effects in parallel-sided slabs and in thin films we have so far taken account of only two reflected beams, and ignored any multiple reflections. Unless special measures are taken to increase them, such reflections are very weak, but if appropriate steps are taken so that they are enhanced, allowing perhaps 10 or more beams to be combined, the fringes change their character and become very much sharper. By suitably coating the faces of a slab with a thin metallic film it is possible to increase the reflection coefficient, so that the front face reflects a large fraction of the light incident upon it. The small amount that does get through to the inside of the slab is reflected back and forth many times, a small proportion emerging at each reflection. It is the interference of these *many* emerging rays, rather than just two, that gives multiple interference its special character.

This simple case is illustrated in Fig. 9.8. Most of the incident light S is reflected into the ray R_1, but after that the further rays on the reflection side R_2, R_3 ... are all of similar strength, dying away gradually. Let A be the amplitude of the incident ray. Then if r and t are the reflection and transmission coefficients from the surrounding medium to the slab, and r' and t' the corresponding quantities from slab to medium, we can write down the amplitude of the reflected rays

$$rA, tt'r'A, tt'r'^3A, \ldots \tag{9.12}$$

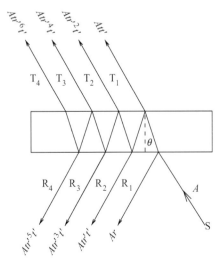

Fig. 9.8   Multiple reflections in a parallel-sided slab.

and of the transmitted rays

$$tt'A, tt'r'^2A, tt'r'^4A, \ldots \tag{9.13}$$

With the exception of the first large reflected ray, these amplitudes go in geometric progression, and the closer to unity that r' is made, the more slowly their size dies away. Their *phase* depends on θ, the angle of refraction inside the plate, its thickness h and its refraction index n. The relative phase of the rays on a plane outside the slab perpendicular to the ray depends on θ as

$$\psi = \frac{4\pi nh \cos \theta}{\lambda}. \tag{9.14}$$

If, now, as in the case of two-beam interference, we provide a lens to bring the transmitted rays to a focus, we can find the conditions for constructive and destructive interference. For constructive interference, all the slowly declining vectors will be in phase if

$$2nh \cos \theta = N\lambda. \tag{9.15}$$

Consequently, with an extended source, fringes of equal inclination, that is to say, circles, will be seen, each corresponding to a particular value of N. It is when we consider the spaces between these fringes that the special character of multiple beam interference emerges. The angle between *each* of the many phasors is given by ψ in Eq. (9.14); a change in ψ of 2π takes us to the next maximum, but a change from 2π by only a very small amount causes the long string of nearly equal phasors to curl up into a near circle. For this reason, the fringes are very sharp, a plot of the irradiance showing almost no light at all transmitted except close to the maxima.

If p beams are transmitted, the complex amplitude is the sum of the geometric series

$$A(p) = Att'\left(1 + r'^2 \exp(i\psi)\right) + \cdots + r'^{2(p-1)} \exp(i(p-1)\psi). \tag{9.16}$$

The sum of this geometric series is

$$A(p) = Att'\left(\frac{1 - r'^{2p} \exp(ip\psi)}{1 - r'^2 \exp(i\psi)}\right). \tag{9.17}$$

If the number of beams p becomes large so that r'^{2p} becomes very small, we can ignore that term, and calculate the irradiance I as $A(\infty)A^*(\infty)$, giving

$$I = \frac{(Att')^2}{(1 - r'^2 \exp(i\psi))(1 - r'^2 \exp(-i\psi))} \tag{9.18}$$

$$= \frac{(Att')^2}{1 + r'^4 - 2r'^2 \cos \psi}. \tag{9.19}$$

In this expression we can recognize A^2 as the irradiance of the incident wave, I_i. The expression for transmitted irradiance I_t is more neatly expressed in terms of $\sin^2(\psi/2)$:

$$\frac{I_t}{I_i} = \frac{(tt')^2}{(1 - r'^2)^2 + 4r'^2 \sin^2(\psi/2)}. \tag{9.20}$$

Now it may be shown from a study of the Fresnel coefficients in section 5.4 that $tt' = 1 - r'^2$ so that Eq. (9.20) may be further simplified to

$$\boxed{\frac{I_t}{I_i} = \frac{1}{1 + F \sin^2(\psi/2)},} \tag{9.21}$$

where

$$F = \frac{(2r')^2}{(1 - r'^2)^2}. \tag{9.22}$$

Notice that the parameter F becomes very large as r'^2 approaches unity. For example, if $r'^2 = 0.8, F = 80$. The effect of this is to keep the transmitted irradiance (Eq. (9.21)) always very small, except when ψ is close enough to a multiple of 2π for $\sin^2(\psi/2)$ to become less than $1 / F$. The value of I_t/I_i then shoots rapidly up to unity when ψ is a multiple of 2π. Figure 9.9 is a plot of I_t/I_i for several values of F. The sharp fringes of multiple-beam interference are known as Fabry–Perot fringes.

The sharpness of Fabry–Perot fringes is often specified as the *finesse* $\mathcal{F}$, which is the ratio of the separation of adjacent maxima to the half-width of a fringe, defined as the width between points of half-irradiance. From Eq. (9.21) the phase shift $\psi_{1/2}$ for the fringe irradiance to be halved is given by

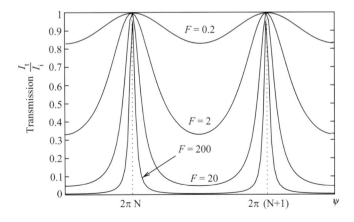

Fig. 9.9   Cross-section of Fabry–Perot fringes for several values of the parameter F.

$$\left(1 + F\sin^2\frac{\psi_{1/2}}{2}\right)^{-1} = \tfrac{1}{2}, \tag{9.23}$$

giving

$$\psi_{1/2} = 2\sin^{-1}(1/\sqrt{F}). \tag{9.24}$$

In practice, F is large, and $\sin^{-1}(1/\sqrt{F}) \approx 1/\sqrt{F}$. Since the phase difference between two adjacent fringes is 2π, the finesse $\mathcal{F}$ is the ratio $2\pi/2\psi_{1/2}$, giving

$$\mathcal{F} = \frac{\pi\sqrt{F}}{2}. \tag{9.25}$$

The multiple beams of the Fabry–Perot act very like those of a diffraction grating (see Chapter 10): the many phasors arising from the multiple reflections give only a small resultant unless the angle θ at which they traverse the slab is just right to put them all in phase. So the slab is a device that will allow light of any fixed wavelength to traverse it only at certain extremely well-defined angles. It can therefore be used as a filter transmitting only selected narrow wavelength ranges, or as a spectrometer.

9.6   THE FABRY–PEROT INTERFEROMETER

This instrument, illustrated in Fig. 9.10, uses the effect discussed in the previous section to produce circular, sharply defined interference fringes from the light from an extended source. The rings on the plate P are images of those points on the source producing (with the help of the first lens) light going in suitable directions between the lenses. The central cavity is usually made of two glass plates, with their inner surfaces coated with partially transparent films of high reflectivity. These plates are held apart by an optically worked spacer made of invar or silica, to which they are pressed by springs. This device is called a *Fabry–Perot etalon.*[1]

If the light from the source is not monochromatic, but contains two spectral components, the ring system is doubled. It is possible by this means to distinguish optically, or to 'resolve', even the hyperfine structure of spectral lines directly. Since its introduction by Fabry and Perot in 1899 the instrument has dominated the field of high-resolution spectrometry. We refer again to the Fabry–Perot interferometer and its use in high-resolution spectrometry in Chapter 13.

---

[1] From étalon, a standard; it can be used as a standard of length calibrated in terms of the wavelength of spectral lines.

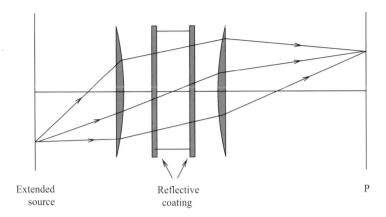

Fig. 9.10 A simple Fabry–Perot interferometer. Without the cavity or etalon an image of the broad source is formed on P. When the etalon is in place, only those parts of the image corresponding to allowable angles through the etalon are transmitted, giving the extremely well-defined ring system.

9.7 INTERFERENCE FILTERS

A parallel-sided glass plate, or a cavity between two parallel glass plates, acts as a spectral filter on light that falls on it at any given angle. Figure 9.9 shows how the irradiance of the transmitted light varies with the phase difference ψ, which is inversely proportional to wavelength. The curve in Fig. 9.9 is therefore the *transmission characteristic* of the filter, showing its relative transmission properties over a range of wavelengths.

A combination of two or more such filters, using plates or cavities with different thicknesses, can be arranged to transmit only one narrow spectral band. Filters transmitting a band only 1 nm wide at optical wavelengths can be made in this way. They can even be made tunable by making the thickness variable; a convenient way of doing this is to move one of the reflecting surfaces by attaching a piezoelectric transducer in which an applied electric field induces a small mechanical movement. An alternative is to vary the optical thickness of the cavity by changing the gas pressure within the cavity, which varies the refractive index.

A very high reflection coefficient, giving a high finesse, is essential for high resolution in interferometers and in interferometric filters. Since thin metal films absorb rather too much light, it is preferable to use dielectric coatings on glass. The reflection coefficient then depends on the step of the refractive index at the interface; it can be increased by using a series of layers of dielectric, with alternate high and low refractive indices.

Fabry–Perot etalons often form an essential component of lasers (Chapter 15), where they are referred to as resonant cavities. A helium–neon gas laser, for example, has a cavity formed by two mirrors enclosing a gas-filled tube; the cavity resonator determines the wavelength of the laser action in the gas. The wavelength of light from semiconductor lasers is similarly determined by a resonant cavity, which is formed by the polished faces of the semiconductor material.

9.8   FURTHER READING

H. A. Macleod, *Thin-film Optical Filters*. Hilger, 1989.

NUMERICAL EXAMPLES 9

9.1 An air-filled wedge between two plane glass plates is illuminated by a diffuse source of light, wavelength 600 nm. Fringes are seen in the light reflected by the air wedge, spaced 5 mm apart. Find the angle α between the glass plates.

9.2 Newton's rings are formed by a lens face with radius of curvature 1 m in contact with a plane surface, using sodium light with wavelength 589 nm. Find the radii of the first and second bright rings.

9.3 Newton's rings are formed using the bright sodium spectral line, which is a doublet with wavelengths 589.0 nm and 589.6 nm. Find the order of rings where the bright ring of one component of the doublet falls on a dark ring of the other.

PROBLEMS 9

9.1 Young's double-slit fringes are formed by two side-by-side coherent sources. Consider, in contrast, the fringe pattern formed by two coherent point sources one in front of the other, and relate it to the pattern from a single source seen reflected in a parallel-sided slab.

9.2 The colours in a soap bubble often fade just before the bubble bursts, and the film becomes dark. Estimate the film thickness at this stage.

9.3 *When two object-glasses are laid upon one another, so as to make the rings of the colours appear, though with my naked eye I could not discern above eight or nine of those rings, yet by viewing them through a prism I could see a far greater multitude, insomuch that I could number more than forty . . . But it was on but one side of these rings . . .*
Newton, *Optics*. Explain!

9.4 The centre of Newton's rings is usually dark. Is this the same phenomenon as in the soap film of Problem 9.2? What happens if oil with a refractive index between those of the lens and the flat surface fills the air space between them?

9.5 Show that, contrary to expectation, a transparent film on a perfectly reflecting surface does not show interference fringes in reflected light. This will require analysis following the lines of section 9.5, noting the change of phase at internal reflection at the top face.

9.6 What happens to the fringes in the air-filled wedge of Numerical Example 9.1 if a liquid with refractive index n fills the wedge? What happens if n approaches the refractive index of the glass?

9.7 Find the highest-order fringe visible in a Fabry–Perot etalon with spacing 1 cm, for light with wavelength 600.000 nm. What is the highest order for a wavelength 600.010 nm? What are the angular radii of the highest-order fringes for these wavelengths?

9.8 The structure of a doublet spectral line is to be examined in a Fabry–Perot spectrometer. If the separation of the doublet is 0.0043 nm at a wavelength of 475 nm, what is the spacing of the etalon that places the nth order of one component on top of the $(n + 1)$th order of the other?

9.9 If the reflectance (see section 5.4) in a Fabry–Perot etalon is 60%, find the ratio of the irradiance at maximum to that halfway between maxima.

9.10 A parallel beam of white light is passed through a Fabry–Perot etalon and focused on the slit of a spectrograph. Describe the appearance of the spectrum. The slit is illuminated also by mercury light. If 200 bands are seen in the spectrum between the blue and green mercury lines (wavelengths 546 nm and 436 nm), what is the spacing of the etalon?

10

The diffraction grating and its applications

In 1912 Laue (1879–1960) had the inspiration to think of using a crystal as a grating.
S. G. Lipson and H. Lipson, *Optical Physics*. Cambridge University Press, 1969.

In Chapter 7 the Fraunhofer diffraction pattern in intensity from two slits illuminated by a plane wavefront was shown to be a set of equally spaced $\cos^2$ fringes. We now discuss the more general problem of how such a system performs when it has a large number of slits instead of just two. This provides a description of an important optical element, the diffraction grating. The general solution for any grating is to evaluate the Fourier transform of the aperture function. However, much physical insight can be gained by using a phasor approach.

In this chapter we use phasor diagrams to illustrate the mathematics of diffraction by gratings, and develop the relation between the grating function and its transform, which is the relation between the properties of the grating and its diffraction pattern. We give examples of diffraction theory applied to radio antenna theory and to X-ray crystal diffraction.

10.1 THE DIFFRACTION GRATING

Consider first the simple case of five slits illustrated in Fig. 10.1, each separated by d from its neighbours. With two slits, maxima were produced by beams from *both* slits being in phase, which occurred when $d \sin \theta = \pm N\lambda$. Clearly, we can have a situation when beams from *all five* slits are in phase, and this will again be when

$$\boxed{d \sin \theta = \pm N\lambda.}$$

(10.1)

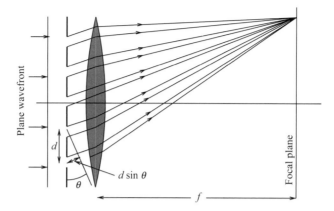

Fig. 10.1   A diffraction grating with five slits. All the light diffracted into the direction θ is brought to one point in the focal plane of the lens where the Fraunhofer diffraction pattern may be seen.

This condition means that the path difference from the reference plane to each slit is in all cases an integral multiple of the wavelength. The phasors lie on a straight line and add up to the same value of amplitude as when there is no path difference between the slits at sin $\theta = 0$. Note that the condition Eq. (10.1) applies for a grating with any number of slits: it is worth remembering as the simplest form of the *grating equation*. The successive maxima in the diffraction pattern of a grating are called its *orders*, first, second, third, etc. according to the value of N. The *central* or *zeroth* order is the one for $N = 0$.

The phasor amplitude pattern for the five-slit grating as sin θ moves away from zero is illustrated in Fig. 10.2; for comparison, the amplitude pattern for a pair of slits at spacing $5d/2$ is also shown. The remarkable thing is that in the

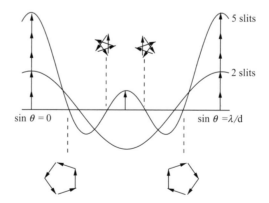

Fig. 10.2   Amplitude diffraction patterns for five slits d apart and two slits $5d/2$ apart. Phase is in each case referred to the centre of the slit pattern. Notice that in the five-slit case the phasor contributed by the central slit remains unchanged throughout.

five-slit case the light has been diffracted mainly into the strong maxima at the several orders, with only weak maxima coming between them. The first zeros on each side of an order are $\pm\lambda/5d$ apart. The orders then are approximately the angular width we should expect for the whole diffraction pattern of a slit as wide as the whole grating; that is to say, $5d$ in this case. Also, we can narrow the intensity distribution of the orders by adding *more slits* at the same spacing.

Diffraction gratings in use at optical wavelengths often have many thousand slits per centimetre, or *lines* as they are more usually called. In practice, gratings are often used in *reflection* rather than *transmission*. The discussion will be continued in terms of transmission, which is probably easier to follow, but reference will be made to reflection gratings where necessary. Figure 10.3, showing the geometric relations of a reflection grating, has been arranged so that the discussion that follows can easily be related to it. In general, a grating is not illuminated with a wavefront arriving exactly parallel to its plane, but with one at an angle θ_0 as measured in a plane perpendicular to the grating's lines. We shall continue in this analysis to assume the lines are narrow enough for each to be considered as the source of a single cylindrical wavelet. Take the first

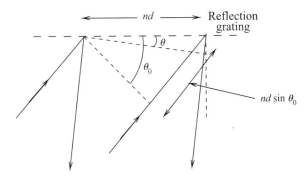

Fig. 10.3   Geometry for a reflection grating.

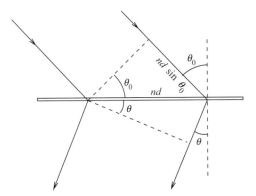

Fig. 10.4   Geometry for a transmission grating.

line as the phase reference and let there be n lines, so that the optical path difference across the width of the plane wavefront incident to the reference plane for emergence at θ in Fig. 10.4 is

$$nd(\sin \theta_0 + \sin \theta). \tag{10.2}$$

It will be convenient to write the phase difference between light paths that pass through two adjacent slits as ψ. Then the phase of light that has gone through the nth line is (with respect to the first)

$$\frac{2\pi nd}{\lambda}(\sin \theta_0 + \sin \theta) = n\psi. \tag{10.3}$$

The complex amplitude obtained as the sum of all the light leaving the grating in the direction θ is then

$$A(\theta, \theta_0) = A_0\{1 + \exp(i\psi) + \exp(2i\psi) + \ldots + \exp((n-1)i\psi)\}. \tag{10.4}$$

This geometrical series may be summed in the usual way, giving

$$A(\theta, \theta_0) = A_0 \left\{ \frac{1 - \exp(in\psi)}{1 - \exp(i\psi)} \right\} \tag{10.5}$$

$$= A_0 \exp\left\{ \frac{i(n-1)\psi}{2} \right\} \frac{\sin(n\psi/2)}{\sin(\psi/2)}. \tag{10.6}$$

The exponential term gives the phase of the resultant $A(\theta, \theta_0)$ relative to the zero of phase that was taken to be the first line. If, instead, we take the centre of the grating (be it line or non-line) as our phase reference, the exponential term becomes unity and we see once again that the amplitude remains real, although the phase can be reversed.

If we are interested in the pattern as a function of θ for a fixed θ_0 it is convenient to regard the inclination of the illuminating wavefront as putting a linear phase shift across the grating. This may be conveniently designated by δ radians per line. Then

$$\delta = \frac{2\pi d}{\lambda}\sin \theta_0. \tag{10.7}$$

Rewriting Eq. (10.6) in terms of θ and θ_0 and with the grating centre as phase reference gives

$$\boxed{A(\theta, \delta) = A_0 \frac{\sin(n\pi d/\lambda \sin \theta + n\delta/2)}{\sin(\pi d/\lambda \sin \theta + \delta/2)}.} \tag{10.8}$$

This is an important general expression for the diffraction pattern from n narrow lines d apart.

10.2   DIFFRACTION PATTERN OF THE GRATING

For most purposes it may be sufficient to remember the basic equation for the diffraction maxima at normal incidence

$$d \sin \theta = N\lambda \qquad (10.9)$$

and for incidence at angle θ_0

$$d(\sin \theta + \sin \theta_0) = N\lambda. \qquad (10.10)$$

We may also need to know the width and shape of the diffraction maxima, for which we need the general diffraction pattern of Eq. (10.8). This is illustrated in Fig. 10.5 where the intensity is plotted for a grating with six lines. It has several important properties:

(i) The orders are equally spaced in $\sin \theta$, and occur whenever the phase difference between adjacent lines is an integral multiple of λ. This occurs when the numerator and denominator of Eq. (10.8) both tend to zero together. This will happen if

$$\frac{\pi d}{\lambda} \sin \theta + \frac{\delta}{2} = N\pi, \qquad (10.11)$$

or

$$\sin \theta = -\frac{\delta \lambda}{2\pi d} + \frac{N\lambda}{d}. \qquad (10.12)$$

Here, N is the diffraction grating order number. So δ, the phase shift per line caused by the angle of the illumination, determines the position of the

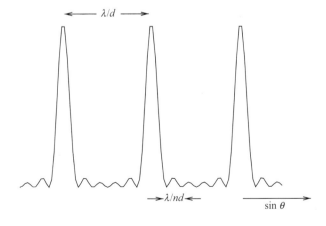

Fig. 10.5   General grating pattern for a grating with six very narrow lines.

orders. Hence, changing δ by altering θ_0 shifts the pattern so that $(\sin \theta + \sin \theta_0)$ remains constant.

(ii) On the other hand, the separation of the orders in $\sin \theta$ is independent of δ. Thus

$$(\sin \theta)_N - (\sin \theta)_{N-1} = \frac{\lambda}{d}, \tag{10.13}$$

and the orders are equally spaced in $\sin \theta$, this spacing depending only on the separation of the lines d and the wavelength.

(iii) Zeros are given by the numerator of Eq. (10.8) being zero when the denominator is not. That is to say, when

$$\frac{n\pi d}{\lambda} \sin \theta + \frac{n\delta}{2} = p\pi \text{ (unless } p/n \text{ is an integer)} \tag{10.14}$$

or, with the same restriction

$$\sin \theta = -\frac{\delta\lambda}{2\pi d} + \frac{p\lambda}{nd}. \tag{10.15}$$

This is similar to the expression in Eq. (10.12), i.e. the condition for orders, except for the restriction that p/n is not integral. So to sum up: from Eqs. (10.12) and (10.15) we see that the n phasors produced by the n lines of the grating form a closed polygon at each zero; zeros thus appear whenever the phase shift across the whole grating is a multiple of 2π. However, in the rare cases where this condition is satisfied, but also the phase shift between *each line* is a multiple of 2π, the phasor diagram is not a polygon at all, but a straight line, and instead of a zero an order is produced.

10.3   THE EFFECT OF SLIT WIDTH AND SHAPE

So far the diffraction grating has been taken to consist of n slits so narrow that the phase change across each could be neglected. This condition is very restrictive as lines of actual gratings may be several wavelengths wide. To analyse the situation where this restriction is not valid, consider first the case of a grating that is opaque except for lines of width w spaced as before d apart. Then the diffraction at a single slit follows the analysis of section 7.3, restricting the light to a range of angles depending on the width of the slit. The amplitude contributed by an individual line to the light transmitted in any direction by the whole grating is governed by the diffraction pattern of that line itself. Hence, the resultant diffraction pattern from the grating is the product of the intensity pattern of a single line with the intensity pattern of Fig. 10.5 for the ideal grating. This is illustrated in Fig. 10.6.

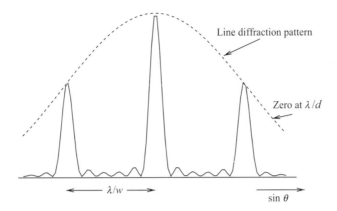

Fig. 10.6 Grating pattern for a grating with lines of width w. The line diffraction pattern envelopes the general pattern for a grating with very narrow lines.

10.4   FOURIER TRANSFORMS IN GRATING THEORY

We pointed out in Chapter 7 that the Fraunhofer diffraction pattern of an aperture is the Fourier transform of the amplitude distribution across the aperture. An idealized grating has the aperture distribution shown in Fig. 10.7(a); this equidistant series of lines, or delta-functions, is known as the grating function. Its Fourier transform is also a grating function, as shown in Fig. 10.7(b). Following section 7.4, the transform applies to the diffraction grating if the scales are in terms of wavelengths (for the aperture distribution) and in terms of direction cosines (for the angular distribution). As expected, the closer the lines of the grating, the wider apart in angle are the diffracted beams of successive orders.

The grating function represents an idealized grating, an infinitely wide grating with infinitely narrow lines. Practical gratings, with finite overall width and with lines of finite width, are also very conveniently analysed by Fourier theory; the theory becomes essential for the more complex case of three-dimensional diffraction encountered in X-ray crystallography.

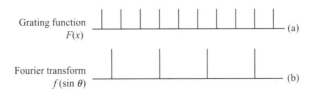

Fig. 10.7 The Fourier transform of a grating function is another grating function. The two scales are inversely proportional.

We now develop the Fourier transform approach. As we will see, this easily extends to cover the general case of arbitrary slit structure and of arbitrary distribution of illumination over the grating. Mathematically, the aperture distributions are constructed as products and convolutions (see section 4.13) of various functions with the grating function. These cases develop as follows:

(i) The grating function: an infinite array of delta functions, spacing d, with the form

$$\psi(x) = \sum \delta(x - nd). \qquad (10.16)$$

(ii) Repeated line structure, i.e. an infinite array of elements each with the form $f(x)$. The overall aperture distribution $F(x)$ is then a convolution of $\psi(x)$ with $f(x)$:

$$F(x) = f(x) * \psi(x). \qquad (10.17)$$

(iii) Finite grating length, i.e. a grating with narrow (delta-function) lines, and limited in extent by a function $H(x)$, where $H(x) = 1$ for $|x| < L$ and $H(x) = 0$ for $|x| > L$. Then $F(x)$ is the product

$$F(x) = \psi(x) \cdot H(x). \qquad (10.18)$$

(iv) Finite array of structured lines, i.e. a combination of (ii) and (iii) above. Then

$$F(x) = (f(x) * \psi(x))H(x). \qquad (10.19)$$

Recalling from section 4.13 that

- *the Fourier transform of the convolution of two functions is the product of their individual Fourier transforms*, and correspondingly
- *the Fourier transform of the product of two functions is the convolution of their individual Fourier transforms*,

we find the required transforms, i.e. the diffraction patterns, of the four cases as follows:

(i) The ideal grating $\phi(x)$ transforms into an ideal angular distribution $A(l)$, consisting of diffracted beams at equal intervals of the direction cosine l.
(ii) The grating with finite line width transforms into the product of the transforms of the grating function and the line structure, as in the envelope curve of Fig. 10.6.
(iii) The grating with finite length transforms into the convolution of the transforms of the grating function and the overall illumination function of the grating, as in Fig. 10.5.

(iv) The general case is a combination of the above, as in Fig. 10.6. In this example the complete line structure of the whole grating is a 'top-hat' function, which transforms into a wide sinc function forming the overall envelope; the illumination of the grating is also a top-hat function, which transforms into a narrow sinc function that is convolved with the grating function to produce diffracted beams with finite width.

Notice also the ideal case of a sinusoidal grating, such as may be produced holographically (Chapter 19), in which $\psi(x) = \sin 2\pi(x/a)$. Here, the diffraction pattern consists simply of single sharp (delta-function-like) diffraction maxima at $\theta = \sin^{-1}(\lambda/a)$. It is often valuable to consider complicated diffraction problems, such as those encountered in determining crystal structures (section 10.9) or in holography (Chapter 19), as the sum of diffraction effects by many simple components that each give single diffraction maxima.

10.5   MISSING ORDERS AND BLAZED GRATINGS

The combination of the individual line pattern and the grating pattern can produce the effect of *missing orders*. Suppose the line width w is commensurate with d; that is to say, that d/w is a rational fraction. Then a zero of the individual line diffraction pattern will fall on an order of the grating pattern, so that this order will not appear.

This effect is shown in Fig. 10.8 for $w = d/2$. So it is possible to remove orders by the skilful choice of the diffraction pattern of the lines. More important, particularly in reflection gratings, it is possible to arrange the

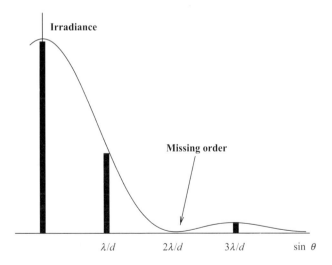

Fig. 10.8   Missing orders. A zero of the line diffraction pattern can suppress an order entirely.

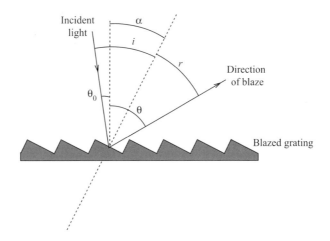

Fig. 10.9   A blazed grating with blaze angle α.

grating so that most of the light goes into one particular order. Such a grating, illustrated in Fig. 10.9, is called a *blazed* grating. The lines are ruled so that each reflects specularly in the direction of the desired order. Another way of looking at this is to observe that the angle of illumination and the tilt of the reflecting lines is such that each line has across it the appropriate phase shift to make it (as a single slit) diffract in the direction of the required order.

It is amusing to notice that a plane mirror can be regarded as a limiting case of a diffraction grating, in which the separation d is equal to the line width w. A little thought shows that now every order but the zero order is missing, since it is suppressed by the zeros of the line pattern. Only the zeroth order (which is simply specular reflection) is produced, its width depending upon the width of the whole grating; that is to say, upon the diffraction of the whole mirror as an aperture.

10.6   MAKING GRATINGS

The main grating effects are easy to observe, and do not require very fine gratings. If a handkerchief is held in front of the eye and one looks through it at a well-defined edge (such as a distant roof against the sky) it is found that as well as the actual edge at least two more can be seen. These, which are displaced by equal increments of angle, correspond to the first and second orders of the grating formed by the threads of the handkerchief and are spaced at λ/d. The human eye can resolve angles down to about 1 arcmin, or 0.0003 rad. With $\lambda = 500$ nm, this means that grating effects can just be detected if a grating of spacing 1 mm is held in front of the eye. The millimetre graduations on a transparent ruler will serve, but only just. A better grating to look through may be made by ruling a 10 cm × 10 cm square with 100 lines 1 mm apart. Photographic reduction to produce a negative 5 mm × 5 mm then gives a grating with a line spacing of 0.005 cm, in which the order separation is 0.01 rad or about 34 arcmin. The Sun viewed through this shows a spectacular and colourful series of orders, overlapping more and more as higher orders are reached. Careful; avoid looking directly at the sun.

Fraunhofer made his first grating in 1819 by winding fine wire between two screws. Later, he made them by ruling with the help of a ruling machine, in which a ruling point was advanced between lines by means of a screw. The ruling was either of a gold film deposited on glass, or etched directly onto glass with a diamond point. Later in the nineteenth century Rowland improved the design of ruling machines and was able to rule 14 000 lines to the inch on gratings as much as 6 in wide. He also invented the *concave* grating that diffracts and focuses the light at the same time; this dispenses with the need for a lens, which allows spectra to be made of ultraviolet light that would be absorbed by glass.

Excellent gratings can be made by exposing a photographic plate to the interference pattern made by two crossing plane waves, as in Fig. 7.1. The two waves must be essentially monochromatic, so that high-order interference fringes still have full visibility; this means in practice that they are both derived from a single laser source. The process is an elementary form of *holography* (see Chapter 19). Holographic gratings are used in most modern optical spectrographs (Chapter 12).

If a grating is ruled by a machine that is not perfect, confusing effects are observed that make its use for spectroscopy difficult. Each single spectral line is seen with several equally spaced and dimmer lines on each side of it. These are called *ghosts* and in a complicated spectrum of many lines may be difficult to distinguish from genuine lines. They arise from *periodic errors* in the ruling. Suppose that the machine's error was such that the depth of the ruling varied so as to go through a cycle of deep, shallow, deep every m lines, and that the transmission of the lines was proportional to their depth. Then the grating would be like a perfect grating with another perfect grating m times as coarse in front of it. When illuminated by monochromatic light, each *order* of the perfect grating would be further split into orders separated by λ/md caused by the coarse grating. It is these satellite orders that are the ghosts. In fact, any type of imperfection with a periodicity every m lines causes such ghosts spaced at λ/md, whether it be of amplitude or *phase*. The case we considered was of amplitude, in which the ability to transmit light was periodically variable. In a phase variation the *spacing* of the lines, whilst remaining on the average d, is periodically variable. In this case, the lines are first too close, then too far apart, repeating this cyclically. This is like phase-modulation of a carrier wave in radio engineering. The spectrum produced has numerous ghosts spaced at multiples of λ/md and of various intensities.

10.7   RADIO ANTENNA ARRAYS

From metre wavelengths to centimetre wavelengths it is often convenient to construct large antennas or aerials from many similar radiating or receiving elements, arranged on a one- or two-dimensional grid. Such an arrangement is called an array. The radiating elements do not here concern us: they may, for example, be half-wave dipoles. In the present discussion they are considered to be identical so that they all have the same polar diagram. The *power* polar diagram used in radio engineering is simply the angular pattern of intensity produced at a large distance from the antenna, often normalized to a fraction of the maximum of intensity. It is a Fraunhofer diffraction pattern. Similarly, the less-familiar *voltage* polar diagram is the complex amplitude. Further nomenclature that is usual in antenna work is that the main maximum or maxima of a polar diagram are called the *main beam* or beams. Subsidiary maxima are referred to as *side-lobes*.

Consider first a one-dimensional array of elements uniformly spaced d apart. The elements can all be excited separately, using suitable lengths of transmission line from the transmitter. New possibilities now arise as compared with the optical diffraction grating, since the phase of each element can be separately controlled. The main beam can be directed at any angle to the line of elements, as in the following examples, illustrated in Fig. 10.10.

End-fire array shooting equally in both directions

An end-fire array means an array with a polar diagram having equal main beams directed each way along its length. To achieve this it must be arranged that one order appears at $\sin\theta = \pm 1$, with no order in between. To make the distance in $\sin\theta$ between orders equal to 2,

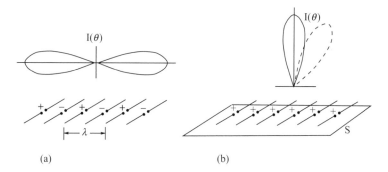

(a)   (b)

Fig. 10.10   Directive radio antennas made from separately excited dipole elements spaced $\lambda/2$ apart. The polar diagrams show the main beam. In the end-fire array (a) the phase is reversed in alternate elements. The broadside array (b) is constructed a short distance (typically $\lambda/8$) above a reflecting sheet S. A progressive phase difference between elements changes the direction of the main beam, as shown in the broken line polar diagram.

$$\frac{\lambda}{d} = 2 \quad \text{so} \quad d = \frac{\lambda}{2}. \tag{10.20}$$

To put a main beam at $\sin\theta = -1$ it can be seen from Eq. (10.12)

$$-\frac{\delta}{2\pi d/\lambda} = -1 \quad \text{so} \quad \delta = \pi. \tag{10.21}$$

So an array of spacing $\lambda/2$ between (say) dipole elements phased alternately positive and negative would have the required property. It is easy to see that in either direction along the array the contributions from each dipole would be in phase, the delay in space from dipole to dipole being matched by the shift $\delta = \pi$ between the dipoles. On the other hand, in the direction at right angles to the array (for example) the contributions from alternate dipoles would cancel each other.

End-fire array shooting in only one direction

Here we must put an order at $\sin\theta = -1$, but not another anywhere. To ensure the last condition

$$\frac{\lambda}{d} > 2 \quad \text{or} \quad d < \frac{\lambda}{2} \tag{10.22}$$

and to put an order at $\sin\theta = -1$

$$-\frac{\delta}{2\pi d/\lambda} = -1 \quad \text{so} \quad \delta = \frac{2\pi d}{\lambda}. \tag{10.23}$$

A reasonable arrangement now might be to make $d = \lambda/4$ and hence $\delta = \pi/2$. Seen from one end of the array the delay in space would now be compensated for by the phase shift of $\pi/2$ between elements. Seen from the other end the dipoles' contributions would cancel in pairs.

The technical problem of obtaining the required phase shift is not here our concern. In practice, because of the mutual coupling between elements, these sorts of arrays are difficult to realize. The familiar television Yagi aerial is a realization of the end-fire array in which a single 'driven' element couples with a reflector and several 'director' elements to approximate to the correct phase conditions.

The broadside array

In the broadside array a single main beam is required, emerging perpendicular to the array at $\sin\theta = 0$. The single main beam with low side-lobe level elsewhere is ensured if

$$\frac{\lambda}{2} = 2 \quad \text{or} \quad d = \lambda/2. \tag{10.24}$$

The side-lobes along the array at $\sin\theta = \pm 1$ will then be those midway between two orders of the general grating pattern and therefore they will be low.

The main beam will emerge perpendicular to the array if

$$-\frac{\delta}{2\pi d/\lambda} = 0 \quad \text{or} \quad \delta = 0. \tag{10.25}$$

So in a broadside array all the elements must be fed in phase. Suppose we do not make $\delta = 0$. Then the beam will emerge at an angle determined by

$$\sin\theta = -\frac{\delta}{2\pi d/\lambda}. \tag{10.26}$$

Evidently with a fixed broadside array the direction of the main beam can be steered simply by altering the phase shift from element to element at the rate given by Eq. (10.26). This principle is used in many large fixed arrays for beam swinging.

Two-dimensional broadside arrays

In the discussions above of one-dimensional arrays we have considered only the plane of the elements. Considering now all three dimensions the one-dimensional broadside array above would have a main beam rather like a pancake with its plane perpendicular to the line of the array. In almost all applications a beam narrow in both dimensions is required, and this may be achieved by making the array two-dimensional. As in the rectangular aperture treated in section 7.5 the two-dimensional polar diagram is now given by the product of the two grating patterns appropriate to each dimension, and the resultant pencil beam may be steered independently in each direction by applying suitable phase shifts.

10.8   X-RAY DIFFRACTION WITH A RULED GRATING

Diffraction gratings are expected to work well when the line spacing is a few wavelengths. If the spacing is very many wavelengths, the diffraction angles will be small and hard to measure. Diffraction of X-rays by a ruled grating is therefore rather difficult: furthermore, X-rays are not easily reflected or absorbed, so that no ordinary grating can be used.

Compton solved these difficulties very neatly by using a metal grating at nearly glancing incidence, when X-rays are reflected quite well. For X-rays, the refractive index of all substances is less than unity by a few parts in 10^6 and hence the phenomenon of total internal reflection is observed when X-rays attempt to pass from free space into a medium rather than the other way round. As the angle of incidence of a beam of X-rays on a highly polished surface approaches $90°$ there is a critical *glancing* angle $\theta_R (= 90° - i)$ at which reflection is observed; since the refractive index n is near unity we write:

$$\cos \theta_R = n = 1 - \delta, \; \sin \theta_R = \theta_R = \sqrt{2\delta}. \tag{10.27}$$

A typical example is copper, for which $\theta_R = 20'24''$ at a wavelength of 0.1537 nm, giving $\delta = 17.7 \times 10^{-6}$. The observation of θ_R is not a very accurate method of determining δ because absorption and other unwanted effects make the transition from non-reflection to reflection gradual rather than sharp. But the ability to reflect X-rays at once leads to the possibility of measuring their wavelength with a grating. Compton and Doan in 1925 successfully achieved this measurement. Using a grating of speculum metal with 50 lines per millimetre they found the wavelength of the K_α line of molybdenum to be 0.0707 ± 0.0003 nm.

In Fig. 10.11 the relationships of the zeroth and first order are shown. For the Nth order

$$BD - AC = d(\cos \alpha - \cos(\alpha + \beta)) = N\lambda. \tag{10.28}$$

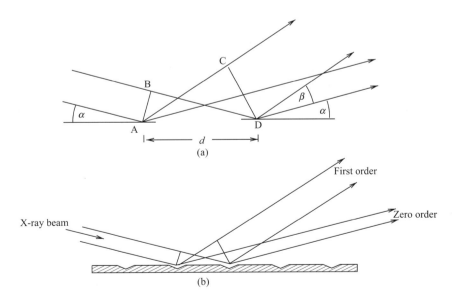

Fig. 10.11   Diffraction of X-rays at glancing incidence on a ruled grating: (a) the geometry of the reflected and diffracted beams; (b) zero-order (reflected) and first-order diffracted beams.

Since α and β are small this may be simplified to

$$d\left(\alpha\beta + \frac{\beta^2}{2}\right) = N\lambda. \qquad (10.29)$$

Notice that there is no difficulty in knowing which order the observed diffraction is at; the orders starting at zero simply come out in series as β increases.

10.9 DIFFRACTION BY A CRYSTAL LATTICE

The short wavelengths of X-rays are evidently not well suited for diffraction by ruled gratings. They are, however, conveniently close to the spacings of atoms in crystal lattices, which therefore provide excellent three-dimensional diffraction gratings for X-rays. The diffraction pattern is intimately connected with the arrangement and spacing of atoms within a crystal, so that X-rays can be used for determining the lattice structure.

It was von Laue who first suggested that a crystal might behave towards a beam of X-rays rather as does a ruled diffraction grating to ordinary light. At the time it was not certain either that crystals really were such regular arrangements, or that X-rays were short-wavelength electromagnetic radiation. In 1912, Friedrich, Knipping and von Laue performed the experiment illustrated in Fig. 10.12. X-rays from a metallic target bombarded by an electron beam were collimated by passing through holes in screens S_1 and S_2 and fell on a single crystal of zinc blende, passing through to a photographic plate. When the plate was developed, as well as the central spot at O where the beam struck the plate there were also present many fainter discrete spots. This showed that there were a few directions into which the three-dimensional array of atoms in the crystal selectively diffracted the X-rays. These depended on the orientation of the single crystal used. How could the significance of these directions be understood, when the three-dimensional and doubtless very complicated crystal structure was not known?

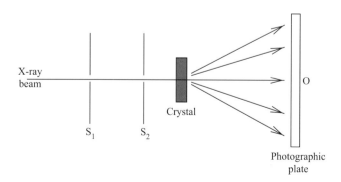

Fig. 10.12 The X-ray diffraction experiment of Friedrich, Knipping and von Laue.

The answer was given by W. L. Bragg in 1912 and is so simple that (in hindsight) it may seem completely obvious. Bragg pointed out that whatever the structure of a crystal, so long as it is *repetitive* in three dimensions, it is possible to draw sets of parallel planes on each of which the arrangements of atoms will be the same. Such planes are called *Bragg* planes and their separations *Bragg* spacings. In any crystal structure it is possible to draw many such sets of planes, but the numbers of atoms on each will vary very much. If plane monochromatic X-rays fall upon the atoms of a Bragg plane, each atom acts as a scatterer and a secondary wavelet spreads out in all directions. If we consider a *single* Bragg plane these wavelets will combine in phase in the undeviated direction, which is uninteresting, but also in a direction corresponding to ordinary reflection just like a mirror. Now add all the other Bragg planes parallel to the first. The specularly reflected waves from the various planes will, in general, be out of phase and will interfere destructively. They will all combine in phase, however, if a rather simple condition involving the glancing angle θ is satisfied. In Fig. 10.13 it is easy to see that the paths R_1A and R_1C are identical, so that the extra path in reflection number 2 is $CR_2 + R_2B$. The conditions for reinforcement are therefore the simple law of reflection, $\theta = \theta'$, plus

$$2d \sin \theta = N\lambda. \qquad (10.30)$$

These two statements together form Bragg's law for X-ray reflection from a crystal. Note that θ is the angle of the incident ray to the crystal plane.

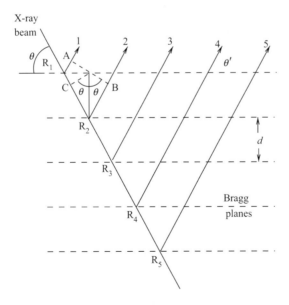

Fig. 10.13   Diffraction of X-rays at the Bragg planes in a crystal. The extra path of ray 2 is the distance $CR_2B = 2d \sin \theta$.

The spots found in the original experiment of Fig. 10.12 represented reflection at Bragg planes. The X-ray source contained a wide range of wavelengths, so that the Bragg law was obeyed for each spot by a wavelength within the range. Modern X-ray diffraction techniques use monochromatic sources; since λ is fixed a spot can then only be found by rotating the crystal until a suitable value of θ occurs. An alternative to rotating a single crystal is to use a powder containing many small crystals at all possible random orientations.

The importance of the many methods of investigation of crystal and molecular structures that sprang from these fundamental discoveries of Laue and Bragg is hard to over-emphasize. The full analysis of an X-ray diffraction pattern requires measurement not only of spot positions but also of their relative intensities, since these reveal the internal structure of the repeating elements in the crystal lattice. This may not, however, be sufficient to determine the structure of large and complex biochemical molecules, even when they can be prepared and analysed in crystalline form; here, a Fourier transformation of a diffraction pattern requires a determination of *phase* as well as intensity. The solution to this problem is found by labelling a particular site in the molecule by adding chemically an atom of a heavy metal. This atom produces a diffraction pattern of comparable intensity to the rest of the molecule, adding to or subtracting from the intensity of the elements of the diffraction pattern according to the relative phase. The heavy molecule acts as a phase reference, allowing a full Fourier transform to be made. The structure of very complex molecules can be completely determined, even if they contain some tens or hundreds of thousands of atoms, as in a protein molecule.

10.10   FURTHER READING

M. C. Hutley, *Diffraction Gratings*, Academic Press, 1982.
Loewen, E. G. Diffraction gratings, ruled and holographic, in R. P. Shannon and J. C. Wyant, eds., *Applied Optics and Optical Engineering*, Vol IX. Academic Press, 1983.

NUMERICAL EXAMPLES 10

10.1 Sodium chloride, NaCl, has a face-centred cubic crystal lattice with Na and Cl atoms alternately spaced equidistant along perpendicular x, y, and z directions. The spacing of the atoms is 2.8×10^{-10} m. At what angles would you expect X-rays of wavelength 1.54×10^{-10} m to be reflected by (a) planes of atoms perpendicular to the z direction and (b) by planes of atoms parallel to the z axis but at $45°$ to the x and y axes?

10.2 A diffraction grating consists of just 10 narrow lines separated by a spacing of 0.0001 cm is used at normal incidence. At what angles are the first and second orders for light of wavelength 500 nm? What is the angular width of the diffraction pattern for these two orders? What wavelengths might therefore be separated in a spectrograph using this grating?

PROBLEMS 10

10.1 Sketch the diffraction pattern (in intensity) of a grating of total width D, in which the slit width is w, and the spacing is d, when the grating is illuminated normally by a plane monochromatic wave.

Indicate how this pattern would be modified if: (a) the illumination is reduced at the edges of the grating; (b) alternate lines are blacked out; (c) the amplitude in alternate lines is reduced by a grey filter to a fraction α.

10.2 How is the pattern of the grating in Problem 10.1 modified if transparent material that retards the wave by half a wavelength covers (a) alternate slits; (b) every third slit; (c) one half of the grating?

10.3 A plane diffraction grating consists of alternate transparent and opaque lines with width a, b, respectively. Babinet's principle shows that the Fraunhofer diffraction pattern is the same for a grating with widths a and b interchanged, apart from the central maximum. Confirm this by means of Fourier analysis.

10.4 Derive the Fraunhofer diffraction pattern of a set of three equally spaced equal width slits by adding the transform of the outer pair of slits to that of the single central slit. Repeat the exercise for a four-slit grating, taking the inner and outer pairs separately.

10.5 The line spacing s in a badly ruled grating increases linearly with distance x across the grating so that

$$s^{-1} = a + bx. \tag{10.31}$$

Show that the nth-order diffraction maximum at angle θ, formed by plane-incident light, does not emerge from the grating as a plane wave but converges on a focus at a distance $\cos\theta/(bn\lambda)$ from the grating.

10.6 The transparency of the lines in a badly ruled grating varies periodically across the grating, so that the pth line has transparency proportional to $1 + a\sin(2\pi p/q)$, where a is small and the pattern of transparency repeats every q lines. Use the convolution theorem (section 4.13) to find the diffraction pattern of the grating.

11

Interferometry: length and angle

Following a method suggested by Fizeau in 1868, Professor Michelson has...produced what is perhaps the most ingenious and sensational instrument in the service of astronomy – the interferometer.
Sir James Jeans, *The Universe Around Us*. Cambridge University Press, 1930.

Optical interferometers, such as the Michelson and the Fabry–Perot (Chapter 9), can be used to measure distances in terms of the wavelength of light. Most of the interferometers for this purpose are two-beam interferometers, using amplitude division. Their performance may often be improved by using multi-beams, as in the Fabry–Perot; the principle is the same, but the fringes are sharper.

Interferometers can also be used to measure angular distributions of brightness across sources with small angular diameters. Michelson was again the pioneer, with his stellar interferometer that made the first measurements of the angular diameters of stars.

In this chapter we describe measurements of lengths ranging from a small fraction of a wavelength, in which the purpose may be to test the optical quality of a surface, to some tens or hundreds of metres, where the objective may be to measure the stability of a large structure or to test the theory of relativity. Interferometers measure the optical path along a light beam, i.e. the product of geometric path and refractive index; by comparing the optical paths of two light beams, very small differences in physical length or refractive index can be measured.

We continue with the measurement of the angular size of light sources, which is achieved by interferometers using elements separated not along but across a wavefront.

11.1   THE RAYLEIGH REFRACTOMETER

We start with the simplest of refractive index measurements. In any two-path interferometer there is a comparison of the optical length of two separate paths. Rayleigh put this to use in measuring the refractive index of a gas. His refractometer (Fig. 11.1) was based on Young's double slits, although evidently any other two-beam interferometer could be used. The tubes T_1 and T_2 are in the separate light paths from the slits S_1 and S_2, illuminated coherently from a single source. When the pressure of gas is changed in one of the tubes, the fringe system, viewed by an eyepiece E at the focus of a long-focus lens, moves across the field of view. A count of the fringes (N) as they move provides a direct measurement of the change in optical path through the tube, and hence the change in refractive index δn as the amount of gas changes. For a tube length l

$$N = \frac{l\delta n}{\lambda}. \tag{11.1}$$

For a dilute gas the refractive index differs from unity by an amount proportional to density, so that for a fixed temperature $n-1$ is proportional to pressure. The refractive index obtained from a single measurement can then be used to calculate the value for any other pressure by simple proportion.

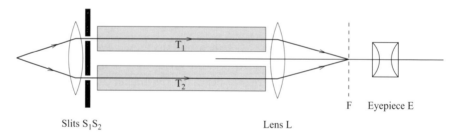

Slits S_1S_2 Lens L

Fig. 11.1   Rayleigh refractometer. S_1, S_2 are illuminated by a common source of light. Interference fringes are formed at the focal plane F of the lens L, and viewed with an eyepiece E. The fringes move across the field of view when the gas pressure is changed in one of the tubes T_1, T_2.

11.2   WEDGE FRINGES AND END GAUGES

Interferometers measuring physical length in terms of light wavelength are used to establish and compare standards of length in the form of *end gauges*. These are metal bars with polished ends that can be used as reflecting mirrors in interferometers.

The simplest comparison that can be made is between two end gauges that are nominally of the same length. They can be placed together on a flat surface, and a partially silvered optically flat glass plate placed on top (Fig. 11.2).

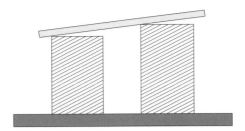

Fig. 11.2 Comparison of the lengths of two end gauges. Wedge fringes are formed in the air gap between a glass plate and the ends of the gauges. The spacing of the fringes, which are viewed from above the plate, is a measure of the wedge angle.

Thin-film fringes (Fizeau fringes) are then seen in the wedge cavity over each of the gauges, with a spacing depending on the angle between the gauge and the glass plate; the difference in height of the two gauges can easily be found by counting the fringes. Irregularities in the reflecting surfaces are seen as deviations from straight lines in the reflected wedge fringes. The partial silvering of the glass plate gives a multi-beam interference effect, like the transmission fringes in the Fabry–Perot interferometer. These fringes are so sharp that surface discontinuities down to 0.3 nm can be detected.

Interferometers for measuring larger differences in light paths are usually based on Michelson's interferometer. An example of the direct measurement of the length of an end gauge is shown in Fig. 11.3. Here, the end gauge $G_1 G_2$ is placed firmly in contact with an optically flat reflector M_1, so that M_1 and the end of the gauge G_1 can both reflect light in the field of the interferometer. The reflector M_2 is inclined at a small angle, so that the field of view is crossed with parallel fringes, as in Fig. 11.3. Part of the field shows fringes from G_1 and part from M_1.

The distance $G_1 G_2$ is measured as a shift of the fringe pattern in wavelengths of the particular light in use. Only the fractional part can be measured, however. The whole number of fringes can be found by repeating the measurement with other wavelengths of light; for example, using four different wavelengths of light from a cadmium lamp, each of which is known to about 1 part in 10^8. When the fractional parts are all known, the whole numbers can be found by computation (see Problem 11.5).

11.3   THE TWYMAN AND GREEN INTERFEROMETER

The fringes shown in Fig. 11.3 are straight and evenly spaced. This would only be so if the surfaces were precisely flat: any departures from flatness would be seen as distortions in the fringe pattern. Twin-beam interferometers can evidently be used to measure the flatness of a reflecting surface. The Twyman and Green interferometer (Fig. 11.4(a)) is a twin-beam interferometer designed for this purpose. It can also be used to test the transmission properties of transparent optical components.

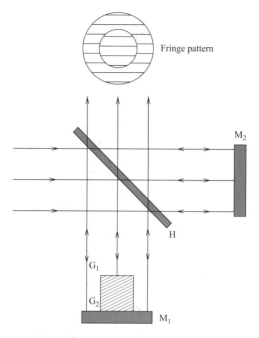

Fig. 11.3   End gauge interferometer. This is similar to a Michelson interferometer, but with fixed mirrors. The gauge G_1G_2 is in contact with M_1, so that the centre of the field of view, seen from above, shows fringes from G_1, while the outer part shows fringes from M_1, as shown in the inset diagram.

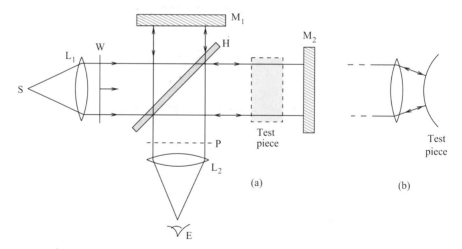

Fig. 11.4   Twyman and Green interferometer: (a) the half-silvered plate H and the two mirrors M_1, M_2 are arranged as in a Michelson interferometer, but the source of light is a coherent plane wavefront W, and the fringes are recorded either directly on a photographic plate at P or viewed at the focus of the lens L_2. The field of view is uniformly bright if M_1 and M_2 are exactly aligned. Imperfections in optical path, as, for example, through a glass slab at G, show as variations of brightness; (b) M_2 can be replaced by a spherical mirror when a converging lens is to be tested.

The Twyman and Green interferometer is a development of the Michelson interferometer in which the source of light is a plane wavefront coherent over the whole field rather than a broad incoherent source. Originally, this was achieved by the use of a pinhole source at the focus of a high-quality lens, but now a suitable source is a laser beam. With this arrangement, light paths can be compared over the whole field. For example, the arrangement of Fig. 11.4(a) may be used to compare the surfaces of two mirrors, while in (b) the optical path through a lens is measured across the whole field. As with the measurement of end gauges in Fig. 11.3, the mirror M_1 is slightly tilted, so that a perfect optical system yields a system of parallel fringes across the field of view; imperfections then show as deviations from straightness.

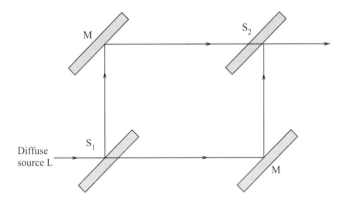

Fig. 11.5 Ray paths in the Mach–Zehnder interferometer. The beam-splitter S_1 and recombiner S_2 are half-reflecting mirrors. The mirrors M are fully reflecting. An extended light source L may be used.

The Mach–Zehnder interferometer, shown in Fig. 11.5, is another amplitude-splitting device intended for comparing optical paths in separated beams. The separation of the beams may be large, so that one beam may for, example, traverse a wind tunnel in the region of a shock wave (Fig. 11.6), or it may traverse a plasma cloud in a thermonuclear test reactor. Variations of refractive index across the field are seen as deviations from straightness in a set of plane-parallel fringes.

11.4   THE STANDARD OF LENGTH

The internationally defined standard of length is the distance travelled in vacuo by light in unit time.[1] Before 1983 the metre was defined as a number of

---

[1] The definition (by the Conference Gènèrale des Poids et Mesures, 1983) is: 'the metre is the length of the path travelled by light in vacuo during a time interval of 1/299 792 458 of a second'. The standard of time, the second, is defined using precision atomic clocks as the duration of 99 162 631 770 periods of radiation between two hyperfine levels of the caesium atom.

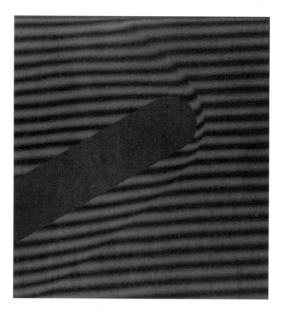

Fig. 11.6 Interference fringes obtained in a Mach–Zehnder interferometer, showing variations of refractive index around a wind-tunnel model. (Andrew Kennaugh, University of Manchester.)

wavelengths of a narrow spectral line in krypton, but because of the finite width of this line this standard was reproducible only to about 3 parts in 10^8. In contrast, the standard of time can be reproduced with an accuracy of 1 part in 10^{13}; this is achieved by linking a clock such as a quartz crystal oscillator through a series of harmonic generators to the caesium standard.

The velocity of light, $c = 299\ 792\ 458\ \text{m s}^{-1}$, is therefore a defined constant that was chosen in 1983 to give agreement, to the best possible accuracy, between the new and old definitions of the unit of length. If the frequency of a narrow-bandwidth laser can be measured in terms of the unit of time, then the wavelength is known in standard units, and it can be used to calibrate a secondary standard such as a metre-long end gauge at the wavelength of a chosen spectral line. (In fact, the previous length definition was in terms of a specific spectral line wavelength.)

Michelson made the first measurements of the metre in terms of wavelengths in a different era, when the metre was defined by a mechanical standard; he was therefore measuring the wavelength of light rather than measuring the metre. The moving mirror of a Michelson interferometer was mounted on a carriage on accurate sliding ways alongside the standard, and with a microscope attached for setting on the fiducial marks of the metre. A direct count of fringes as the carriage traversed the metre would involve some millions of fringes; furthermore, the spectral line sources available to Michelson were not sufficiently narrow to allow the use of path difference as large as a metre. He therefore used an intermediate-sized standard of length called an etalon, and built up the full length by adding a series of etalons.

11.5 THE MICHELSON–MORLEY EXPERIMENT

This classic experiment, which is now regarded as a test of special relativity, was originally devised as a measurement of the velocity of the Earth through space. If light could be regarded as a wave in a medium, called the ether, through which the Earth was moving, then the velocity of light as measured on Earth would depend on its direction of travel. A Michelson interferometer, with one light path along this direction and the other at right angles, would show the effect as a fringe shift. The shift could be detected by rotating the interferometer so as to interchange the optical paths (Fig. 11.7).

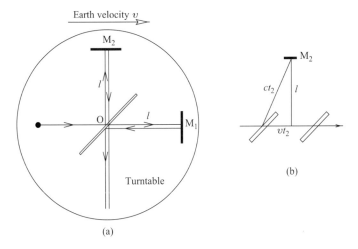

(a)

(b)

Fig. 11.7 (a) The Michelson–Morley experiment. The path to mirror M_1 is initially along the direction of the Earth's orbital motion; the whole interferometer can be rotated to bring the path to M_2 into this direction. (b) The path followed by beam OM_2, according to classical theory

In the Michelson–Morley experiment both light paths were increased to 11 m by folding them between a series of mirrors. The expected effect, according to the ether theory, of the Earth's orbital velocity was nevertheless only 0.4 of a fringe. This was to be detected by rotating the whole apparatus by 90°, so interchanging the arms parallel and perpendicular to the Earth's motion. To preserve the stability of the interferometer, it was mounted on a stone bed floated in mercury.

The path difference expected from simple 'classical' arguments is found from the times of travel t_1 and t_2 along the two beams, assigning a velocity v to the ether drift. For an observer moving with the velocity of the Earth in its orbit, the ether should be moving past the interferometer with the same velocity v, so that the velocity of light along and against the ether drift would be $c + v$ and $c - v$. For a light path l the time t_1 along and against the ether flow is

$$t_1 = \frac{l}{c-v} + \frac{l}{c+v}$$

$$= \frac{2l}{c}\beta^2, \tag{11.2}$$

where $\beta = \left(1 - v^2/c^2\right)^{-1/2}$.

The time t_2 transverse to the flow is increased by the movement of the mirror with velocity v, as shown in Fig. 11.7, so that

$$c^2 t_2^2 = v^2 t_2^2 + (2l)^2, \tag{11.3}$$

giving

$$t_2 = \frac{2l}{c}\beta \tag{11.4}$$

$$t_1 - t_2 = \frac{2l}{c}\left(\beta^2 - \beta\right). \tag{11.5}$$

Expanding each term in β into a series gives a good approximation

$$t_1 - t_2 \approx \frac{l}{c}\left(\frac{v^2}{c^2}\right). \tag{11.6}$$

The expected fringe shift would be $c(t_1 - t_2)/\lambda$, which would reverse when the interferometer was rotated through 90°. For a path $l = 11$ m, wavelength $\lambda = 550$ nm and an ether velocity $v = 30$ km s^{-1}, which is the Earth's orbital speed, there would be a shift in the fringe pattern by 0.4 fringe on rotation. No such fringe shift was detected. Just in case the Earth happened to be moving at the same velocity as the ether, the experiment was repeated six months later when the Earth's velocity was reversed. Again there was no fringe shift.

We should realize that this result was a great surprise when it was first obtained in 1887, which was 18 years before Einstein published his theory of special relativity. The first reasonable explanation came from Lorentz, who suggested that moving bodies contract in the direction of motion by the factor β in our analysis above. Poicaré commented that this was a conspiracy of nature that made it impossible to detect an ether wind by *any* experiment! The null result of the Michelson–Morley experiment is, of course, now seen to be entirely in accordance with the special theory of relativity, in which the velocity of light is invariant between any pair of frames of reference in relative linear motion.

11.6   THE RING INTERFEROMETER

In 1911 Sagnac constructed an interferometer in which the two beams follow the same optical path but in opposite directions (Fig. 11.8). The ring may have

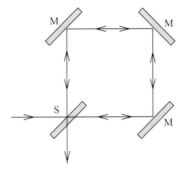

Fig. 11.8   Ray paths in the Sagnac interferometer. The rays that recombine at the beam-splitter S traverse the circuit in opposite directions, reflected by the three mirrors M.

three or four mirrors; a small angular misalignment of any of them will produce a parallel fringe pattern.

The ring interferometer is insensitive to any effect apart from a continuous rotation about an axis perpendicular to the ring. In such a rotation one optical path is effectively shortened in comparison with the other. Special relativity in no way suggests that rotation cannot be detected, in contrast to the linear velocity that was the objective of the Michelson–Morley experiment.

The following simplified analysis should be regarded with some suspicion, since it appears not to take proper account of relativity. We assume simply that the time for light to travel once around the ring from a given point P is increased or decreased by the movement of that point during the circuit. If the ring has a mean radius R, then without rotation the time is $2\pi R/c$. If, however, the ring is rotating with angular velocity Ω the point P moves in that time a distance $\delta s = (2\pi R/c)\Omega R$. Light in one direction therefore arrives at P late by $\delta s/c$, while in the other direction it arrives early by $\delta s/c$. The difference Δt in travel time is therefore

$$\Delta t = 2\frac{\delta s}{c} = 4\frac{\pi R^2 \Omega}{c^2} = 4\frac{A\Omega}{c^2}, \tag{11.7}$$

where A is the area of the ring.[2]

---

[2] The special theory of relativity shows that, as measured in the laboratory frame, the velocity of light in the two directions around a ring with refractive index μ whose peripheral velocity is v is

$$c^+ = \left(\frac{c}{\mu} + v\right)\left(1 + \frac{v}{\mu c}\right),$$

$$c^- = \left(\frac{c}{\mu} - v\right)\left(1 - \frac{v}{\mu c}\right).$$

The result, Eq. (11.7), is obtained for $\mu = 1$ and $R\Omega \ll c$.

The ring interferometer can therefore be used to measure the rate of rotation about the axis of the ring, since the displacement of the fringe pattern is proportional to Ω. In 1925 Michelson and Gale[3] used such a system to measure the rate of rotation of the Earth, so demonstrating the contrast between linear and rotational motion in relativity.

11.7   OPTICAL FIBRES IN INTERFEROMETERS

In the twin-beam interferometers that we have described so far, two light beams from a single source are recombined after travelling different paths in air or blocks of glass. We will see in Chapter 18 that light can be guided down long lengths of glass fibre with very little loss, so that some remarkably simple and stable interferometers can be constructed using long lengths of glass fibre. Light from a compact semiconductor laser (Chapter 16) can be split between two fibre paths, and recombined in a simple diode detector (Chapter 21). As in the original Rayleigh interferometer, the sum is dependent on the relative phase of the two beams, allowing measurement of differences in length or refractive index and their dependence on other parameters in the two light paths.

Figure 11.9 shows a basic system in which light from a laser is split between two fibre paths, usually of nearly the same length, and recombined in a detector. The relative phase is detected by incorporating a modulator, which periodically reverses the phase in one arm; the detector output is then observed to be modulated at the switch frequency, and the modulation depth depends on the phase difference between the two beams. The output is selectively amplified

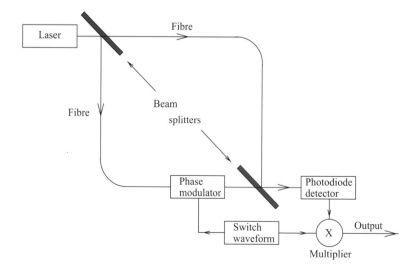

Fig. 11.9   Optical fibre interferometer.

[3] Michelson and Gale, *Astrophys J.*, **61**, 140, 1925.

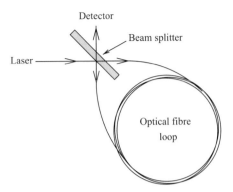

Fig. 11.10   The Sagnac interferometer, used as a gyroscope.

at the switch frequency. This is the fibre equivalent of the Mach–Zehnder interferometer. It may be adapted for measurements of any parameter that affects the phase path differently in the two arms, such as temperature differences, or mechanical strain.

The Sagnac ring interferometer has been developed into a useful gyroscope, one form of which uses a fibre optic ring. The system shown in Fig. 11.10 uses only one length of fibre, but compares the phase of light signals sent round a long coil in two opposite directions. This system is sensitive to continuous rotation about the axis of the coil (Eq. (11.7)). The difference in transit times for light travelling in clockwise and anticlockwise directions gives rise to an optical path difference $c\Delta t$. This may be detected as a relative phase difference ϕ_s proportional to the rotation rate, the area A and the number of turns:

$$\phi_S = \frac{8\pi\Omega NA}{\lambda c}. \tag{11.8}$$

For example, consider the measurement of the rotation of the Earth (15°/h). For a ring with $R = 0.2$ m, made of 1000 turns of optical fibre and operating with an He–Ne laser at 632.8 nm, a phase shift of 1.2×10^{-3} rad is required to be measured.

In another form of the Sagnac gyroscope, the ring laser gyroscope, the laser source is contained within the ring, which acts as a laser cavity (see Chapter 15) producing clockwise and counterclockwise travelling laser waves. The rotation of the ring leads to a difference $\Delta\nu$ between the two resonant laser frequencies given simply by

$$\Delta\nu = \frac{2\Omega R}{\lambda}. \tag{11.9}$$

(This result can be obtained from Eq. (11.7) by substituting $\Delta\nu/\nu = \Delta t/t$, where $t = 2\pi R/c$ is the time for light to travel around the ring.) For a ring with $R = 0.5$ m operating on an He–Ne laser at 632.8 nm, Earth rotation gives a frequency difference $\Delta\nu = 115$ Hz. This can be detected as a beat between the two laser frequencies.

11.8 DETECTING GRAVITATIONAL WAVES BY INTERFEROMETRY

Although the existence of gravitational waves has been demonstrated by the loss of radiated gravitational energy in a binary pulsar system, their direct detection on Earth is extremely difficult. A detectable gravitational wave might be radiated from a catastrophic event such as the coalescence of a binary pair of stars, but the amplitude at the Earth would correspond to a transient or low-frequency change in length scale by a factor of only 10^{-21} or less. Gravitational wave detectors attempt to obtain such a sensitivity by using a laser interferometer on a monumentally large scale, as shown in Plate 11.1.

The interferometer used in the gravitational wave detector follows the Michelson principle in comparing the length of the two arms, which are each typically several kilometres long. The distant mirrors are mounted on large suspended masses, which will move differentially depending on the direction of the gravitational waves. There are (only!) 10^{10} wavelengths of visible light (wavelength 600 nm) in a double journey along one beam, so the requirement is to measure a fringe shift of 3×10^{-12} fringe! Obviously, the Fabry–Perot technique must be employed, so obtaining very sharp fringes, and a very stable and high-powered laser must be used to allow the positions of the edges of the fringe profile to be measured with great accuracy. The achievement of a positive detection must be rated as one of the greatest challenges in the measurement of lengths by optical interferometry.

11.9 DOUBLE SOURCE INTERFEROMETERS

Interference and diffraction theory has in previous chapters considered waves originating from an idealized point source, or from an ideal plane wavefront. In the second part of this chapter we are concerned with the effect on the interference fringes when the source has a finite size; in general, this reduces the *visibility* of the fringes. The reduction of visibility enables a measurement to be made of the size of the source; more precisely, the measurement is of the angular spread of plane waves entering the interferometer. The most important example of such a measurement is Michelson's stellar interferometer.

We first consider Young's double source interferometer, which was introduced in Chapter 7. This is an example of an interferometer using *division of wavefront*, in contrast to *division of amplitude*, as discussed in Chapter 9. Figure 11.11 shows two pinholes or slits S_1, S_2 illuminated by a point source of monochromatic light L. Light from S_1 and S_2 spreads and overlaps in the shaded region: throughout this region there is interference between the two sets of waves, and interference fringes would appear on a screen or photographic plate placed anywhere in the region. (The interference fringes are said to be *non-localized*, in contrast to the *localized* fringes seen in the thin films that we discussed in Chapter 9). We must distinguish between the effects of an

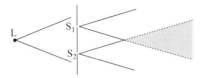

Fig. 11.11   Young's double slit. Overlapping beams from two slits, S_1, S_2, illuminated by the same source L. Interference occurs in the whole of the shaded area.

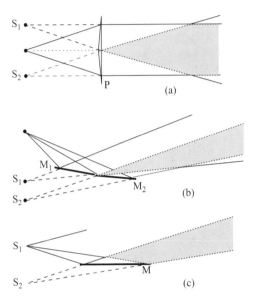

Fig. 11.12   (a) Fresnel's biprism; (b) Fresnel's double mirror; (c) Lloyd's mirror. In each arrangement twin beams from the same source overlap in the shaded area, appearing to diverge from the twin sources S_1, S_2.

extended, rather than a point, source, and the effect of slits with a finite width. The same arguments will apply to many related types of interferometer that involve two overlapping light beams from a single small source. Fresnel used the thin biprism of Fig. 11.12(a) and the nearly coplanar mirrors of Fig. 11.12(b). A particularly simple arrangement is Lloyd's mirror (Fig. 11.12(c): here, the two sources are slit S_1 and its image S_2). An interesting feature of Lloyd's mirror is that the light reflected at grazing incidence to form S_2 suffers a phase reversal at reflection (see Chapter 5), so that the interference fringes are exactly out of step with those of the double slit.

Similar arrangements can be devised for the much longer wavelength radio waves. A classic example is the equivalent of Lloyd's mirror (Fig. 11.13(a)) used in early radioastronomical observations in Sydney, Australia. The radio telescope, mounted on a cliff overlooking the sea, received radio waves of around $1\frac{1}{2}$ m wavelength (frequency 200 MHz) from the Sun and other celestial radio

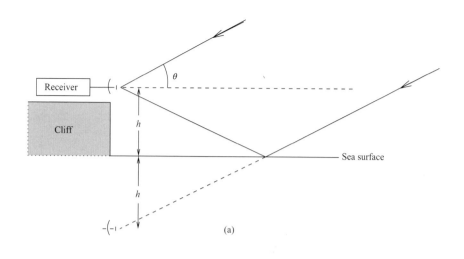

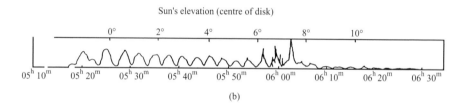

Fig. 11.13   Lloyd's mirror in radioastronomy: (a) the radio telescope receives radio waves both directly and indirectly from the sea; (b) the recorded radiation from the Sun as it rises above the horizon. The interference fringes are disturbed by refraction near the horizon, and by solar outbursts.

sources as they rose above the horizon. Both direct and reflected radio waves were received; as the Sun rises the path difference between them changes and produces a set of interference fringes. A typical trace of the interference fringes is shown in Fig. 11.13(b); this was recorded at a time of strong and variable radio emission from above a sunspot.

11.10   THE EFFECT OF SLIT WIDTH

So far each slit in Young's experiment has been supposed to be the source of a uniform cylindrical wave, which for an ideal narrow slit covers 180°. We now recall the analysis of Chapter 7, in which the angular distribution of amplitude in the wave spreading from each slit by Fraunhofer diffraction was related to the slit width by Fourier theory. The simplest way to think of what happens is to notice in Fig. 11.11 that fringes cannot be observed in any particular direction θ unless each slit, acting as a diffracting aperture, sends some light

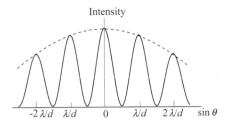

Intensity

$-2\lambda/d$ λ/d 0 λ/d $2\lambda/d$ $\sin\theta$

Fig. 11.14   Young's fringes with slits of finite width w and separation d. The dotted envelope of the fringes is the diffraction pattern of each slit and would reach its first zero at $\pm\lambda/w$.

that way. Each slit contributes according to its own Fraunhofer diffraction pattern, while the relative phase of the two contributions is $(2\pi d/\lambda)\sin\theta$, where d is the spacing of the slits. Young's fringes are then observed within the intensity envelope of the diffraction pattern of a single slit, as shown in Fig. 11.14.

Combining Eqs. (7.18) and (7.21), we can immediately describe the fringe pattern as the product of a sinc function due to the slit width w and a cosine fringe system due to the spacing d. Using the direction cosine notation $l = \sin\theta$, Eq. (7.23) may be written:

$$A(l) = \frac{\sin \pi wl/\lambda}{\pi wl/\lambda} \cos\left(\frac{\pi \, dl}{\lambda}\right). \tag{11.10}$$

The same result may be reached via the convolution theorem in Fourier transforms (Chapter 4). In this equivalent approach the twin slits are described as the convolution of a top hat function, width w, with a pair of delta functions, spaced d apart; the Fourier transform, which is the angular distribution of diffracted plane waves, is the product of the two separate transforms.

11.11   SOURCE SIZE AND COHERENCE

So far the wavefronts leaving the two apertures have been considered as parts of a single plane monochromatic wavefront. No wavefront is ideally plane and no wave is perfectly monochromatic, although laser light can approach the ideal very nearly. We now investigate interference between non-ideal wavefronts, such as those obtained from a source of finite size.

In any ordinary light source, such as an incandescent filament or a sodium lamp, the output of the lamp is made up of the sum of the light waves produced by a very large number of individual atoms. Their phases are unrelated, so that only their intensities add. Light derived from different parts of the source cannot interfere. Interference is only observable if the two interfering beams

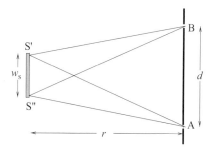

Fig. 11.15 A source of finite size illuminating Young's slits. Fringes of high visibility can only be seen if the path difference AS − BS only changes by a small fraction of a wavelength for all positions of S from S' to S".

are both derived from the same region of the source, when the two beams are said to be *mutually coherent*.[4] Two beams derived from any other part of the source may also interfere, but the interference patterns from different parts of the source may not coincide in time and space. But if the interference patterns made by each component of an extended source are identical, then they add to produce the same pattern as that from a point source.

Let us look at the problem in the case of Young's slits. Suppose that the plane wave we have so far considered is replaced by the wavefront from a source S'S" of finite width w_s at a large distance r from the slits, as in Fig. 11.15. Each component of the source produces interference fringes with the same spacing, but the position of the interference fringes depends on the relative phase of the wave as it arrives at the slits. This relative phase depends on the difference in distance between the source and the slits; if this difference is nearly the same for all components, then the fringe patterns will coincide and add to give the normal point source fringe pattern.

The interference patterns will coincide nearly enough to give a fringe pattern with high visibility if the relative path length from each end of the source is the same within a small fraction of a wavelength. Seen from the slits this means that the directions of S' and S" must be the same within an angle of λ/d. The waves at A and B can then again be regarded as coherent, as they were when the light came from a single point source. This gives us a condition for coherence between the light at A and B:

$$\frac{w_s}{r} \ll \frac{\lambda}{d}. \qquad (11.11)$$

A perfect point source would have the property that all pairs of points on its wavefront would be coherent. The important result of Eq. (11.11) gives the condition under which a finite source may be regarded as a point source as far

---

[4] The concept of coherence is discussed in more detail in Chapter 14.

as a particular diffracting system is concerned. Looking back at the source from the diffracting system, Eq. (11.11) may be restated as

$$\boxed{\text{angle subtended by source} \ll \lambda/\text{linear size of diffracting system.}} \qquad (11.12)$$

11.12   MICHELSON'S STELLAR INTERFEROMETER

In the previous section it was shown that one condition for fringes to be produced by Young's slits was Eq. (11.12), a restriction on the angular size of the source seen from the slits. This suggests that such a system might be used to *measure* the angular size of a source by altering the slit separation until fringes could not be seen. This is the principle of *Michelson's stellar interferometer*. Some modification of the simple Young system is needed to make an interferometer capable of producing fringes by the light of a star; this concerns the method of combining the light that comes through the two slits. In Young's system, diffraction at the slits is relied upon to make part of the emergent light from each slit reach the same region so that interference can take place. This means that the slits have to be narrow. In Michelson's system (Fig. 11.16) the slits are replaced by two large plane mirrors M_1 and M_2 a distance d_M apart. Each reflects light inwards to two further inclined mirrors M_3 and M_4. These reflect two parallel beams of light into a telescope objective (which may be reflecting or refracting). Each beam forms an image of the star in the focal plane F, and fringes are seen crossing the diffraction disc of the combined

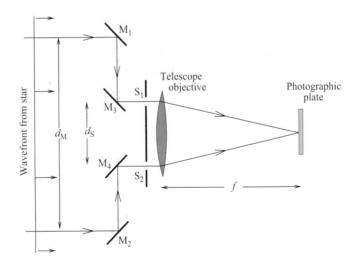

Fig. 11.16   Michelson's stellar interferometer. Light from parts of the star's wavefront d_M apart is made to interfere in the focal plane of the telescope. The visibility of the resultant fringes as d_M is varied allows some estimate to be made of the star's angular diameter.

image. The two apertures S_1 and S_2 in front of the objective, which limit the size of the beams, act like Young's double slits with a spacing d_S.

Suppose first that a star is observed that has so small an angular diameter ϕ_0 that the inequality (11.12) is satisfied, i.e. $\phi_0 \ll \lambda/d_M$. Then the wavefronts falling on M_1 and M_2 are coherent and hence (if the mirrors are perfectly adjusted) S_1 and S_2 are illuminated by coherent wavefronts. In the focal plane F the interference pattern of the slits is observed; it consists of intensity fringes with a spacing $(\lambda/d_S)f$, where f is the focal length of the telescope. Notice that the fringe *spacing* is independent of the separation d_M of the outer mirrors M_1 and M_2. The extent of the area over which fringes can be observed is limited by diffraction at the individual apertures, width w_S, and is therefore of the order of $(\lambda/w_S)f$; this is the approximate width of the central maximum of the diffraction pattern of each aperture.

Now consider the effect of observing a star of finite angular diameter ϕ_0. Suppose it to be divided up into very narrow strips of width $\delta\phi$, each of which satisfies the condition Eq. (11.12), which in this case is

$$\delta\phi \ll \frac{\lambda}{d_M}. \tag{11.13}$$

Then each such strip produces a set of fringes, but each set is non-coherent with any other set, originating as it does from a different part of the source. These sets of fringes then add non-coherently; that is to say, their *intensities* add. Each elementary strip on the star produces its own fringe system, which overlaps the others to a greater or lesser degree according to the star's angular diameter. Note that there is no *interference* between fringes from different parts of the

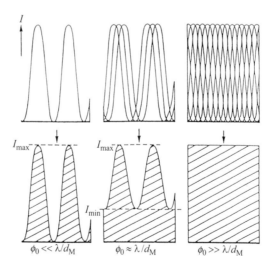

Fig. 11.17 Fringes in three cases with the Michelson interferometer. The angular diameter of the observed star is ϕ_0.

star (which are non-coherent), only addition of intensity. Three cases can easily be distinguished, and are illustrated in Fig. 11.17, where only the central portion of the pattern is considered, so that the $\sin \psi/\psi$ term in the single slit pattern, which gradually modulates the cosine fringes, is usually taken as unity. The sets of fringes from different parts of the star will be shifted sideways in y, the total shift in y being $\phi_0 f$, as is easily shown by geometrical considerations. An angular movement ϕ amounting to λ/d_M causes a difference of path of one wavelength for the light going by the two routes. Examining the three cases in turn shows that more or less blurring of the fringes occurs.

1. $\phi_0 \ll \lambda/d_M$. The condition (11.12) is satisfied and the transverse shift of the fringes from different parts of the source is slight. Completely dark minima will be seen.

2. $\phi_0 \simeq \lambda/d_M$. The transverse shift of the fringes from different parts of the source is the same order as their spacing λ/d_S. There will be no completely dark minima, but a sinusoidal variation of the intensity will be seen.

3. $\phi_0 \gg \lambda/d_M$. The overlapping of the sets of fringes from different parts of the source is complete; no variation of the intensity will be seen.
   The *fringe visibility* $V(d_M)$, given by

$$V = \frac{I_{max} - I_{min}}{I_{max} + I_{min}} \tag{11.14}$$

is used in the measurement of source diameter and the brightness distribution across the source. The maximum and minimum intensities used here are illustrated in Fig. 11.17. To see how V varies quantitatively with d_M it is necessary to define the brightness distribution of the source as a function of ϕ. Let this be $B(\phi)$, and let $B(\phi)$ be symmetrical about the centre of the source. Then each elementary strip of the source, $d\phi$ wide and at angle ϕ from the centre of the source, produces a fringe system with intensity proportional to $B(\phi)$ and displaced by angle ϕ from the centre of the fringe system. The intensity of a fringe maximum now becomes the integral

$$I_{max} = a \int_{source} B(\phi) \cos^2 \frac{\pi d_M \phi}{\lambda} \, d\phi, \tag{11.15}$$

$$= \frac{a}{2} \int_{source} B(\phi) \left(1 + \cos \frac{2\pi d_M \phi}{\lambda}\right) d\phi, \tag{11.16}$$

where a is a constant. Similarly, the minimum intensity becomes

$$I_{min} = \frac{a}{2} \int_{source} B(\phi) \left(1 - \cos \frac{2\pi d_M \phi}{\lambda}\right) d\phi. \tag{11.17}$$

The fringe visibility is therefore

$$V = \frac{\int B(\phi)\cos(2\pi\, d_M\phi/\lambda)\mathrm{d}\phi}{\int B(\phi)\mathrm{d}\phi}. \tag{11.18}$$

For a source of uniform brightness $B(\phi)$ over an angular width ϕ_0

$$V = \frac{\sin(\pi\, d_M\phi_0/\lambda)}{\pi\, d_M\phi_0/\lambda}. \tag{11.19}$$

This is a sinc function similar to that representing the Fraunhofer diffraction pattern of a single slit. There is, in fact, a close relation between the fringe visibility function, Eq. (11.18), and the diffraction formula, Eq. (7.20), which can be traced through the fact that the numerator of Eq. (11.18) is the cosine Fourier transform of the brightness distribution across the source. Equation (11.19) may be taken to represent the variation of visibility with ϕ_0 at a fixed spacing d_M; alternatively, if d_M can be varied the fall of visibility can be used to measure ϕ_0 for a given star.

Michelson in the early 1920s set up such an interferometer at Mount Wilson on the 100-in telescope, in which the maximum spacing of the outer mirrors was about 6 m. The angular separation of the interference fringes at this spacing was 10^{-7} rad, or 0.02 arcsec. Unfortunately, this is large compared with the angular diameters of most nearby stars, and thermal instabilities in the air paths of the interferometer did not allow precise measurements of fringe visibility. Nevertheless, the diameters of a small number of red giants could be measured. The first of these was Betelgeuse, which was found to have an angular diameter of 0.047 arcsec. From this measurement and the distance (known from measurements of parallax) the linear size turned out to be three hundred times that of the sun, and large enough to enclose the earth's orbit! Only a few such stars could be measured with the 6-m spacing. Michelson attempted to use larger spacings, but thermal instabilities in the separated light paths made the interference fringes fluctuate rapidly and impossible to observe. Larger spacings can, however, be used if a detector with sufficiently rapid response is used, as has been demonstrated more recently by J. Baldwin in Cambridge and J. Davis in Sydney, Australia.

11.13   VERY LONG BASELINE INTERFEROMETRY

Interferometers with very much longer separations can be used at radio wavelengths, where the effects of random refraction in the atmosphere are negligible. Radio telescopes up to 200 km apart in the MERLIN array centred on Jodrell Bank are connected by radio links in a system closely analogous to Michelson's stellar interferometer. At this spacing a radio wavelength of 2 cm gives the same fringe spacing of 0.02 arcsec, although the system is usually operated at longer wavelengths.

Longer baselines with direct connections between the separate radio telescopes are impractical, but radio interferometry is nevertheless routinely used with baselines extending over some thousands of kilometres. Instead of directly transmitting the radio signals to a common receiver, they are recorded on magnetic tapes that are subsequently transported and replayed into the common receiver. The relative phase of the signals must be preserved in this operation; this is achieved by using very stable oscillators as phase references at the separated receivers. Very long baseline interferometry (VLBI) with a baseline of 1000 km and a wavelength of 1 cm has a fringe spacing of 2 micro-arcsec, giving a very much greater angular resolution than any optical interferometer. Surprisingly, there are some distant and powerful radio sources, the quasars, that demand still longer baselines; this has been achieved by using a radio telescope in an orbiting satellite as one element of a VLBI system, giving baselines of up to 6000 km.

11.14   THE INTENSITY INTERFEROMETER

The difficulties of achieving stable interference fringes in Michelson's stellar interferometer were overcome in a remarkable way by the intensity interferometer of Hanbury Brown and Twiss. This was originally conceived by Hanbury Brown at Jodrell Bank for use at radio wavelengths. Longer baselines of radio versions of Michelson's interferometer were required so as to increase their resolving power, but the links that conveyed the radio frequency signals to the central station where they could interfere and produce fringes did not have the necessary phase stability.

Hanbury Brown's suggestion was that instead of conveying back the amplitude of the radio waves received at each end from the radio source, the intensity only need be conveyed. This suggestion may seem ridiculous, since the intensity does not depend on phase; however, the signals from radio sources, like other naturally occurring radio signals, are characteristically noisy and have fluctuations in intensity. According to Hanbury Brown these fluctuations in intensity would correlate if the two stations were close, but as the baseline was increased the correlation would fall off in a way that would allow the angular diameter of the source to be determined. We discuss this concept in more detail in Chapter 14.

These two radio versions of Michelson's interferometer are shown in Fig. 11.18. The easing of the problem of conveying the information to the central station this allowed was enormous. The radio frequencies occupying a bandwidth of several megahertz must be transmitted in a way that preserves phase; the intensity fluctuations are, instead, at the low frequencies normally carried by telephone lines, and can be transmitted without loss over large distances. The system was used first by Jennison and das Gupta at Jodrell Bank in 1951 in

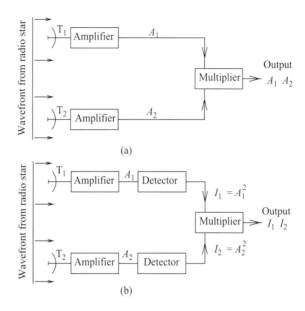

Fig. 11.18 Radio interferometers: (a) a conventional form similar to Michelson's. The chief difficulty is in conveying the amplitudes A_1 and A_2 from the radio telescopes T_1 and T_2, which may be intercontinental distances apart; (b) the intensity interferometer where the much more easily handled intensity is brought in, the amplitude being squared (in detection) at each end.

a measurement of the angular diameter of the second most powerful radio source in the sky, that in Cygnus. Encouraged by its success at radio wavelengths, Hanbury Brown then proceeded to apply the same technique at optical wavelengths. Here, there seemed to be a fundamental question: could an optical photon detector be used to measure correlation between two light beams? An initial laboratory scale demonstration[5] using a mercury lamp showed that it could, so that the optical stellar interferometer should work.

The equipment for the new intensity interferometer consisted of two searchlight mirrors focusing the light of the observed star onto the cathodes of two photomultipliers. The observed fluctuations in current were then correlated in an electronic multiplying circuit. The equipment was used to observe Sirius on every possible night in the winter of 1955–1956, when a total of only 18 hours observing was achieved. As the two ends of the interferometer were moved apart to a final spacing of over 9 m, there indeed was a fall-off in correlation, giving an angular diameter for Sirius of 0.0069 ± 0.0004 arcsec.

This result and the laboratory experiments in intensity interferometry that had preceded it started a storm of controversy amongst theoretical physicists. The conventional explanation of the intensity fluctuations was that the *numbers*

[5] R. Hanbury Brown and R. Q. Twiss, Correlations between photons in two coherent beams, *Nature* **127**, 27, 1956.

of photons arriving at each photocathode had statistical variations; how could the separate photons caught by searchlight mirrors several metres apart have correlated fluctuations? This is, of course, a problem that applies to all interferometers, but it was brought into sharp focus by the measurement of intensity at the two telescopes, which was related more obviously to a flux of photons than in the more conventional interferometers.

The work of Hanbury Brown and Twiss epitomizes the dual character of light: a wave nature (which makes interferometers work), and a quantum nature (which is clearly demonstrated in the detection of individual photons). The correct approach to the problem is to apply wave theory to the propagation, diffraction, and interference of the light waves, and quantum theory to the interaction of the light waves with the detectors. We consider this problem further in Chapter 14. The wave theory shows what correlation exists between the waves at the detectors, while the quantum theory shows how this correlation may become masked by a statistical 'photon noise'.

Moving to clearer skies than those of Cheshire, Hanbury Brown set up at Narrabri in Australia a very large version of his intensity interferometer. The advantage over the Michelson technique was the measurement of correlation over very much larger baselines, giving a crucially important improvement in angular resolution. With the Narrabri interferometer Hanbury Brown measured the diameters of several hundred of the brightest stars, the first direct measurement of the diameters of any stars smaller than the red giants.

11.15 FURTHER READING

P. Hariharan, *Optical Interferometry*. Academic Press, 1985.
R. K. Tyson, *Principles of Adaptive Optics*. Academic Press, 1991.

NUMERICAL EXAMPLES 11

11.1 In a Michelson interferometer used to measure the wavelength of monochromatic light, 185 fringes crossed the field of view when the mirror was moved by 50 μm. What was the wavelength of the light?

11.2 A simple double slit with separation 1 mm is held immediately in front of the eye, and a distant sodium street lamp is observed. If the lamp is 10 cm across, how far away must it be for clear interference fringes to be observed?

11.3 In an experiment to demonstrate Young's fringes, light from a source slit falls on two narrow slits 1 mm apart and 100 mm from a slit source; the fringes are observed on a screen 1 m away. The source is white light filtered so that only the wavelength band from 480 nm to 520 nm is used. (i) What is the separation of the fringes? (ii) Approximately how many fringes will be clearly visible? (iii) How wide can the source slit be made without seriously reducing the fringe visibility?

PROBLEMS 11

11.1 A Fresnel biprism (Fig. 11.12 (a)) with wedge angles α and refractive index n_1 is at a large distance from a point source of light with wavelength α. It forms fringes on a screen at a distance d from the light source. Show that the spacing of the fringes is $\lambda d/a$, where $a = 2D(n_1 - 1)\alpha$, where D is the distance between the source and the prism. What is the spacing if the whole system is immersed in liquid with refractive index n_2?

11.2 Hanbury Brown's development of the Michelson interferometer operated at optical wavelengths with spacings up to 50 m between the two mirrors. Calculate the diameter of an object just resolvable by this instrument at a distance equal to the Earth's diameter.

11.3 A Rayleigh refractometer is used to measure the refractive index of hydrogen gas. The tubes are 100 cm long, and a pressure change of 50 cm mercury gives a count of 145.7 fringes at $\lambda = 589.3$ nm. What is the refractive index at normal atmospheric pressure?

Starting with both tubes at atmospheric pressure, what pressure change will give a minimum fringe visibility due to the pair of sodium D lines at 589.0 and 589.6 nm?

11.4 The Jamin interferometer shown in Fig. 11.19 uses two parallel-sided glass plates to form separated but identical optical paths, like the twin paths of the Rayleigh refractometer. The plates are set at $\theta = 45°$ to the light path, and one is tilted by a small angle $\Delta\theta$ to produce a set of interference fringes. The plates have refractive index n and thickness h. Find the phase difference $\Delta\phi$ between the paths at the centre of the field, using Eq. (9.9) to show that

$$\frac{\Delta\phi}{\Delta\theta} = \frac{4\pi h}{\lambda}\cos\theta\sin\theta\left(n^2 - \sin^2\theta\right)^{1/2}. \tag{11.20}$$

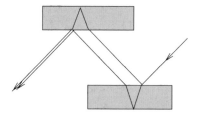

Fig. 11.19   The Jamin interferometer.

11.5 An etalon used as a length standard known to be near 5 mm was measured in terms of three wavelengths of cadmium light. Only the fractional parts of the fringe numbers were known, as follows:

Wavelength (nm)	Fraction
479.992	0.150
508.582	0.495
643.847	0.800

What was the spacing of the etalon?

12

Prism and grating spectrometers

One important object of this original spectroscopic investigation of the light of the stars and other celestial bodies, namely to discover whether the same chemical elements as those of our Earth are present throughout the universe, was most satisfactorily settled in the affirmative; a common chemistry, it was shown, exists throughout the universe.

The scientific papers of Sir William Huggins.

Spectroscopy, the study of spectra, scored a spectacular success at the solar eclipse of 1868, when Lockyer[1] noted that a bright yellow-orange line emitted by solar prominences corresponded to no known element. He named it helium; it was isolated from the terrestrial atmosphere by Ramsey in 1895. Spectroscopy is still the general term for the study of both absorption and emission spectra, but we may distinguish between a *spectroscope* for direct visual use, a *spectrograph* providing simultaneous measurement over a wide spectral range, and a *spectrometer*,[2] which measures and records photoelectrically details of spectral lines at any part of the electromagnetic spectrum.

In this chapter we describe the simplest and most universally used spectrometers, and analyse their *resolving power*, i.e. their ability to distinguish between light of closely adjacent wavelengths. Spectroscopy at higher resolving power is the subject of Chapter 13, and a more general treatment of resolving power will be given in Chapter 14.

12.1 THE PRISM SPECTROMETER

The simplest way of examining a spectrum is to use a prism to spread a light beam in angle, following the example of Newton, who placed a prism in a beam

---

[1] Sir Norman Lockyer (1836–1920), founded and edited for 50 years the journal *Nature*.
[2] The term *spectrometer* was originally used for an instrument to measure the refractive index.

of sunlight and showed how the light was split into a spectrum of colours. The geometry is best understood from the thin prism, as presented in Chapter 2, where the angles are small. The prism deviates a wavefront through an angle θ given by

$$\theta = (n - 1)\alpha, \tag{12.1}$$

where α is the apex angle of the prism and n its refractive index. Since the refractive index n varies with wavelength λ, different colours emerge as wavefronts with different angles when a single superposed plane wavefront is incident on the prism. The angular separation $\delta\theta$ of two wavefronts with wavelengths separated by a small amount $\delta\lambda$ depends on $dn/d\lambda$, and differentiation of Eq. (12.1) leads to

$$\delta\theta = \alpha(dn/d\lambda)\delta\lambda. \tag{12.2}$$

The *angular dispersive power* of the prism $d\theta/d\lambda$ is therefore directly proportional to the angle α and to $dn/d\lambda$, the rate of variation of the refractive index with wavelength in the prism. As we saw when discussing chromatic aberration in a lens (Chapter 2), the *dispersive power* of a particular type of glass is often quoted in terms of its refractive indices at specific wavelengths; for example, a typical flint glass has refractive indices $n_F = 1.632$ at $\lambda = 486$ nm, $n_D = 1.620$ at $\lambda = 588$ nm and $n_C = 1.616$ at $\lambda = 656$ nm. The conventional dispersive power is then $(1.632 - 1.616)/1.62 - 1) = 1/39$. The differential $dn/d\lambda$ is approximately 10^{-4} nm^{-1}, varying slowly over the wavelength range.

The proper design of a practical spectrometer requires the angular spread of the spectrum to be large enough for an adequate separation of components with different wavelengths; for the prism this requires glass with a large dispersive power and a large prism angle. Light with a single wavelength will however spread over a range of angles, so that light with two closely spaced wavelengths may overlap and not be clearly *resolved*. This limit on the possible resolution of closely spaced wavelengths is due to diffraction at the whole aperture. As we shall see later in this chapter, the best possible *spectral resolving power* of a prism depends mainly on its overall dimensions rather than on its apex angle or refractive index.

Equation (12.2) may be stated exactly for any size of prism, and for any configuration, whether symmetrical or not, if the path p in the prism is used as a parameter (Fig. 12.1). It is the change $p\delta n$ in the optical length of this path due to the change $\delta\lambda$ that causes the angular dispersion $\delta\theta$. This change affects the emergent wavefront along a width w, so that the angle $\delta\theta$ is now given by

$$\delta\theta = \frac{p\,dn}{w\,d\lambda}\delta\lambda. \tag{12.3}$$

A practical prism spectrometer is now seen to require in essence an incident plane wavefront, and a means of recording separately the plane wavefronts that

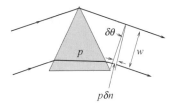

Fig. 12.1   The dispersion of a prism. The optical length of the path p changes by $p\delta n$ with wavelength change $\delta\lambda$. The change $p\delta n$ over the width w deviates the emergent wavefront by the angle $p\delta n/w$.

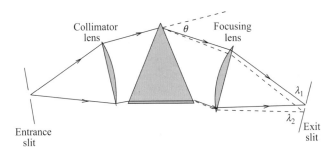

Fig. 12.2   Optical system of a prism spectrometer. The entrance slit is at the focus of the collimating lens; the exit slit with its focusing lens forms a telescope that is moved to collect plane wavefronts over a range of angles.

emerge in different directions. The incident plane wave may be obtained from a source at a great distance, or from a point source of diverging waves which are made plane by a lens with a positive power, such as a simple biconvex lens. The emergent wavefronts may be examined by eye, since the eye is designed to sort out waves travelling in different directions, or by the eye with the help of a telescope, or by a form of camera, as shown in Fig. 12.2. A narrow source of light, usually an *entrance slit*, is needed. A large part of the spectrum may be seen or recorded simultaneously on a photographic or photodiode array detector such as a charge-coupled device (CCD) (see Chapter 21); the instrument may then be called a *spectrograph*. A spectrograph using a linear array of diode detectors may be termed an optical spectrum analyser (OSA) or an optical multichannel analyser (OMA). Alternatively, a narrow range may be selected by an exit slit as a source of monochromatic light; this is then a *monochromator*. If the selected wavelength range is recorded in a photoelectric detector, then this becomes a *spectrometer*. (Although this is the formal definition, we shall follow common practice and use the term spectrometer in a more general sense.)

It will be seen that for each wavelength the exit slit accepts a monochromatic image of the original point source, formed by the two lenses, referred to as the *collimating* and the *focusing* lenses. If the source is made a line, such as a slit in

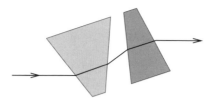

Fig. 12.3 Direct vision prism for a central wavelength λ.

front of a flame or a discharge tube, then a line will appear for each colour of the spectrum. The width of the line depends on the width of the slit; a wide slit admits more light, but produces wider final images and thus gives poorer resolution. The line from a narrow slit will be broadened by diffraction, which limits the resolving power of the spectrometer.

As shown in Fig. 12.2, the focusing lens images the entrance slit onto the plane of the exit slit. If the focal length of the focusing lens is f, then the distance between the images for λ and $\lambda + \delta\lambda$ is

$$\Delta x = f\Delta\theta = f\frac{\mathrm{d}\theta}{\mathrm{d}\lambda}\Delta\lambda = \frac{\mathrm{d}x}{\mathrm{d}\lambda}\Delta\lambda. \tag{12.4}$$

The quantity $\mathrm{d}x/\mathrm{d}\lambda$ is the *linear dispersion* of the spectrometer when used at wavelength λ.

A small hand-held prism spectrometer, which is useful for detection of elements in a flame or discharge tube, would probably use the *direct vision prism* of Fig. 12.3. Here there are two prisms made of glasses with different dispersive power, deviating in opposite directions. The prism angles are chosen to give zero deviation at a central wavelength.

12.2   THE GRATING SPECTROMETER

As we saw in Chapter 10, a diffraction grating can act like a prism in deviating a wavefront through an angle which depends on wavelength; a diffraction grating could therefore be substituted for the prism in Fig. 12.2. The diffraction grating is, however, usually used in reflection, since it is often advantageous to avoid transmission through glass which loses light either by absorption or partial reflection. If collimation and focusing are achieved with concave mirrors instead of lenses, then the spectrometer can work not only at visible wavelengths but also at ultraviolet wavelengths. The Czerny–Turner spectrometer of Fig. 12.4 is a common arrangement; here the scanning in wavelength is accomplished by rotating the plane grating.

The essential geometry of diffraction at a grating, which applies equally in transmission and reflection, is shown in Fig. 12.5. Here θ_I is the angle of

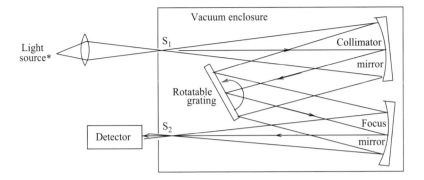

Fig. 12.4   The Czerny–Turner spectrometer. Only reflecting optical components are used, allowing operation at ultraviolet wavelengths.

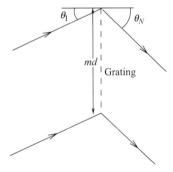

Fig. 12.5   A diffraction grating showing the angles of incident and emerging light. The total number of lines in the grating is m.

incidence, and θ_N is the angle of emergence for wavelength λ in the Nth order. The grating equation, from Chapter 10, is

$$d(\sin \theta_N + \sin \theta_I) = N\lambda, \tag{12.5}$$

where d is the line spacing and N is the order of the diffraction.

   The *angular dispersive power* of the grating used at order N is the rate of change of the angle of emergence against the wavelength. For a given angle of incidence, we obtain from Eq. (12.5)

$$\frac{d\theta_N}{d\lambda} = \frac{N}{d \cos \theta_N}. \tag{12.6}$$

Thus for any given order a grating may be used just like a prism, except that it is possible to get a much greater angular dispersion with a grating.

A new problem arises, however, when a wide range of wavelengths is being observed: the spectra in different orders may overlap. If a range from λ_1 to λ_2 is observed in the Nth and $(N + 1)$th order, than there is an overlap if

$$N\lambda_2 > (N + 1)\lambda_1. \qquad (12.7)$$

The wavelength range between the overlapping orders is the *free spectral range*. If the spectrometer is set for operation at wavelength λ_1 in order N, then it will pass $N\lambda_1$ in first order, $N\lambda_1/2$ in second order, and so on. For a grating used at normal incidence, and with the diffracted beam in the Nth order at angle θ, the overlap occurs when

$$\lambda_2 = d \sin \theta/N, \qquad \lambda_1 = d \sin \theta/(N + 1). \qquad (12.8)$$

The free spectral range $(\lambda_2 - \lambda_1)$ for order N is then

$$\delta\lambda_{FSR} = \frac{\lambda_1}{N}. \qquad (12.9)$$

The confusion this may cause when observing a spectral range greater than $\delta\lambda_{FSR}$ may be avoided by using a filter to restrict the wavelength range of the incident light, or by adding a *cross-dispersing* device such as a second grating or prism which spreads the spectrum in an orthogonal direction. Note that a prism used alone concentrates light into a single spectrum, with no overlapping orders; for this reason astronomical telescopes may use a large thin prism in front of the objective lens or mirror to spread each star image over a large angular field into small spectra.

12.3   RESOLVING POWER IN WAVELENGTH

The purpose of a spectrometer is to distinguish between light waves separated by a small wavelength difference $\delta\lambda$. The prism and the grating spectrometers change the *wavelength* difference $\delta\lambda$ into a difference of *emergence angle* $\delta\theta$ in the wavefronts at the two wavelengths. The relation between $\delta\lambda$ and $\delta\theta$ is determined for a prism by the geometry of the prism and the dispersive power of its material, and for the grating by the line spacing and the order of diffraction. Light from a single wavelength will, however, emerge over a spread of angles, so that there is a limit to the possibility of distinguishing two spectral lines closely spaced in a wavelength.

Consider the problem of distinguishing two adjacent spectral lines, such as the two sodium D-lines at 589.0 and 589.6 nm. Even if the entrance slit of the spectrometer is made very narrow, the exit slit will be scanning across two diffraction images, as in Fig. 12.6(a). If these are well separated, the spectral lines are resolved. If they are so close as to merge into a single image, as in Fig. 12.6(b), then they are unresolved. In Fig. 12.6(c) the separation is such that

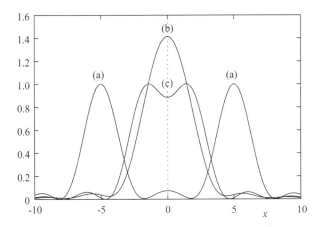

Fig. 12.6 Diffraction images of adjacent spectral lines in a spectrograph. In (a) the lines are clearly resolved, while in (b) they merge and are unresolved. The separation in (c) illustrates Rayleigh's criterion for resolution.

the first diffraction zero of one image falls on the maximum of the other, giving an obviously double line. This is known as *Rayleigh's criterion* for the limit of resolution of the spectrometer.

The quantity $\lambda/\delta\lambda$ is obviously a useful measure of the power of any device to distinguish different wavelengths and is called the *chromatic resolving power* R of the spectrometer:

$$R = \frac{\lambda}{\delta\lambda} = \frac{\nu}{\delta\nu}. \tag{12.10}$$

For example, a resolving power greater than 1000 is needed to resolve the two sodium D-lines.

12.4 RESOLVING POWER: PRISM SPECTROMETERS

Following a similar argument to the discussion of diffraction in Chapter 7, we note that the minimum width (at a given wavelength) of the image at the exit slit in Fig. 12.2 is due to diffraction in the limited width of the wavefront emerging from the prism. The angular spread is determined by the ratio of the wavelength to the width w of the wavefront. This angular width $\delta\theta$ is λ/w, measured from the line centre to the first minimum. Using the thin prism approximation, Eq. (12.2), this is related to the dispersion in the prism by

$$\delta\theta = \frac{\lambda}{w} = \alpha\frac{\mathrm{d}n}{\mathrm{d}\lambda}\delta\lambda, \tag{12.11}$$

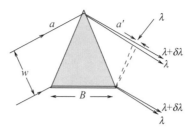

Fig. 12.7   Geometry for the chromatic resolving power of a prism.

giving the criterion for resolving the two spectral lines:

$$\frac{\lambda}{\delta\lambda} = w\alpha\left(\frac{dn}{d\lambda}\right). \tag{12.12}$$

Instead of extending the simple geometry applicable to Eq. (12.12) from *thin* to the geometrically more complicated case of *thick* prisms, we choose to derive an interesting simple expression for the chromatic resolving power of any prism spectrometer by a direct consideration of optical paths and Rayleigh's criterion (section 12.3 above). In Fig. 12.7 a thick prism is shown at the position of minimum deviation showing the approximate paths for plane waves of light with wavelengths λ and $\lambda + \delta\lambda$. Now for the diffraction maximum of one emerging wavefront to lie on the minimum of the other there must be one wavelength difference between them at the top of the wavefront emerging from the prism (see, for example, the way the phasors curl up in Figure 7.8). So for light of wavelength λ, equating the optical path lengths for the extreme rays in air and in the prism

$$2a = nB, \tag{12.13}$$

where n is the refractive index of the prism at wavelength λ and B is its baselength. For wavelength $\lambda + \delta\lambda$ the refractive index is $n + \delta n$. The plane waves for λ and $\lambda + \delta\lambda$ and at the resolved wavelength $\lambda + \delta\lambda$ are separated by a small angle because of the extra optical path in the prism, so that

$$2a - \lambda = (n - \delta n)B, \tag{12.14}$$

giving

$$\lambda = \delta nB, \tag{12.15}$$

which may be written as

$$\boxed{R = \frac{\lambda}{\delta\lambda} = B\left(\frac{dn}{d\lambda}\right).} \tag{12.16}$$

At minimum deviation the resolving power of the prism spectrometer depends on the baselength and the spectral dispersion of the material of the prism. Equation (12.16) for the chromatic resolving power of a thick prism shows that the *angle* of a prism is unimportant; what matters is the distance B traversed in the prism by the extreme ray, and the value of $dn/d\lambda$ for the material of the prism. For a heavy flint glass $dn/d\lambda$ can be about $10^{-4}\,\mathrm{nm}^{-1}$ so that for a wavelength of 500 nm and a large prism with $B = 10\,\mathrm{cm}$ we have

$$\frac{\lambda}{\delta\lambda} = 10^4 \quad \text{and} \quad \delta\lambda = \frac{5 \times 10^{-7}}{10^4} = 0.05\,\mathrm{nm}, \qquad (12.17)$$

which is adequate for the resolution of the two sodium D-lines but insufficient for detailed measurement of the structure of each line.[3]

12.5 RESOLVING POWER: GRATING SPECTROMETERS

We have seen that the chromatic resolving power of a prism is related to its overall size. The same arguments applied to the grating spectrometer give a similar result: the resolving power is again related to its overall size. Figure 12.8 shows diffracted wavefronts for two wavelengths λ and $\lambda + \delta\lambda$ emerging from a grating. The angular distribution of irradiance in these two spectral components is shown for a separation $\delta\lambda$, where they are just distinguishable. Again following Rayleigh's criterion, the maximum of one diffraction image falls on the first zero of the other.

The diffraction angle θ for wavelength λ at normal incidence is given by the grating equation

$$d \sin \theta = N\lambda, \qquad (12.18)$$

where d is the line spacing of the grating and N is the order of diffraction. For a grating of width w and a total number of lines m we can write $w \sin \theta = mN\lambda$. Across the emerging wavefront there is a difference in path $mN\lambda$. We now concentrate on the irradiance in this one direction as the wavelength is changed by a small amount $\delta\lambda$. Light of wavelength $\lambda + \delta\lambda$ has its principal maximum at the same angle as the first minimum for light of wavelength λ. If the extra path $d \sin \theta$ changes by one wavelength, than the irradiance will fall to zero. So for two adjacent spectral lines to be distinguished the criterion is

---

[3] The concept of spectral *lines* is very deep in the language: we talk of atoms having emission lines, and of the 21-cm hydrogen line in the radio spectrum, and so on. But of course the atoms do not have lines, they emit or absorb at certain wavelengths. It is the spectrograph that displays the different wavelengths in the light presented to it as a series of lines, each of which is an image of the slit, each at a different wavelength. So when we speak of lines in an X-ray spectrum, or of the emission lines of molecules in millimetre wave astronomy, we are using a word that is an interesting fossil originating in the simplest and oldest technique of spectral analysis, the prism spectrometer.

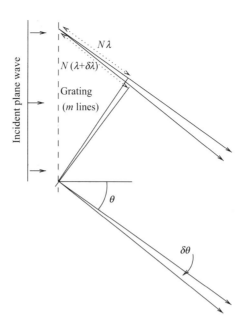

Fig. 12.8 Chromatic resolution in a grating spectrometer with m lines, used in the Nth order, showing the separation of two components of a plane wavefront.

$$mN\lambda + \lambda = mN(\lambda + \delta\lambda) \tag{12.19}$$

or

$$\boxed{R = \frac{\lambda}{\delta\lambda} = mN.} \tag{12.20}$$

That is to say, the chromatic resolving power of a grating is the product of the total number of lines across it and the order in which it is used. The order acts like a kind of gearing: in the third order a given change of wavelength $\delta\lambda$ changes the path difference between adjacent lines by three times as much as it does in the first order, giving three times the resolution.

We can also derive an expression for the angular dispersion of the grating, the rate of change of angle of emergence against wavelength for an order. From Eq. (12.5) this is

$$\frac{\mathrm{d}\theta}{\mathrm{d}\lambda} = \frac{N}{d\cos\theta_N}. \tag{12.21}$$

Thus for any given order a grating may be used just like a prism, except that it is possible to obtain much higher resolving powers with a grating.

Note that the dispersion of the grating is related to the line spacing, while the resolving power is related to the number of wavelengths N in the extra path labelled $N\lambda$ in Fig. 12.8. To obtain the same resolving power with a grating in the second order as for the large prism in section 12.4 above would need 5000 lines across it. So a grating on the scale of the prism that was 10 cm across would need only 500 lines per centimetre. Fraunhofer, Rowland and Michelson all improved techniques for ruling conventional gratings, Michelson eventually producing gratings more than 15 cm across giving resolving powers of 4×10^5. Modern gratings are produced by a simple form of holography (Chapter 20), in which two crossing beams of monochromatic laser light form an interference pattern on a photographic plate. The plate surface is a film of photoresist which is subsequently etched to leave lines of clear glass on which a metallic coating is deposited. Holographic gratings can be made with up to 6000 lines per milli-metre; furthermore, they are very uniform, avoiding the periodic errors that produce 'ghosts' (Chapter 10).

A further development of holographic gratings is volume phase holography (VPH), which gives gratings in which none of the light is lost at a partially reflecting surface. The crossed laser beams used for making holographic grat-ings on a surface will make a three-dimensional pattern in a thicker film of gelatin; this pattern can be preserved as a three-dimensional pattern of changed refractive index in a completely transparent film. The refractive index changes are produced by a hardenening process in the gelatin, in which the collagen molecules become cross-linked when exposed to blue light. The result is a grating that behaves partly like a crystal in X-ray diffraction (Chapter 10). It can be used in reflection or transmission, and has the normal dispersive power of a plane grating. The Bragg wavelength, which is the centre of the envelope of efficiently reflected wavelengths, can be tuned by tilting the grating. The width of the envelope is related to the thickness of the gelatin film.

12.6   CONCAVE GRATINGS

In an ordinary spectrograph a grating is usually illuminated by a plane wave, requiring a collimator lens or mirror with the light source at the focus. The diffracted spectrum is then focused onto a detector, so that two lenses or mirrors are needed; these may introduce losses and aberrations, especially for infrared and ultraviolet light. The difficulty may be avoided by using an arrangement due to Rowland in which a concave grating is itself used for focusing. In this grating the lines are ruled on the surface of a concave mirror. An interesting piece of geometry shows that if the slit is located on a circle tangential to the mirror and with *diameter* equal to its radius of curvature, then the several orders of diffraction are also in focus along this circle.

In Fig. 12.9 S is the slit and C the centre of curvature of the grating. Then all rays from S that are reflected from the grating at R have the same angle of incidence $\alpha = $ SRC because CR is normal to the grating. The directly reflected ray SQP crosses the circle again at P. Other rays such as SR are very nearly also focused on the same point P; if R were on the circle, then the angle SRP $= 2\alpha$ would be independent of the position of R, so that all rays from S would pass

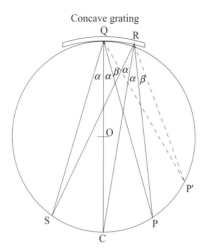

Fig. 12.9   The geometry of the Rowland circle.

through P. In fact if the size of the mirror is small compared with the diameter of the circle, then this is true enough. Now if one wavelength in one of the orders is diffracted by an angle β more than the direct reflection, so that it cuts the circle at P$'$, then the same argument applies to angle SRP$'$ which is constant at $2\alpha + \beta$. Hence if the slit, the grating and the photographic plate are all located on this *Rowland circle*, then sharp spectra may be recorded without the intervention of further optics.

12.7   BLAZED, ECHELLETTE, ECHELLE AND ECHELON GRATINGS

We have seen that for a grating the resolving power is mN, the product of the number of lines and the order of the diffraction. A *blazed* grating (Fig. 12.10) is a reflection grating with tilted reflection faces, so that light is diffracted predominantly in one of the higher orders, with the advantage of greater resolution at a high light level. The angle α between the normal to the grating and the normal to the grooves is called the *blaze angle*. The diffracted light satisfies the grating equation $d \sin \theta = N\lambda$ and also $\theta = 2\alpha$.

   To obtain still higher resolution from gratings it is easier to use fewer lines but increase the order of the diffraction. By setting $\theta_I = \theta_N = 90°$ in Eq. (12.5) it can readily be seen that for a conventional plane grating the order cannot be higher than $2d/\lambda$, the number of wavelengths in the space between lines, so that close ruling does not permit the use of high orders. For example, if a grating with 5000 lines per centimetre was to be used at high resolution at wavelength 500 nm, the highest order it could possibly be used in would be

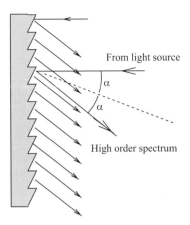

Fig. 12.10   A blazed reflection grating, with blaze angle α, illuminated at normal incidence. The diffracted light is concentrated in the direction θ.

$$N = \frac{2d}{\lambda} = \frac{2}{5 \times 10^{-5}} \frac{1}{5 \times 10^{3}} = 8, \qquad (12.22)$$

and to realize this extreme case the light would be at grazing incidence (at $\pi/2$, parallel with the surface) and be diffracted back through π. High-order diffraction is achieved in practice by the use of blazed gratings, and the *echelette, echelle* and *echelon* gratings. The idea of all these basically similar systems is to separate the fixed relationship between the line spacing and the order by making the grating not flat but rather like a flight of stairs viewed from a distance. The riser of the stairs corresponds to the line, and the tread to a displacement backwards of each line. The lines have thus become reflecting surfaces, each one displaced backwards from the previous one to give a high order of interference. The angle of these reflecting surfaces can now be adjusted to reflect light in the direction in which it is desired to observe spectra.

An echelle grating about 25 cm across with 10^4 steps or grooves can be used in the 1000th order for visible light with a resolving power of 10^6. The echelle grating is often used as a tuning element in lasers, since it gives high angular dispersion and high efficiency. For use at longer wavelengths, into the far infrared, a small number of grooves may be ruled directly onto metal: these are called echellettes, meaning 'little ladders'. Similar systems due to Michelson, called echelons, consist of a pile of glass plates arranged like a flight of stairs, which may be used in either transmission or reflection at orders as high as 20 000. The difficulties of realizing high resolution in this system become very great, and Michelson never in fact perfected the reflection echelon, though he made transmission echelons (Fig. 12.11) successfully with some tens of plates.

Blazed gratings are often used in an arrangement due to Littrow (Fig. 12.12(a)) in which the diffracted light returns almost along the incident path. This allows the same lens to be used as a collimator and for focusing.

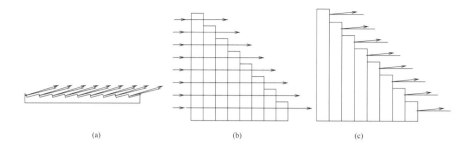

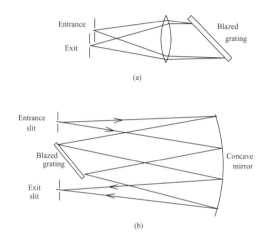

Fig. 12.11   (a) Echellette; (b) echelle; and (c) echelon gratings.

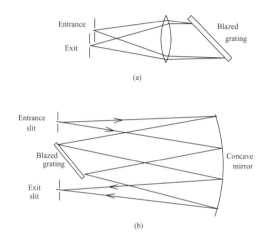

Fig. 12.12   Mountings for blazed gratings: (a) Littrow; (b) Ebert.

A similar arrangement due to Ebert is shown in Figure 12.12(b); here the collimator and focusing elements are combined in a single concave mirror, avoiding the losses inherent in lens systems.

With the very high orders of interference obtained in these devices the problem of overlapping orders becomes extreme. Overlapping orders may however be dealt with by crossing any high-resolution spectrometer with a low-resolution spectrometer, such as a prism, whose resolution is in a perpendicular direction. The various orders are then separated in a two-dimensional format. An example is shown in Fig. 12.13, where a prism is used as a cross-disperser for a Fabry–Perot spectrometer.

The combination of a grating and a prism, often called a 'grism', has another advantage when it is used in reflection (Fig. 12.14). As we have seen, the resolving power of a grating is related to the line spacing and the wavelength. If the grating is bonded to, or etched into, the glass of the prism, then the wavelength is reduced by the refractive index of the glass, and the resolving power may be increased in proportion by using a grating with a smaller line spacing.

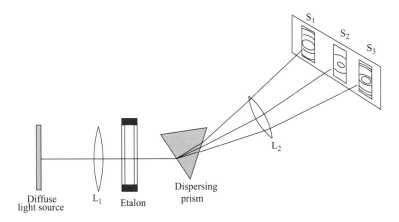

Fig. 12.13   Cross-dispersion with low-resolution prism and high-resolution Fabry–Peron etalon spectrometer. The spectral lines S_1, S_2, S_3 are images of the source, dispersed by the prism P. The etalon produces a high dispersion spectrum within each of the spectral lines.

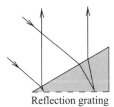

Fig. 12.14   A combination of a blazed grating and a prism, used in reflection. This 'grism' has a higher resolution than a grating in air with the same geometry

12.8   EFFICIENCY OF SPECTROGRAPHS

Our discussion of spectrometers has concentrated on their resolving power, and the various optical arrangements that allow their use over a wide wavelength range. Two other factors need to be considered in any practical design. First, the detector system need not be a single element behind a single exit slit: it may be, and often is, a multi-element detector such as a photoelectric array (see Chapter 21). The detector system then becomes a camera, which may have to work efficiently over a wide angle. The overall efficiency, or speed of operation, is proportional to the number of detectors which are simultaneously in use.

Second, it is usually important to ensure that as much light as possible reaches the detector. In Fig. 12.15 light from a source S is focused on a spectrometer slit by a lens, so that as it leaves the slit it exactly fills the area of the grating. The useful light is proportional to the area A of the slit and the solid angle Ω. The product $A\Omega$ is referred to as the *étendue* of the spectrometer. The term refers to the amount of light that passes through the spectrometer;

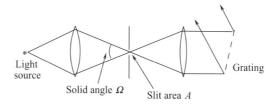

Light source

Solid angle Ω

Slit area A

Grating

Fig. 12.15 The étendue of a spectrometer. The light entering the spectrometer and falling on the grating is proportional to the solid angle Ω and the slit area A.

synonomous terms are *luminosity, throughput* and *light-gathering power*. Note that widening the slit to increase the étendue may reduce the resolution by blurring the spectrum.

These two requirements are crucial in spectrographs designed to be used at a large telescope for measuring the spectra of very faint objects. Cameras with small *F*-numbers are required, but without using multiple optical elements which may cause an unacceptable loss of light from reflections at the surfaces. The Schmidt telescope described in Chapter 3 is often used as a spectrograph camera for this reason.

NUMERICAL EXAMPLES 12

12.1 A spectrometer uses a prism with base width 5 cm and apex angle 11.5°, i.e. 0.2 rad., made of glass with refractive index $n = 1.70$ at $\lambda = 650$ nm and 1.72 at $\lambda = 590$ nm. Calculate the resolving power, using Eq. (12.16), and the angular separation of the two sodium lines at 589.0 nm and 589.6 nm. Will this spectrometer resolve the hydrogen doublet at 656.272, 656.285 nm?

12.2 In a high resolution spectrograph three prisms are arranged with their bases on a semicircle with diameter 20 cm as in Fig. 12.16, so as to deflect light through 180°. Show that for refractive index 1.5 the prism angle must be approximately 82°. Find the resolving power if

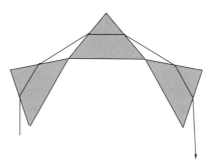

Fig. 12.16 A prism spectrometer with increased dispersion and resolving power.

$$\frac{dn}{d\lambda} = 5 \times 10^4 \text{ m}^{-1}.$$

12.3 Calculate the spectral resolving power for wavelengths near 500 nm of the following spectrometers: (i) a glass prism, baselength 4 cm, with refractive index varying linearly between $n = 1.5477$ at $\lambda = 546$ nm and $n = 1.5537$ at $\lambda = 486$ nm; (ii) a grating 4 cm across with 1500 lines per cm, used in the third order; (iii) a Fabry–Perot interferometer in which $F = 40$, and with a spacing 4 cm between the plates.

PROBLEMS 12

12.1 Light falls normally on a reflection echelon grating in which the step height is h and the step width is w. Show that the path difference between light beams reflected in the direction θ to the normal from corresponding points on adjacent step faces is

$$h(1 + \cos \theta) - w \sin \theta. \tag{12.23}$$

For small θ the mth order then emerges at

$$\theta = \frac{2h - m\lambda}{w}. \tag{12.24}$$

For an echelon with $h = 1$ cm and $w = 0.1$ cm, and with 40 such steps, for wavelengths near 500 nm find:

(i) the order m for θ near zero
(ii) the angular separation of orders
(iii) the resolving power.

12.2 The resolving power $\lambda/\delta\lambda$ of a grating spectrograph is the difference between extreme optical paths measured in wavelengths. Show that the resolvable frequency difference $\delta\nu$ is related to the difference τ in light travel times in the extreme paths by

$$\delta\nu = \frac{1}{\tau}. \tag{12.25}$$

12.3 For a Lummer plate (Fig. 12.17), which produces a fringe pattern with high resolution, show that the resolving power at the grazing emergence angle is approximately

$$\frac{\lambda}{\delta\lambda} = \frac{L}{\lambda}\left(n^2 - 1\right). \tag{12.26}$$

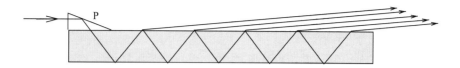

Fig. 12.17   The Lummer plate.

Note that the plate is used with the emergent beams at a very small angle to the surface of the plate. The light beam enters the plate via prism P; the refractive index is n and the length is L.

13

High resolution spectrometry

Modern improvements in optical methods lend additional interest to an examination of the causes which interfere with the absolute homogeneity of spectral lines.

Lord Rayleigh, 1915.

We may conveniently distinguish three ranges of wavelength resolution in spectrometers, both from the point of view of technique and of application. We have already described the simplest, using prisms and gratings, in Chapter 12; these are useful for distinguishing the various spectral lines of a complex source and deducing its atomic or molecular content. The higher resolution demanded for measuring the detailed shape of spectral lines usually demands an interferometric technique, as described in Chapter 9. Finally, the very narrow lines in scattered laser light may be resolved by a totally different technique in which fluctuations of intensity, measured through optical mixing spectroscopy, are related to the spectrum.

In this chapter we start by outlining the main factors that determine the width and shape of spectral lines. We show how *Fourier transform spectrometry* has developed from the interferometer techniques that are needed for the resolution of narrow spectral lines. We then describe the *intensity fluctuation* or *photon correlation* spectrometry which is used to examine very narrow spectral lines, notably those of scattered laser light.

13.1 THE SHAPE AND BROADENING OF SPECTRAL LINES

Strictly speaking, there is no such thing as monochromatic radiation. A wave train must begin and end; if it lasts a time τ, than it will have a frequency bandwidth of order $1/\tau$. The ratio of frequency ν to bandwidth $\delta\nu$, or of wavelength λ to line width $\delta\lambda$, is often referred to as the quality factor.[1]

---

[1] The quality factor Q is familiar in electrical engineering as a measure of the sharpness of resonance in an electrical circuit Q. Values of Q over the whole field of spectrometry extend from 10^2 to 10^{14} or more; for atomic spectral lines in the optical range we most frequently encounter values of 10^4 to 10^6.

Spectral lines are broadened by various processes. Transitions between energy levels in individual atoms or molecules have a natural spread in energy, giving a natural linewidth. This may be increased by interactions affecting all the emitters equally, due to collisions or electrostatic forces between neighbours; this is known as *homogeneous broadening*. Inhomogeneous broadening applies when the individual members of a species have a distribution of broadening characteristics; this applies to Doppler broadening due to thermal velocities in a gas, when there is a distribution of Doppler shifts among the emitters.

The effects on spectral line shape of these two types of broadening are similar at first sight, but the difference in origin has important consequences. As we will see in Chapters 15, 16 and 17, the distinction is particularly important in the action of lasers.

13.2 NATURAL LINEWIDTHS

The mean lifetime of photon emission from an atom τ leads, from the Heisenberg uncertainty principle, to an uncertainty $\Delta\varepsilon$ in the energy ε of a transition from an excited state given by

$$\Delta\varepsilon \approx \frac{\hbar}{\tau}, \tag{13.1}$$

where $\hbar = h/2\pi$. A group of excited atoms will lose energy exponentially with the time constant τ. Although energy is lost in discrete quanta from each atom, the overall effect is statistically the same as an exponential decay of energy in each individual atom, which is simply represented by writing the electric field $e(t)$ radiated by each atom as

$$e(t) = E_0 \exp(-t/2\tau) \exp(i\omega_0 t). \tag{13.2}$$

Note the factor 2 in the exponential; the decay in intensity follows $\exp(-t/\tau)$, but it is the sum of the *squares* of the fields that gives the intensity. The frequency distribution is found by taking the Fourier transform of (13.2):

$$E(\omega) = \frac{1}{2\pi} \int_{-\infty}^{\infty} e(t) \exp(-i\omega t) dt \tag{13.3}$$

$$= \frac{E_0}{4\pi} \left[\frac{1}{\left(\omega_0 - \omega + \frac{i}{2\tau}\right)} - \frac{1}{\left(\omega_0 + \omega - \frac{i}{2\tau}\right)} \right]. \tag{13.4}$$

The intensity of the emitted radiation is

$$I(\omega) \propto |E(\omega)|^2 \propto \frac{1}{(\omega - \omega_0)^2 + (1/2\tau)^2} \tag{13.5}$$

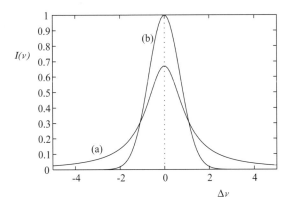

Fig. 13.1 The Lorentzian (a) and Gaussian (b) functions, describing natural lineshape and Doppler lineshape.

and in frequency terms

$$I(\nu) = \frac{1}{(\nu - \nu_0)^2 + (1/4\pi\tau)^2}. \tag{13.6}$$

This function is the *Lorentzian* shape shown in Fig. 13.1(a). The full width at half maximum (FWHM) $\Delta\nu$ is

$$\Delta\nu = \frac{1}{2\pi\tau}, \tag{13.7}$$

and the spectral line shape is

$$I(\nu) \propto \frac{1}{(\nu - \nu_0)^2 + (\Delta\nu/2)^2}. \tag{13.8}$$

Note the greater extension of the Lorentzian curve in comparison with the Gaussian curve of Fig. 13.1(b), which has the same FWHM.

13.3 PRESSURE BROADENING

The Lorentzian function describes all types of homogeneous line broadening as well as natural line shapes. Homogeneous broadening also occurs in solid state media (in which the collisions of phonons perturb the phase of an excited emitting species) and in gases (by pressure broadening due to the collisonal interaction of an emitting particle with its neighbours). Pressure broadening is also referred to as collisional broadening.

The importance of collisions in an emitting gas depends on the average time between collisions compared with the natural lifetime τ. A collision disturbs the wavetrain from an atom, and there is a discontinuity of phase at average intervals of the time τ_c between collisions. A Fourier analysis of the effect of these discontinuities on the line shape must take account of the random nature of the collisions, and we quote only the result that the lineshape is Lorentzian. The line widths add simply. Equation (13.7) gave the width due to a lifetime τ. If the transition is between energy levels with natural lifetimes τ_1 and τ_2, then the linewidth is

$$\Delta\nu = \frac{1}{2\pi}\left(\frac{1}{\tau_1} + \frac{1}{\tau_2}\right) \tag{13.9}$$

and the width including collisions at an average time interval τ_c is

$$\Delta\nu = \frac{1}{2\pi}\left(\frac{1}{\tau_1} + \frac{1}{\tau_2} + \frac{2}{\tau_c}\right). \tag{13.10}$$

For gases at normal temperature and pressure (300 K and 10^5 Pa) $\tau_c \approx 3 \times 10^{-11}$ s, so that for visible light where $\nu_0 \approx 5 \times 10^{14}$ s^{-1} the wavetrain completes on average 15 000 cycles between collisions. The resultant linewidth in this example is comparable with the Doppler linewidth, and is usually much larger than the natural linewidth.

13.4   DOPPLER BROADENING

Inhomogeneous broadening occurs when the individual emitters experience different environments leading to different perturbations or frequency shifts. The most important case is the *Doppler broadening* of a spectral line due to thermal motion. In low density gases the Doppler broadening is usually the dominant factor determining the spectral linewidth.

In a gas there is a random distribution of particle velocities; as seen by a stationary observer, this leads to a distribution in the emission centre frequency of different emitting particles. The normalized Maxwell–Boltzmann distribution of atomic velocities for atoms of mass M at temperature T is

$$f(v_x, v_y, v_z) = \left(\frac{M}{2\pi kT}\right)^{3/2} \exp\left[-\frac{M}{2kT}\left(v_x^2 + v_y^2 + v_z^2\right)\right]. \tag{13.11}$$

For N atoms per unit volume, the number of atoms per unit volume that have velocities in the range $v_x \to v_x + \mathrm{d}v_x$, $v_y \to v_y + \mathrm{d}v_y$ and $v_z \to v_z + \mathrm{d}v_z$ is $Nf(v_x, v_y, v_z)\mathrm{d}v_x\mathrm{d}v_y\mathrm{d}v_z$. The normalized one-dimensional velocity distribution, giving the probability that the velocity of a particle towards the observer is in the range v_x to $v_x + \mathrm{d}v_x$, is

$$f(v_x) = \left(\frac{M}{2\pi kT}\right)^{1/2} \exp\left(-\frac{Mv_x^2}{2kT}\right). \tag{13.12}$$

The observed frequency of a transition for an atom whose stationary centre frequency is ν_0 and whose component of velocity towards the observer is v_x is

$$\nu = \nu_0 + \frac{v_x}{c} \cdot \nu_0, \tag{13.13}$$

where c is the velocity of light in the gas. The distribution of emitted frequencies of the Doppler broadened transition is therefore a Gaussian lineshape:

$$g(\nu) = \frac{c}{\nu_0} \left(\frac{M}{2\pi kT}\right)^{1/2} \exp\left[\left(-\frac{M}{2kT}\right)\left(\frac{c^2}{\nu_0^2}\right)(\nu - \nu_0)^2\right]. \tag{13.14}$$

The FWHM linewidth is

$$\Delta\nu_D = 2\nu_0 \left(\frac{2kT \ln 2}{Mc^2}\right)^{1/2}. \tag{13.15}$$

In terms of linewidth the Doppler broadened lineshape function may be written

$$g(\nu) = \frac{2}{\Delta\nu_D} \left(\frac{\ln 2}{\pi}\right)^{1/2} \exp\left[-\left\{\frac{2(\nu - \nu_0)}{\Delta\nu_D}\right\}^2 \ln 2\right]. \tag{13.16}$$

This Gaussian lineshape is shown in Fig. 13.1(b). Note the contrast with the Lorentzian shape of natural broadening in Fig. 13.1(a), in which there is a sharp peak and more extended wings to the line profile.

An approximation to $\Delta\nu_D$ can be readily derived by equating the particle kinetic energy to the mean thermal energy:

$$\frac{1}{2}Mv^2 = \frac{3}{2}kT \tag{13.17}$$

from which $v = (3kT/M)^{1/2}$. Assuming $\Delta\nu/\nu_0 \sim v/c$ gives

$$\Delta\nu_D \simeq \frac{\nu_0}{c} \left(\frac{3kT}{M}\right)^{1/2}. \tag{13.18}$$

This is close to the value given in Eq. (13.15).

In gases, inhomogeneous broadening through the mechanism of Doppler broadening usually dominates over all other mechanisms. An extreme example is the radio frequency spectral line at 1420 MHz (wavelength 21 cm) emitted by the very tenuous neutral hydrogen gas in interstellar space. The transition involved is between two states in which the spins of the neutron and the electron are parallel and antiparallel; it has a very low transition probability (the Einstein coefficient $A = \frac{1}{\tau} = 2.85 \times 10^{-15}\,\text{s}^{-1}$), so that a hydrogen atom excited to the higher energy level might wait many million years before spontaneously decaying. The collision rate is also very low, so that the linewidth is determined by Doppler shifts, partly due to thermal broadening and partly due to differential velocities in the spiral structure of the Galaxy.

Occasionally a lineshape profile must be described by a combination of a Gaussian and a Lorentzian; such intermediate lineshapes are known as *Voigt* profiles. Laser light has quite different characteristics: the line centre frequencies and widths in lasers are determined by a resonant cavity more than by the resonance of the lasing medium, and the widths can be very much smaller than those of normal spectral lines (Chapter 15).

13.5 TWIN-BEAM SPECTROMETRY: FOURIER TRANSFORM SPECTROMETRY

We now turn to interferometric methods of spectrometry which extend the resolution by many orders of magnitude beyond that available from grating spectrometers. We have so far described the performance of a twin-beam interferometer in terms of an ideally narrow spectral line. The next step is to consider its action with a single spectral line with finite width or structure, and then to generalize for a wide spectral range and complex spectrum. The twin-beam interferometer, in its many and varied forms and in many ranges of wavelength, will then be seen to be a very powerful spectrometer, capable of resolving and measuring the width and shape of narrow line profiles.

Suppose that sodium light provides the illumination in a twin-beam interferometer, such as the Michelson, seen in outline in Fig. 13.2. As is known from examination in any optical spectroscope that can resolve wavelengths separated by a few ångström units ($1\,\text{Å} = 10^{-10}\,\text{m}$), the prominent yellow light from sodium is made up of the two D-lines, at approximately $\lambda = 5890\,\text{Å}$ and $5896\,\text{Å}$ (i.e. at 589.0 nm and 589.6 nm) of almost equal intensity. These wavelengths differ by just over 0.1%. As a first approximation consider them as very narrow compared with their separation. Suppose that the interferometer is first set up with one mirror M_1 and the image M_2 of the other in coincidence at the centre and at a slight angle so that vertical fringes of near zero order are seen. Then as M_1 is moved farther away the fringes move sideways, and as each crosses the centre of the field of view it indicates a change of one wavelength in the optical paths between the two arms; that is to say, a movement of $\lambda/2$ in the position of M_1. As M_1 moves further and the order of the interference increases, the fringes become less and less visible. This is because the two sets of fringes from the two wavelength components get progressively out of phase until a point is reached when they are exactly in antiphase and give a uniform intensity. The condition for this is that the mirror M_1 moves a distance d_1 where

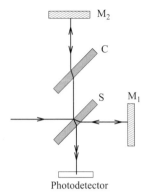

Fig. 13.2   Outline of a Michelson interferometer.

$$2d_1 = N_1\lambda_1 = \left(N_1 - \frac{1}{2}\right)\lambda_2. \tag{13.19}$$

We must be clear that there is no interference between the two sets of fringes, as light on different frequencies cannot be coherent. It is simply that the addition in *intensity* of two nearly equal but antiphase sine waves gives a more or less uniform intensity. Increasing d still further, the visibility improves and the fringes become sharp again when

$$2d_2 = N_2\lambda_1 = (N_2 - 1)\lambda_2. \tag{13.20}$$

Changing d by about 3 cm allows about 100 such cycles of visibility variation to be counted, each with about 1000 fringes between them. Such an observation allows the separation of the two lines to be accurately determined.

In general when light from a spectral line with any structure is examined in the twin-beam interferometer it is found to give high-visibility fringes at zero path difference d, which decrease in visibility as d is increased, and finally disappear. (See section 11.12 for the definition of fringe visibility.) A recording of the fringe visibility as d is varied is an *interferogram*. In the example of sodium light above it is easy to see that the form of the interferogram implies the spectrum of the light causing it. Michelson realized this and pointed out that the two quantities were related as a Fourier transform pair. He was then able to use a twin-beam interferometer to find the shape and structure of a single spectral line, and discovered the hyperfine structure of many spectral lines previously regarded as monochromatic.

The Fourier relationship is easily demonstrated by considering first a single spectral component with wavenumber k (where $k = 1/\lambda$), and extending to a multi-component spectrum. Two waves of equal amplitude from the single component arrive at a field point along path lengths x_1 and x_2, and add to give an intensity I

$$I = \langle \{E_0 \exp i(kx_1 - \omega t) + E_0 \exp i(kx_2 - \omega t)\}^2 \rangle, \qquad (13.21)$$

where $\langle \rangle$ indicates a time average.

At a path difference $x = x_1 - x_2$ the intensity is

$$I_x = 2I_0(1 + \cos kx). \qquad (13.22)$$

This is the familiar interference pattern of cosine fringes. Extending to a spectrum $I(k)$ with finite width (expressed for convenience in wavenumber k), the interferogram becomes

$$I(x) = \int_{-\infty}^{\infty} I(k) \cos(kx)dk, \qquad (13.23)$$

which is the Fourier transform of the spectrum $I(k)$ (apart from a factor $2/\pi$ which occurs in the cosine Fourier transform). A more general formulation will be found in the next chapter, but the general result is that the measurement of fringe visibility as a function of interference order gives the profile of a spectral line via a Fourier transform. The resolving power is equal to the order of interference reached in the measurement.

As an example of Fourier transform spectrometry we show in Fig. 13.3 the fringe visibility functions for the two types of line broadening, Lorentzian and Gaussian. The fringe visibility is labelled g_{12}, anticipating the discussion in the next chapter; this is formally the correlation between the fields at points 1 and 2, separated by a time delay τ. A comparison with Fig. 13.1 shows that the sharp peak of the Lorentzian corresponds to the extension of the visibility function to larger delays τ.

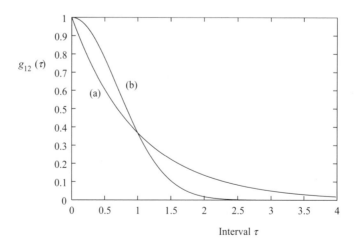

Fig. 13.3   The fringe visibility functions $g_{12}(\tau)$ for (a) Lorentzian and (b) Gaussian line profiles.

13.6   PRACTICAL FOURIER SPECTROMETRY

Twin-beam interferometers such as the Michelson of Fig. 13.2 have two fundamental advantages in sensitivity over a conventional grating or prism interferometer. First, an extended source can be used, instead of a narrow slit. This
is known as the Jacquinot advantage. Second, a single detector can be used to
record light from the whole spectrum simultaneously, in contrast to a detector
scanning a narrow part of a dispersed spectrum. The second, known as the
Fellgett advantage, is less important in the visible spectrum where a photographic plate or a multiple-element detector such as a charge-coupled device
(CCD) (see Chapter 21) is available, but it is vital for efficient measurements in
the far infrared where multiple element detector arrays are not available. The
only loss of light occurs at the arrangement for splitting the beam, but even this
can be avoided by systems such as those of Fig. 13.4. In (a) double mirrors are
used in a Michelson interferometer to allow both beams to be detected at D_1
and D_2, while in (b), due to Strong, an ingenious interleaved mirror reflects all
the light into a single detector.

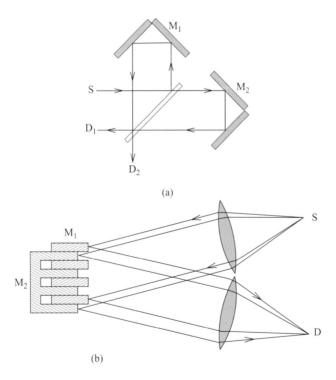

Fig. 13.4   Examples of efficient twin-beam interferometers: (a) the double mirrors in a
Michelson interferometer allow the returning beam to be detected at D_1 in addition to
the beam at D_2; (b) all the light from the source S reaches a single detector in this
arrangement due to Strong. The path difference is changed by moving the multiple
mirror M_2, which interleaves with a fixed mirror M_1.

Starting in the 1950s, the speed and sensitivity of Fourier transform spectro-scopy revolutionized infrared astronomy; for example, observations of the spectra of planetary atmospheres could be made in a single night which pre-viously would have required many years to complete.

13.7　INTENSITY, OR PHOTON CORRELATION SPECTROSCOPY

When the width of a spectral line is so small that the path difference in a twin-beam interferometer with sufficient resolving power becomes impracticably long, a different technique becomes available for measuring spectral lineshapes. As we shall see in the next chapter, the twin-beam interferometer which is used in Fourier transform spectrometry is measuring the correlation between the *ampli-tudes* of light as the path difference between the two beams is changed. In contrast, the technique of intensity fluctuation spectrometry is concerned with fluctuations of *intensity* at a single point. These fluctuations are usually on a very short time-scale, and are averaged out in most photometric and interferometric measurements. The intensity of light from a spectral line with width $\delta\nu$ fluctuates only on a time-scale of order $1/\delta\nu$, which is so small that it is usually unresolv-able, and the fluctuations are unnoticed. But if they can be resolved, using techniques with a time resolution better than $1/\delta\nu$, the frequency spectrum of the intensity fluctuations can be related to the spectral lineshape and linewidth through a Fourier transform similar to that of Fourier transform spectrometry.

We now consider the amplitude and intensity of light from a number of atoms radiating independently, so that their phases are randomly distributed. (This is an example of *chaotic light*, as contrasted with laser light; see Chapters 14 and 17.) The radiation from each atom is coherent for a time τ; then the phase changes discontinuously by a random amount, as might occur at a collision in a gas. We add the contributions of a large number n of atoms to the observed instantaneous intensity. Assuming the contributions all have equal amplitudes, the sum contains the resultant of the individual phases as an amplitude factor $a(t)$ and the intensity averaged over a long time is proportional to

$$\overline{a(t)^2} = |\exp i\phi_1 t + \exp i\phi_2 t + \ldots + \exp i\phi_n|^2. \tag{13.24}$$

The average intensity is, as expected, simply n times the intensity from an individual atom.

Instantaneously, however, the intensity may be very different. The sum of many amplitude contributions with random phase is shown in Fig. 13.5. After a time greater than τ the phases change and the sum will change unpredictably. The *probability distribution P (I)* of the intensity follows a statistical law familiar in the theory of the random walk:

$$P(I)dI = \overline{I}^{-1} \exp\left(-\frac{I}{\overline{I}}\right) dI. \tag{13.25}$$

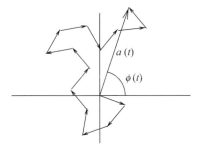

Fig. 13.5   The sum of many vectors with random phases, as in a random walk. This is a phasor diagram for chaotic light.

The average amplitude of the intensity fluctuations given by the difference ΔI between the instantaneous intensity and the mean is

$$\left[\overline{(\Delta I)^2}\right]^{1/2} = \left[\overline{I^2} - \overline{I}^2\right]^{1/2} = \overline{I}, \tag{13.26}$$

so that the root mean square fluctuations equal the mean intensity itself.

   Figure 13.6 shows an example of the form of fluctuations in the intensity of chaotic light, on a time-scale comparable with the coherence time τ. The rate of fluctuation is inversely proportional to the coherence time, and in more detail the spectrum of the fluctuations in intensity is related to the shape and width of the spectral line by a Fourier transform; however, the information about the line is not as comprehensive as in normal Fourier transform spectrometry. We briefly set out the mathematical relations involved in amplitude and intensity transformations.

   In Fourier transform spectrometry the detector measures an intensity obtained by adding two fields with a time lag δt corresponding to a path difference $c\delta t$. For equal amplitudes this gives

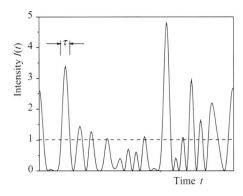

Fig. 13.6   An example of the fluctuations in intensity of chaotic light.

$$\bar{I}(\delta t) = \langle E^*(t)E(t+\delta t)\rangle = \frac{I}{T} \lim_{T\to\infty} \int_0^T E^*(t)E(t+\delta t)\mathrm{d}t. \qquad (13.27)$$

This average is called the *first-order autocorrelation function*, which gives the required spectrum by Fourier transforming the interferogram (see Chapter 4).

The intensity measurements are also separated by an interval δt. Taking the product of the differences from the mean intensity, and averaging over time as in Eq. (13.27):

$$\langle(\bar{I}(t) - \bar{I})(\bar{I}(t+\delta t) - \bar{I})\rangle = \langle\bar{I}(t)\bar{I}(t+\delta t)\rangle - \bar{I}^2 \qquad (13.28)$$

since

$$\langle\bar{I}(t)\rangle = \langle\bar{I}(t+\delta t)\rangle = \bar{I}. \qquad (13.29)$$

The first term of Eq. (13.28) is a *second-order correlation function*.[2] Expanding in terms of the E fields this is

$$\langle|E(t)|^2|E(t+\delta t)|^2\rangle = \langle E^*(t)E^*(t+\delta t)E(t)E(t+\delta t)\rangle. \qquad (13.30)$$

Expanding each term as $E \exp i\omega t$ or $E \exp i\omega(t+\delta t)$, and averaging over times large compared with $1/\delta t$, we find

$$\langle|E(t)|^2|E(t+\delta t)|^2\rangle = |\langle E^*(t)E(t+\delta t)\rangle|^2 + \bar{I}^2. \qquad (13.31)$$

The second-order correlation function is therefore determined by the *magnitude* of the first-order correlation function.

The loss of phase in the correlation function is not usually a serious limitation, since in most applications the spectral line can be assumed to be symmetrical, as in the two broadening mechanisms described in section 13.1. The second-order correlation functions of Lorentzian and Gaussian spectral lines are shown in Fig. 13.7. Note that the correlation in intensity asymptotically approaches unity at large time intervals, when the correlation is between two measures of the average intensity. As in the first-order correlations of Fig. 13.3, the sharper Lorentzian profile is revealed by the extension of the correlation function to larger time intervals.

Intensity fluctuation spectroscopy requires the high time resolution of detectors such as the photomultiplier, which can reach 10^{-9} s. Narrow linewidths are often expressed as a frequency bandwidth; the intensity fluctuation technique therefore applies to bandwidths up to about 10^8 Hz. In the same terms diffraction grating spectroscopy is applicable to bandwidths of 10^{10} Hz *and higher*, while Fabry–Perot interferometry methods are applicable in the range 10^6 to 10^{12} Hz, overlapping with intensity fluctuation and diffraction grating methods.

---

[2] The first- and second-order correlation functions are usually designated $g_{12}^{(1)}$ and $g_{12}^{(2)}$, respectively.

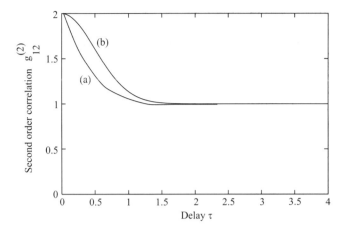

Fig. 13.7   The second-order correlation functions of (a) Lorentzian and (b) Gaussian spectral lines.

The technique is often referred to as *photon correlation spectroscopy*, since the intensity fluctuations can be measured as fluctuations in the rate of arrival of photons at a detector. The photon aspect, however, only becomes important when the flux of photons is insufficient for a large number to be detected in the smallest time interval involved in the measurement. This problem is unimportant if intense sources of light can be used, as in the scattered laser light technique that we now describe.

13.8   SCATTERED LASER LIGHT

The technique of measuring the width and shape of very narrow spectral lines through intensity fluctuations finds its most useful application in the examination of the scattering of laser light in substances such as colloids and polymers. Here a slightly different technique is employed, as shown in Fig. 13.8(a). The light entering the detector is the sum of the scattered light and a reference beam direct from the laser. The fluctuations in intensity are measured by photon counting in time intervals comparable with δt, the inverse of the frequency width of the scattered laser line radiation. The laser light itself contains effectively no fluctuations (see Chapter 17), and the dominant term in the fluctuations is again the first-order correlation of the scattered light as measured by an autocorrelation function such as that in Fig. 13.8(b). The magnitude of the fluctuations, and the ease of measurement, is greatly enhanced if the intensity of the reference beam is much larger than that of the scattered light.

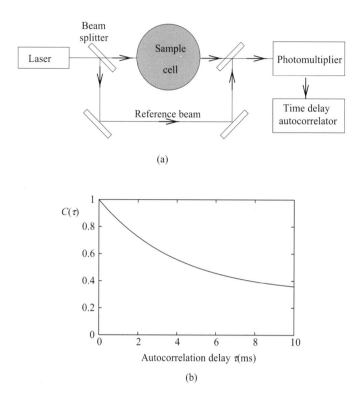

(a)

(b)

Fig. 13.8 Photon correlation heterodyne spectrometry applied to laser light scattered from a colloidal solution: (a) the laser beam is divided by a beam splitter to give a sample beam and a reference beam, which are recombined in the detector photomultiplier; (b) autocorrelation function $C(\tau) = A + B\exp(-\Gamma\tau)$, where A and B are constants and Γ is the linewidth of the scattered light.

13.9 FURTHER READING

R. J. Bell, *Introductory Fourier Transform Spectroscopy*. Academic Press, 1972.
A. Corney, *Atomic and Laser Spectroscopy*. Oxford University Press, 1972.
W. Demtroder, *Laser Spectroscopy*, 2nd edn. Springer-Verlag, 1995.
R. Loudon, *The Quantum Theory of Light*. Clarendon Press, Oxford, 1973.

NUMERICAL EXAMPLES 13

13.1 Find the linewidth in wavelength and frequency terms for a cooled neon (atomic mass number $M = 20$) discharge at 300 K.

13.2 Find the half-width of the $H\alpha$ line ($\lambda = 656$ nm) emitted by atomic hydrogen in an ionized interstellar cloud at a temperature of 10^4 K assuming that the width is entirely due to the thermal Doppler effect. The half-width is $\Delta\lambda = \lambda_0 - \lambda_{1/2}$, where the intensity is half maximum at wavelength $\lambda_{1/2}$.

14

Coherence and correlation

All nature is but art unknown to thee,/ All chance, direction which thou canst not see;/ All discord, harmony not understood.

Pope, *An Essay on Man.*

In much of the discussion of diffraction and interference phenomena in previous chapters we have been concerned with monochromatic light produced by a point source. No actual source is either a point or strictly monochromatic, so that no light has a perfect sinusoidal wavefrom extending indefinitely in space or in time. In practice there is a loss of *coherence* both in space and in time, whose consequences have already been encountered in the two basic types of interferometer, of which Michelson's stellar interferometer and spectral interferometer are examples. The stellar interferometer investigates the waves from a source which is nearly, but not quite, a point, finding that the loss of coherence *across* the wavefront is a measure of the angular diameter of the source. The spectral interferometer investigates the waves from a narrow spectral line by exploring the loss of coherence between two points separated *along* the path of the wave, which is a measure of the coherence in time.

In this chapter we define coherence more precisely, and apply the concepts of coherence and correlation to the practical issues of spatial and temporal coherence, and to angular and spectral resolution in optical instruments. We also discuss the concept of spatial filtering, in which the Fourier components of an object are modified in instruments such as the phase contrast microscope.

14.1 TEMPORAL AND SPECTRAL COHERENCE

The loss of coherence along the path of a wave from a source that is nearly, but not quite, monochromatic can be understood by supposing that the wave is made up of a large number of individual wavetrains of finite length, each produced by a single atom or other emitter, and that a large number of such wavetrains pass a point in the time taken to make an observation of intensity. Light from two points closer together than the length of an individual wavetrain will be coherent and will interfere as for a monochromatic source. Light from two points along the wavetrain separated by more than the length of the wavetrain is incoherent, and cannot show interference effects. (Instantaneously the two samples will add according to their phase relation, but this will change randomly during the observation, since the relative phase of different wave-trains is randomly distributed.) There is a typical *coherence length* in the light beam, which is the length of an elementary wavetrain. There is also a typical *coherence time*, which is the time for the elementary wavetrain to pass any point.

Coherence time is fundamentally related to spectral width, as may be seen from the Fourier analysis of Chapter 4. The precise relation depends on the shape of the spectral line, but it is useful to remember that for a coherence time δt, and an oscillation with angular frequency bandwidth $\delta\omega$, there is a general relation, known as the *Bandwidth Theorem*[1]

$$\delta t \cdot \delta \omega \sim 1. \tag{14.1}$$

An important example is a Gaussian wave group, i.e. a cosine wave modulated by a Gaussian envelope (Fig. 14.1) with spectral width $\Delta\nu(= \Delta\omega/2\pi)$, which has a coherence time t_G where

$$t_G = \frac{1}{\pi\Delta\nu}. \tag{14.2}$$

Correspondingly, for a wave velocity c the coherence length is

$$l_c = ct_G = \frac{2c}{\Delta\omega} = \frac{\lambda^2}{\pi\Delta\lambda}. \tag{14.3}$$

In this example, the decrease in coherence with time and distance follows Gaussians with widths t_G and ct_G.

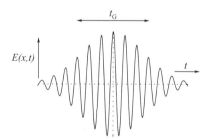

Fig. 14.1   A wave group with Gaussian profile. The spectral width of the group $\Delta\nu$ and the coherence time t_G are related by $t_G = (\Delta\nu)^{-1}$.

---

[1] This is analogous to the Heisenberg uncertainty relation between the momentum p and the position x of a particle

$$\delta p \delta x \geq \hbar.$$

Typical coherence lengths for light are easy to visualize. A colour filter on a white light might isolate a band 50 nm wide at a wavelength of 500 nm; the coherence length is then about 1 μm; a Fabry–Perot filter with a band-pass width of 1 nm increases this to 50 μm. A narrow spectral line from a sodium or mercury lamp can have a coherence length of 1 cm. Light from a carefully constructed laser can have a coherence length of 10 km or more, although very short wavetrains only a few microns long can be also be made by lasers specially designed to operate in a pulsed mode (Chapter 17).

In the discussions of the stellar interferometer (Chapter 11) the condition for obtaining interference between light derived from mirrors transversely separated across the wavefront was expressed as the inequality $\phi_0 \ll \lambda/d_M$, where ϕ_0 is the angle subtended by the source and d_M is the separation of the mirrors. Put in a way more suitable for the present discussion, the maximum distance apart of the mirrors for interference fringes to be observed is of order λ/ϕ_0; this is a measure of the *transverse coherence distance*. The pair of mirrors can be thought of as exploring the degree of coherence across the light wave; full coherence can only be found if the pair are close together, while if they are separated by more than the transverse coherence distance the light at the two mirrors becomes incoherent and no interference can be observed.

Typical transverse coherence widths are also often easy to visualize. Light from a source 1 arcsec across has a coherence width of about 10 cm: light from the nearest large-diameter stars (angular diameter ~ 0.01 arcsec) has a coherence width of 10 m, and light from the sun, which subtends an angle of 30 arcmin at the Earth has a coherence width of only 50 μm.

We are thus led to the idea that around any point in the light field produced by a real source there is a *region of coherence*, with a transverse size governed by the angular diameter of the source, and a longitudinal size governed by the bandwidth of the radiation from the source. Any interferometer that is to produce fringes from the light of the source must derive its two beams from points within this volume. The two sorts of Michelson interferometer we have discussed are the archetypes, the stellar using transverse separation and the spectral using longitudinal separation. We now define coherence more precisely and quantify these relationships.

14.2   CORRELATION AS A MEASURE OF COHERENCE

Let $E_1(t)\exp i\omega t$ and $E_2(t)\exp i\omega t$ be the amplitudes at points P_1 and P_2 in a light field. The irradiances at P_1 and P_2 are then $E_1(t)E_1^*(t)$ and $E_2(t)E_2^*(t)$, where the asterisk indicates the complex conjugate. An interferometer, of any type, combining the light from these two points adds the two amplitudes with, in general, a time delay and measures the square of the sum. The output of the interferometer $I(\tau)$ as a function of τ, the relative delay, is given by

$$I(\tau) = \langle \{E_1(t+\tau) + E_2(t)\}\{E_1^*(t+\tau) + E_2^*(t)\} \rangle. \qquad (14.4)$$

The brackets $\langle\rangle$ denote that a time average is taken. If this expression is multiplied out it gives

$$I(\tau) = \langle E_1(t+\tau)E_1^*(t+\tau)\rangle + \langle E_2(t)E_2^*(t)\rangle + \langle E_1(t+\tau)E_2^*(t)\rangle + \langle E_1^*(t+\tau)E_2(t)\rangle. \tag{14.5}$$

The first two terms are simply the average irradiances at P_1 and P_2, and are not functions of t. The second two terms give the interference fringes (note that they are each other's complex conjugate, so that their real parts are equal). Suppose the field at P_1 and P_2 is from a monochromatic point source of period T. Then when $\tau = NT$ (N is an integer) all four terms are equal and the intensity is four times that at P_1 or P_2. On the other hand, when $\tau = \left(N + \frac{1}{2}\right)T$, the second pair of terms is negative (each being the average of the product of cosines in antiphase) and they exactly cancel the first pair of terms. Thus fringes of 100% visibility are observed.

Evidently it is the second pair of terms that is of interest, expressing the relationship between the complex amplitudes at the two points. As they are each other's conjugate each has the same information as the other and conventionally the first is taken. Mathematically this is the *cross-correlation* of $E_1(t)$ and $E_2(t)$, regarding these as complex functions of time; in optics it is the *mutual coherence* $\Gamma_{12}(\tau)$. Thus

$$\Gamma_{12}(\tau) = \langle E_1(t+\tau)E_2^*(t)\rangle. \tag{14.6}$$

The interferometer output now includes the real part of Γ_{12}:

$$I(\tau) = I_1 + I_2 + 2|\Gamma_{12}(\tau)|. \tag{14.7}$$

Note that when P_1 and P_2 coincide and $\tau = 0$, the mutual coherence reduces to $\langle E_1(t)E_1^*(t)\rangle$, which is simply the intensity.

The *visibility* of a set of interferometer fringes, as in the Young's double slit fringes of Chapter 7, is defined as

$$V = \frac{I_{\max} - I_{\min}}{I_{\max} + I_{\min}}. \tag{14.8}$$

If, as often occurs, $I_1 = I_2$, then the visibility is simply the modulus of the mutual coherence, giving $V = |\Gamma_{12}|$. More generally, $\Gamma_{12}(\tau)$ may be normalized to give the *complex degree of mutual coherence* where $\Gamma_{11} = I_1$, etc.:

$$\gamma_{12}(\tau) = \frac{\Gamma_{12}(\tau)}{\{\Gamma_{11}(0)\Gamma_{22}(0)\}^{1/2}}. \tag{14.9}$$

Then the visibility is

$$V = \frac{2\sqrt{I_1 I_2}}{I_1 + I_2}|\gamma_{12}(\tau)|. \tag{14.10}$$

The function $\gamma_{12}(\tau)$ makes precise the conceptual ideas of the previous section. If we regard P_1 as fixed and P_2 as exploring the space around it, there is in general a complex number $\gamma_{12}(\tau)$ for each position of P_2 and value of τ. The degree of correlation, i.e. the magnitude of γ_{12}, varies between 0 and 1. Most of the interference phenomena we have discussed will be seen to be interpretable in the terms of the complex degree of correlation. For example, in the case of Young's double slit, the delay τ varies across the plane where the fringes are seen. If the light illuminating the slits is monochromatic and the slits are effectively point sources, then the light will be completely coherent and $\gamma_{12}(\tau)$ will be unity everywhere. Fringes of unit visibility result. If a wide source is used, then the light at the two slits is only partially correlated, $\gamma_{12}(\tau)$ is less than unity and so is the visibility. Similarly, if a broad-spectrum source is used, then the fringe on the axis, where $\tau = 0$, will be visible, but those off the axis where $\tau \neq 0$, rapidly decline in visibility. The explanation in Chapter 11 in terms of overlapping of fringes, in the case of the broad source from different *parts* of the source, and in the case of the white light source from different *wavelengths*, is now seen to be more elegantly expressed in terms of coherence.

14.3   AUTOCORRELATION AND COHERENCE

In Chapter 4 we considered autocorrelation in a time-varying quantity $A(t)$. The autocorrelation function is defined as the time average

$$\Gamma(t) = \langle A(t + \tau)A^*(t)\rangle. \tag{14.11}$$

This was shown to be the Fourier transform of the power spectrum of $A(t)$. Comparison with Eq. (14.6) shows that the longitudinal coherence function for a plane wave, where the moduli of E_1 and E_2 are equal, is the autocorrelation function, which is the Fourier transform of the power spectrum. The transverse autocorrelation $\Gamma(x)$ similarly is the Fourier transform of the angular distribution of brightness across the source. Any interferometer that measures coherence along a wavetrain can find $\Gamma(t)$, and hence the spectrum of the wavetrain; any interferometer that measures coherence across an axis x transverse to a wavefront can find $\Gamma(x)$, and hence the brightness distribution across the source.

Fourier transform spectroscopy, as described in Chapter 13, is therefore a process of measuring the autocorrelation along a wavetrain, using an interferometer such as Michelson's spectral interferometer over a range of path differences. The fringe amplitude is measured as a function of path difference, and a Fourier transformation gives the spectrum. The spectrum can be measured with a resolution that depends only on the maximum delay τ between the two beams; the frequency resolution is approximately $1/\tau$.

The extent of the coherence across the wavefront depends on the angular width of the source of light; in more detail there is again a Fourier transform relation between the angular distribution of brightness across the source and the decrease in coherence across the wavefront.

The stellar interferometer samples a wave at two separated points across the wavefront, and measures the coherence between the two samples. We can describe the coherence between the two sample points as a function of distance between them. Similarly, we can sample the wave at two points along its direction of travel, obtaining the coherence as a function of time interval. The whole area of coherence can be delineated by the two sampling processes, one in space transverse to the wavefront, the other in time. The description of the coherence is expressed by two autocorrelation functions, one in space and the other in time. Autocorrelation in space, i.e. transverse to the wave, is related to the angular distribution of the source; autocorrelation in time, i.e. along the wave, is related to its spectrum.

The concept of coherence is also useful in communications, where a narrow bandwidth electrical signal is analogous to an optical spectral line. Any modulation of the signal will give a finite width to the spectrum, and very broad bandwidth electrical noise is analogous to white light. In radio astronomy the spectral lines of interstellar gas, such as hydrogen at 21 cm wavelength and carbon monoxide at 2.7 mm wavelength, have a width that is due to a combination of thermal broadening and Doppler shifts within an interstellar cloud. A typical linewidth might be 1 MHz, when the coherence length would be about 1 μs. The coherence length, and the whole autocorrelation function, can be measured by an autocorrelation technique shown in Fig. 14.2. Here the electrical signal passes through a circuit containing a variable delay (about 1 μs in

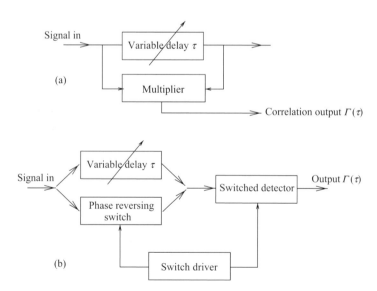

Fig. 14.2 Measuring the spectrum of an electrical signal by autocorrelation. The detector measures the product of the signal with the same signal delayed by a variable amount.

the example above), and the direct and delayed signals are recombined in a detector. The detector multiplies the sum, giving an average product that measures their correlation. The correlation function is obtained by measurements over a range of delays. The spectrum of the signal is then found by a Fourier transform of the autocorrelation function.

The output of the detector in Fig. 14.2(a) also contains the intensity of the input signal as an unwanted component. This can be removed by the switching system of Fig. 14.2(b), where a phase reversing switch has been included in the direct signal path. When this operates the sign of the correlation reverses, while the intensity component is unchanged. The phase switch is operated periodically by a driver which also reverses the output of the detector. The intensity component then averages to zero, leaving only the correlated signal. This technique of *phase switching* also has very many applications in optics.

14.4   TWO-DIMENSIONAL ANGULAR RESOLUTION

We saw in Chapter 11 that coherence across a wavefront is related to the angular distribution of the source of the radiation, so that a measurement of the coherence as a function of distance across a wavefront gives the width and shape of the source. This applies in two dimensions; we now show that a two-dimensional mutual coherence across a wavefront is directly related to the two-dimensional angular distribution of radiance across the source. Exploring the coherence of the wavefront allows a map to be drawn of the angular distribution of brightness across the source of the wavefront.

Suppose that the amplitude of the wavefront at a point in the (x, y) plane is $A(x, y)$, and that at another at a distance X, Y is $A(x + X, y + Y)$. There is no time delay between these samples of the wavefront, so that the two-dimensional mutual coherence is

$$\Gamma(X, Y) = \langle A(x + X, y + Y)A^*(x, y)\rangle. \qquad (14.12)$$

By a similar argument to that in the time-frequency case, the two-dimensional Fourier transform of this turns out to be the two-dimensional distribution of luminance with *angle*. Put in simpler terms, this is the distribution of brightness giving rise to the sampled amplitudes. The coherence $\Gamma(XY)$ is complex; in circumstances where phase as well as amplitude of the correlation can be measured the Fourier transform will give the brightness distribution across the source without any assumptions about symmetry. The resolution in angle is of the order of λ/X and λ/Y in the x and y directions, respectively.

This relationship is the basis of *aperture synthesis* in radio astronomy. At radio wavelengths, typically of order 10 cm, it is difficult to obtain sufficient angular resolution by using a single radio telescope. It is, however, straight-

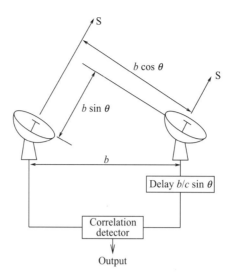

Fig. 14.3 Aperture synthesis in radio astronomy. The interferometer's output is the complex degree of coherence between the two radio telescopes, spaced a distance b apart, which together sample the transverse coherence function. The source under observation may be at an angle θ to the normal to the baseline, so that the correct correlation requires one signal to be delayed by $b \sin \theta$, and the effective baseline length is $b \cos \theta$. Observations at many baselines are combined and transformed to produce maps such as that of the radio galaxy Cygnus A.

forward to measure the correlation between radio waves received by two or more telescopes separated by large distances (Fig. 14.3), even up to some thousands of kilometres (see section 11.13). Through a succession of measurements of the two-dimensional coherence using pairs of telescopes at various spacings and orientations, a sufficient map of $\Gamma(X, Y)$ can be obtained. A map of complex correlation is constructed which, when Fourier transformed, gives a map of the angular distribution of radio brightness across the source.

Radio interferometers using aperture synthesis may require a network of 10 or more radio telescopes operating simultaneously, so that sufficient baselines are available for measuring the distribution of the complex correlation. They can however be operated with baselines up to some thousands of kilometres, using wavelengths of a few centimetres. Since the angular resolution of the resulting source map is of order λ/D, where D is the largest available baseline, it is possible by this method to construct maps of radio brightness with resolutions down to 10^{-3} arcsec (1 milliarcsec). An example is shown in Plate 14.1. It is interesting to note that this angular resolution is two orders of magnitude better than the resolution of the largest optical telescopes, even though the radio wavelength is four orders of magnitude larger than the optical wavelength.

14.5   THE INTENSITY INTERFEROMETER

In Chapter 13 we recounted the history of the intensity interferometer used by Hanbury Brown and Twiss for measuring very small angular diameters of stars. No explanation was given there as to why the intensity fluctuations observed at separated positions should correlate. A full discussion of the intensity interferometer would be out of place here, but it is easy to see why it works in terms of the ideas in the first section of this chapter.

Each atomic emitter in the source gives rise to a finite wavetrain of random phase. We can imagine a multiplicity of spherical waves spreading out from the source. At any point P_1 in space the amplitude at a particular time depends on how many wavetrains are present and how their phases happen to be arranged. Sometimes favourable interference will take place and the amplitude – and hence the intensity – will go up; sometimes destructive interference will make it go down. In these rather oversimplified terms one can see that intensity fluctuations should exist. Now let us consider whether the fluctuations at another point P_2 will be correlated with those at P_1. The same wavetrains reach P_2 as reach P_1: the only difference is in their relative phases caused by the different paths they have travelled.

The condition for identical fluctuations at P_1 and P_2 is the same as that for interference at P_1 and P_2: the relative phases of the wavetrains must be the same, which is to say that the waves are coherent at P_1 and P_2. The phase condition has already been found in section 11.12; it is

$$\phi_0 \ll \frac{\lambda}{d}, \tag{14.13}$$

where ϕ_0 is the angular width of the source and d is the separation between P_1 and P_2. The discussion can be put on a quantitative basis in terms of the measure of coherence Γ_{12}. We have seen that the irradiances at P_1 and P_2 are

$$I_1 = \Gamma_{11} = \langle E_1 E_1^* \rangle, \tag{14.14}$$

$$I_2 = \Gamma_{22} = \langle E_2 E_2^* \rangle. \tag{14.15}$$

The intensity interferometer takes the time average of the product of I_1 and I_2, which is

$$\langle I_1 I_2 \rangle = \langle (\langle E_1 E_1^* \rangle)(\langle E_2 E_2^* \rangle) \rangle \tag{14.16}$$

$$= \langle (\langle E_1 E_2^* \rangle)(\langle E_1^* E_2 \rangle) \rangle \tag{14.17}$$

$$= \langle (\Gamma_{12} \Gamma_{12}^*) \rangle = \langle |\Gamma_{12}|^2 \rangle. \tag{14.18}$$

The crucial step here is that between the first and second lines. Readers should try to convince themselves it is justified. This being done we see that the intensity interferometer measures the *modulus* of the mutual coherence between P_1 and P_2, but not its phase. A variation of the spacing of P_1 and P_2 thus allows $|\Gamma_{12}|$ to be measured over the lateral coherence area of the source. As in the aperture synthesis discussed in the previous section the source brightness distribution must then be obtained by Fourier transformation of $|\Gamma_{12}|$. This cannot be achieved unambiguously without knowledge of the phase of Γ_{12}, but it may be allowable to assume that the source is symmetrical, giving a constant phase at all interferometer spacings. This is at least a reasonable assumption when the diameter of a star is first measured. The optical intensity interferometer was first used in 1956 at Jodrell Bank on the bright star Sirius. Fig. 14.4 shows the measured fall-off of $|\Gamma_{12}^2|$ with baseline increasing up to 9 m. The angular diameter of Sirius is 7×10^{-3} arcsec; a circular disc of this size would give the theoretical variation of $|\Gamma_{12}|$ shown in the figure, agreeing well with the observations.

Note that this account of the intensity interferometer is a purely *wave* explanation. We presented in Chapter 13 a similar discussion of the relation between the shape of a spectral line and intensity fluctuations on a time-scale related to the width of the line. For both some consideration should be given to the photon nature of light. There is no need for this if the flux of photons is large enough for a large number to arrive within a single measurement time, but if the flux is small, then the random variations in photon count become important. These appear as statistical fluctuations in intensity, and the correlator output becomes noisy. This does not change the coherence and correlation, but it reduces the accuracy of the measurements.

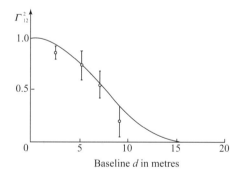

Baseline d in metres

Fig. 14.4   Hanbury Brown and Twiss' results for the variation in intensity fluctuations with baseline for the star Sirius. The full curve is the variation of Γ_{12}^2 with d, calculated from an assumed angular diameter of 0.0069 arcsec. The optical system consisted of two standard army searchlight mirrors 1.56 m in diameter and 0.65 m focal length, capable only of focusing the light from a star into an area 8 mm in diameter. A photomultiplier was mounted at the focus of each mirror and the anode currents were multiplied and integrated to give Γ_{12}^2.

14.6   SPATIAL FILTERING

We have seen that the resolving power of instruments such as the telescope and microscope, and of the Fourier transform spectrometer, is best understood by considering the range of Fourier components that contribute to the output, whether it is an optical image or the shape of a spectral line. This concept can be extended further to consider what happens if instead of a direct reconstruction of the original light source we modify some of the Fourier components before reconstruction. Such a process is familiar in communication engineering, where a signal may be modified by a filter; for example, an unwanted oscillation might be removed by a narrow band filter. The directly analogous process in optics is spectral filtering in the Fourier transform spectrometer (Chapter 13), where the output can be modified by adjusting the amplitude and phase of the measured longitudinal correlation function. In this section we consider the modification of measured transverse correlation functions, and its effects on an optical image. This is the process of *spatial filtering*, shown schematically in Fig. 14.5.

The first application of spatial filtering (although not then described as such) was to the microscope, in Abbe's theory of image formation. Consider the formation by a microscope objective lens of an image of a grating-like object, illuminated by fully coherent light. In Fig. 14.6 the objective, shown as a single

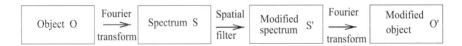

Fig. 14.5 Spatial filtering. Light from the object O is Fourier transformed into the spectrum S. The spectral components are modified by filtering, producing the spectrum S', and an inverse Fourier transform produces the reconstructed object O'.

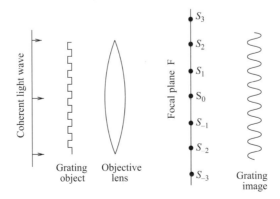

Fig. 14.6 The Abbe theory of microscopy. The object is a grating, coherently illuminated. The diffraction pattern in the focal plane F of the microscope objective is the Fourier transform of the complex amplitude across the object; as the object is periodic, the resulting diffraction pattern comprises discrete components $S_0, S_1, S_2 \cdots S_n, S_{-1}, S_{-2} \cdots S_{-n}$.

lens, collects light leaving the object over a wide range of angles. If we consider this light as an angular spectrum of plane waves, we find that these components are focused on the focal plane F of the lens, which therefore contains the angular spectrum. The plane waves continue beyond this focal plane, forming an image farther away which is then examined by an eyepiece; in this discussion, however, we are only concerned with the angular spectrum in the focal plane F of the objective. How can this be modified, and with what effect on the final image?

As in all Fourier analysis, the finer detail is contained in the highest order components; if these are lost, the resolution is reduced. An object which is a periodic grating will produce a series of components S_0, $S_1 \cdots S_n$, S_{-1}, $S_{-2} \cdots S_{-n}$, as shown in Fig. 14.6. A purely sinusoidal grating would produce only the two first order components S_1, S_{-1}; a grating with sharp narrow lines will produce a series of high order components. If a mask is placed in the focal plane so that only the first-order components are admitted to the rest of the microscope, then a grating with any line shape will be seen in the image plane simply as a sinusoidal grating. The reason for the resolution limit is clear: the objective must accept plane waves leaving the object over a sufficiently wide range of angles. If this range is $\pm i$, and the space between the grating and the objective has refractive index n, then the finest detail of the source that can appear in the final image has a size $d = \lambda/n \sin i$. The spectrum in the focal plane has a zero-order component at S_0 in Figure 14.6, which may be very intense for a nearly transparent object. This provides the first example of spatial filtering: the central zero-order component can be removed by placing a simple mask in the focal plane. This provides *dark-field microscopy*; an example is shown in Plate 14.2.

A more subtle effect, which is also used in Plate 14.2, is obtained by changing the phase of the central component, by using a phase filter, i.e. a transparent mask in which a central zone is thicker. Transparent objects then become visible because of the pattern of phase changes which they impose on the light passing through them. This is often important in biological specimens, which otherwise would have to be stained if they are to be made visible in an ordinary microscope. *Phase-contrast microscopy* was introduced by Zernike, and is often named after him. Consider again a simple grating in the object plane (Fig. 14.6), but a phase changing grating instead of an amplitude grating. The grating has no effect on the modulus of an incident plane wave, but introduces a small phase shift that varies periodically across the object plane. The diffraction pattern in the focal plane F then contains components S_1, $S_2 \cdots S_n$, S_{-1}, $S_{-2} \cdots S_{-n}$ as before, except that the components are in quadrature with the light at S_0, as may be seen by describing the complex amplitude in the object plane as

$$A(x) = A_0 \exp\left(i\phi_0 \cos \frac{2\pi x}{d} \right). \tag{14.19}$$

If ϕ_0 is small, we may write this as

$$A(x) = A_0\left(1 + i\phi_0\cos\frac{2\pi x}{d}\right) \qquad (14.20)$$

and the wave has two components, one with the unchanging amplitude, the other in phase quadrature (because of the i) and with an amplitude varying periodically across the aperture. These produce respectively the central zero order component S_0 and the diffracted components S_1, S_{-1}, etc.

The idea of phase-contrast microscopy is to retard the phase of the large undeviated component S_0 by a quarter wavelength, so as to reproduce the diffraction pattern of an amplitude grating. This may be achieved by inserting in the focal plane F a glass plate with an extra thickness in the central region, so that the light at S_0 is retarded by $\lambda/4$ or $3\lambda/4$. In the first case, regions of the object having greater optical thickness will appear brighter, and in the second darker. These are called bright and dark, or positive and negative phase contrast, respectively.

The undeviated light at S_0 forms an image of the light source. Instead of using a point source it is convenient to use a ring source, as shown in the practical arrangement of Fig. 14.7. The phase-changing plate is then also in the form of a ring, which covers the image of the light source. This arrangement allows a larger light source to be used, giving greater illumination in the image.

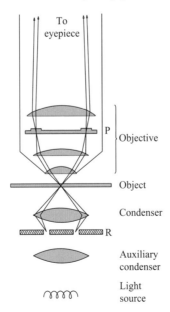

Fig. 14.7 Practical arrangement of a phase-contrast microscope. The ring source of light R is focused on the phase contrast plate P, within the objective lens system. The undeviated light is retarded by the plate, while light diffracted by the object passes through the thinner part of the plate and appears in quadrature in the final image.

14.7   FURTHER READING

R. Hanbury Brown and R. Q. Twiss, Correlations between photons in two beams of light, *Nature* **177**, 27, 1956.

R. Loudon, *The Quantum Theory of Light*. Oxford University Press, 1983.

L. Mandel and E. Wolf, *Optical Coherence and Quantum Optics*. Cambridge University Press, 1995.

G. Troup, *Optical Coherence Theory*. Methuen, 1967.

NUMERICAL EXAMPLES 14

14.1 Calculate the transverse coherence length for sunlight and starlight at the surface of the Earth, given that the Sun subtends an angle of 0.5°, while atmospheric scintillation spreads light from a star typically over 0.5 arcsec.

14.2 Calculate the longitudinal coherence length for laser light with a bandwidth of 60 MHz. What bandwidth $\delta\nu$ and linewidth $\delta\lambda$ would be required in a laser to produce a coherence length of 10 km?

PROBLEMS 14

14.1 A 'cross' type of radio telescope consists of two perpendicular strips of receiving area each with length D and width d. (They may, for example, be large arrays of dipoles or parabolic reflectors.) The radio signals from these two are multiplied together in a receiver which records only their product. What is the angular resolution of the system?

14.2 The wavelength of a beam of particles, mass m and velocity v is given by $\lambda = h/mv$, where h is Planck's constant. (i) Show that the wavelength for electrons accelerated by a field of V volts is approximately $1.23V^{-1/2}$ nm; (ii) calculate the best possible resolving power of an electron microscope with numerical aperture 0.1 and accelerating field 30 000 V.

14.3 A spectrograph used in radio astronomy is required to resolve the structure of the hydrogen spectral line at 1420 MHz, as observed when a radio telescope is receiving radiation from several hydrogen gas clouds moving with different speeds in the line of sight. The spectrograph divides the radio signal into two paths, inserting a variable digital delay into one path, and then recombines them in a multiplier. The smallest increment of delay is τ and the largest total delay is $N\tau$. If gas clouds with velocity differences between 10 km s^{-1} and 1000 km s^{-1} are to be distinguished, what values of τ and N are required?

15

Lasers

...there are certain situations in which the peculiarities of quantum mechanics can come out in a special way on a large scale.

The Feynman Lectures on Physics, vol. III, p. 21.

Lasers, the outcome of elegant physical theory and extensive experimentation, have become a vitally important tool in contemporary research in physics and chemistry, and indeed in all branches of science. Lasers are also used extensively in everyday life, from reading barcodes to playing CD recordings, and in technology, where they have many diverse uses such as optical communication and processing materials, and for many types of measurement.

In this chapter we set out the fundamentals of laser action.[1] The laser produces light in a significantly different way than normal light sources. The essential process of *stimulated emission* is considered along with absorption and spontaneous emission. This leads to the Einstein relations between the rate coefficients for these processes. The creation of *population inversion* is seen to produce *optical gain*. In most lasers the laser medium is inside an *optical resonator* to provide a long gain path length. We look at some of the properties of these resonators and their influence on the output laser radiation.

We describe some of the main types of laser, leaving the all-important semiconductor lasers to a separate chapter. The special characteristics of laser light, such as monochromaticity and directionality, which depend on its high degree of coherence, will be described in Chapter 17, which also deals with the tuning of lasers and the generation of ultrashort duration pulses of laser light.

15.1 STIMULATED EMISSION

Before the invention of the laser the available sources of light were essentially either thermal, such as from a tungsten filament lamp, or spontaneous emission

---

[1] The acronym laser stands for **L**ight **A**mplification by **S**timulated **E**mission of **R**adiation. The device commonly referred to as a laser is usually not an amplifier of light but an oscillator generating light. It is convenient now to refer to the process of light amplification as laser action and reserve the term laser for its common usage for the oscillating device.

from atoms and molecules, as in a gas discharge; in either case their brightness was limited by the temperature of the emitter. The broadband white light of solar radiation, for example, is limited in brightness by the temperature of the photosphere, while the brightness of the spectral lines is limited by the temperature of the gases in the chromosphere or corona above it. Light from a thermal source is *incoherent*; it is the chaotic sum of a disorderly outpouring of photons from individual atoms, radiating at random without any relation to one another. In a laser, however, the emission from individual atoms is synchronized, giving *coherent* radiation with very much higher brightness. The process of synchronization is *stimulated emission*, a concept introduced by Einstein in 1916.

The first use of stimulated emission to achieve a high brightness was in the microwave spectrum; hence the original name *maser* rather than laser. In 1953 Gordon et al.[2] demonstrated stimulated emission between the two lowest levels of ammonia molecules, giving a very narrow emission line at a wavelength of 12.6 mm. For this achievement Townes shared the 1964 Nobel Prize in Physics with Basov and Prokhorov of the USSR. The first laser, originally called the optical maser, followed in 1960, when Maiman produced red light at a wavelength 694.3 nm from the chromium ions in a ruby crystal. The essential effect of stimulated emission in the maser and in the laser is the coherent emission of radiation from excited atoms, adding precisely in phase and with the same direction and polarization. Three components are needed to achieve this in a laser (Fig. 15.1): an active medium with suitable energy levels, the injection of energy so as to provide an excess of atoms in an excited state, and a resonator system in which multiple reflections allow the build up of the coherent laser light.

There are three distinct processes in the absorption and emission of light which are relevant to a laser: absorption of a single photon, the emission of a single photon, and stimulated emission. All three are involved in the basic *three-level laser* shown in Fig. 15.2. Here the atoms in the active medium have

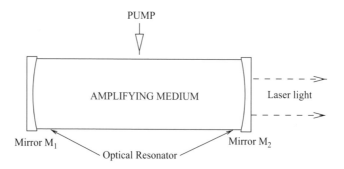

Fig. 15.1   The basic elements of a laser: amplifying medium, pumping energy source, and resonator $M_1 M_2$.

---

[2] J. P. Gordon, H. J. Zeiger and C. H. Townes, *Phys. Rev.* **95**, 282, 1954.

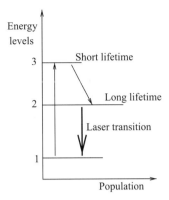

Fig. 15.2 Energy levels and the level populations in a three-level laser. 'Long' and 'short' refer to the lifetimes of the levels.

three energy levels involved in the laser action. Absorption raises the energy from level 1 to level 3 (this process is called *pumping*), spontaneous emission (or a non-radiative transition) reduces the energy to level 2, which is a longer lived or metastable state, and stimulated emission occurs between levels 2 and 1. The accumulation of excited atoms in the metastable state results in an overpopulation, or *population inversion*, in relation to the ground state. Stimulated emission is the release of this accumulated energy; one photon arrives at the excited atom, and two leave, with the same energy, travelling together and in phase. The stimulated photon has the same momentum as the incident photon, and hence travels in the same direction. Both photons can then repeat the process at other excited atoms, and the resulting chain reaction causes the light wave to grow exponentially. The ruby laser is an example of a three-level laser in which the active species is the Cr^{3+} ion rather than a neutral atom.

One further element is needed to make such an amplifier into a self-excited oscillator: the light must be fed back into the laser material. This is achieved by enclosing the lasing material between mirrors, forming a resonant cavity.

15.2 PUMPING: THE ENERGY SOURCE

As shown in Fig. 15.2 the energy that is converted into laser light is injected, or *pumped*, into the laser at a higher photon energy $h\nu_{31}$ than the laser output photons with energy $h\nu_{21}$. The excited ions then lose energy $h\nu_{23}$, falling into the intermediate level 2 which has a longer lifetime. Ions accumulate in this metastable state and are available for the stimulated emission process.

The original ruby laser was pumped by an intense flash of white light, which is selectively absorbed by chromium ions dispersed through the aluminium oxide crystal. Only a small part of the energy in the white light is at the right wavelength to be absorbed and produce the population inversion; this is an inefficient use of light, which is the reason for the use of an intense source of light. Other types of laser use more finely tuned pumping systems; the very common helium–neon gas laser provides a good example.

Energy is fed into the helium–neon laser by mixing the gases in an electrical discharge tube. Both gases are excited and ionized in the discharge. The amplifying medium is neon, which is pumped into a state of population inversion by collision with excited helium atoms; these in turn have been energized by electron collisions in the discharge. This more complex system works because of a close coincidence between energy levels in the excited helium and the upper levels required for the laser action in neon. Fig. 15.3 shows the outline of the helium–neon laser, and the energy levels involved. The coincidence is between the two metastable levels 2^1S_0 and 2^3S_1 in helium and the two metastable levels 5s and 4s in neon.[3] Stimulated emission from the 5s and 4s levels

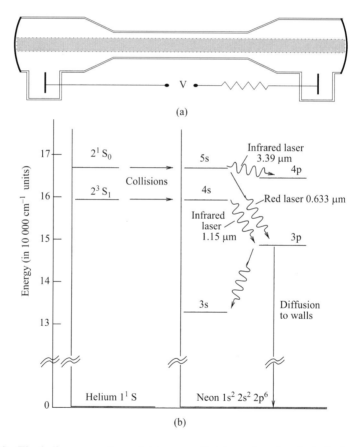

Fig. 15.3 The helium–neon laser: (a) laser excited by a d.c. electrical discharge, with potential V; (b) simplified energy level diagram.

---

[3] The helium levels are described by Russell–Saunders coupling, while for neon the levels are designated as $(1s^2 2s^2 2p^5)3s$, ( )4s, ( )5s, etc.; note that in the older and seldom used Paschen notation the ( )3s configuration is designated 1s and ( )4s is designated 2s, and so on.

can be through transitions to several different energy levels, allowing laser action at 3.39 µm, 1.15 µm, 632.8 nm and 543.5 nm. The familiar red beam of the helium–neon laser is operating on the 632.8 nm transition. Figure 15.3 shows the mirrors that enclose the laser, forming a resonator. As will appear later, a particular laser wavelength can then be selected by a choice of resonator system.

15.3   ABSORPTION AND EMISSION OF RADIATION

We now review the basic theory of the three processes involved in the interaction of radiation and matter, which we introduced briefly in section 1.7.

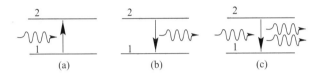

Fig. 15.4   (a) Absorption; (b) spontaneous emission; and (c) stimulated emission in a two-level system.

The processes of absorption, spontaneous emission and stimulated emission are depicted in Fig. 15.4. This illustrates the interaction of radiation with atoms depicted in simplified form as having two energy levels: a ground state energy E_1 and an excited state of energy E_2. These two states are populated with number densities N_1 and N_2. Absorption occurs when radiation of frequency $\nu = (E_2 - E_1)/h$ is incident on the medium, with excitation from the ground state to the excited state. The rate of absorption in which atoms are raised from level 1 to level 2 is

$$\left(\frac{\mathrm{d}N_1}{\mathrm{d}t}\right)_{\mathrm{ab}} = -B_{12}N_1\rho(\nu),\tag{15.1}$$

where N_1 is the population per unit volume in level 1 and $\rho(\nu)$ is the energy density of the incident field (units of energy per unit volume per unit frequency interval, $\mathrm{J\,m^{-3}Hz^{-1}}$). $\rho(\nu)$ is a function of the frequency ν of the radiation field. B_{12} is the *Einstein absorption coefficient*, which is a constant characteristic of the pair of energy levels in the particular type of atom.

Spontaneous emission of a photon occurs with transition of the electron from the excited level 2 to the ground level 1 with the emitted photon energy $h\nu = E_2 - E_1$. The rate of decrease of the population N_2 by spontaneous emission is

$$-\left(\frac{\mathrm{d}N_2}{\mathrm{d}t}\right)_{\mathrm{spon}} = A_{21}N_2.\tag{15.2}$$

The constant A_{21} (unit: s^{-1}) is related to the spontaneous radiative lifetime τ of the excited state as

$$A_{21} = \frac{1}{\tau}. \tag{15.3}$$

In stimulated emission atoms in level 2 are stimulated to make a transition to level 1 by the radiation field itself. The rate at which the transition occurs is proportional to the number of atoms in level 2 and the energy density of the radiation field:

$$\left(\frac{dN_2}{dt}\right)_{stim} = -B_{21}N_2\rho(\nu). \tag{15.4}$$

The constant B_{21} is the Einstein coefficient for stimulated emission from energy level 2 to level 1. Note that the rate of stimulated emission is proportional to the energy density at the resonant frequency $\nu = (E_1 - E_2)/h$, so that for high levels of radiation energy density stimulated emission dominates over spontaneous emission.

The rate of change of population in level 2 is the sum of the effects of spontaneous and stimulated transitions given by Eq. (15.1), (15.2) and (15.4), which yields the rate equation

$$\frac{dN_2}{dt} = -B_{21}\rho(\nu)N_2 + B_{12}\rho(\nu)N_1 - A_{21}N_2. \tag{15.5}$$

The relation between the three Einstein coefficients is found by considering an equilibrium situation, where a collection of atoms within a cavity is in thermal equilibrium with a radiation field. Then the populations of the two levels N_1 and N_2 are constant, so that

$$\frac{dN_2}{dt} = \frac{dN_1}{dt} = 0. \tag{15.6}$$

In thermal equilibrium, when there is detailed balancing between the processes acting to populate and depopulate the energy levels, we obtain

$$A_{21}N_2 + B_{21}\rho(\nu)N_2 = B_{12}\rho(\nu)N_1, \tag{15.7}$$

giving the relation between the values of N_1, N_2 and $\rho(\nu)$ at thermal equilibrium.

We now use two fundamental laws relating $\rho(\nu)$ and the relative populations N_1 and N_2 to the temperature T. These are *Planck's radiation law* for cavity radiation (section 5.7),

$$\rho(\nu) = \frac{8\pi h\nu^3}{c^3} \frac{1}{\exp(h\nu/kT) - 1},$$

(15.8)

and the *Boltzmann distribution* of atoms between the two energy levels:[4]

$$\frac{N_2}{N_1} = \exp\left(-\frac{E_2 - E_1}{kT}\right) = \exp\left(-\frac{h\nu}{kT}\right).$$

(15.9)

From Eqs. (15.7) and (15.9):

$$\rho(\nu) = \frac{A_{21}}{B_{12}(N_1/N_2) - B_{21}}$$

(15.10)

$$= \frac{A_{21}}{B_{12}\exp(h\nu/kT) - B_{21}}.$$

This equation may be combined with Eq. (15.8) to give

$$\frac{A_{21}}{B_{12}\exp(h\nu/kT) - B_{21}} = \frac{8\pi h\nu^3}{c^3}\left(\frac{1}{\exp(h\nu/kT) - 1}\right).$$

(15.12)

Equation (15.12) is satisfied when

$$\frac{A_{21}}{B_{21}} = \frac{8\pi h\nu^3}{c^3}$$

(15.13)

and

$$B_{12} = B_{21}.$$

(15.14)

These are the required relations between the three Einstein coefficients.

A small modification is necessary if either of the levels is *degenerate*, i.e. there is more than one state with the same energy value. For a degeneracy level g the populations are then given by

$$N_1 = g_1 n_1$$

(15.15)

$$N_2 = g_2 n_2$$

(15.16)

with $n_2/n_1 = \exp(-h\nu/kT)$. Then,

$$\frac{N_2}{N_1} = \frac{g_2}{g_1}\exp\left(-\frac{h\nu}{kT}\right)$$

(15.17)

and the Einstein relations for degenerate levels become

$$\frac{A_{21}}{B_{21}} = \frac{8\pi h\nu^3}{c^3}$$

(15.18)

$$g_1 B_{12} = g_2 B_{21}.$$

(15.19)

---

[4] Both levels are assumed here to be non-degenerate.

These equations and the concept of transition probability are fundamental to the theory of exchange of energy between matter and radiation.

The crucial factor for lasers is the ratio between the rates of stimulated and spontaneous emission:

$$\frac{\text{rate of stimulated emissions}}{\text{rate of spontaneous emissions}} = \frac{N_2 B \rho_v}{N_2 A} = \frac{1}{\exp(h\nu/kT) - 1}. \qquad (15.20)$$

The factor $\exp(h\nu/kT)$ shows us that in thermal equilibrium stimulated emission is very unlikely at optical frequencies, and explains why the first successful device was the maser, operating at a much lower radio frequency.

We also note that since the stimulated emission rate is $N_2 B_{21} \rho(\nu)$ we may increase the rate by increasing $\rho(\nu)$, which is achieved in a resonant cavity, and by increasing N_2 in the population inversion resulting from pumping.

15.4 LASER GAIN

We can now consider the growth of a light wave as it passes through an active laser medium, and find the conditions for the wave to grow by stimulated emission. The result, in Eq. (15.29) below, depends on four factors: the population inversion, the spectral lineshape, the frequency, and the transition probability A_{21}. We start by finding the emission and absorption in a small element dz of the path through the laser medium, and then integrating over the whole path, which may involve many to-and-fro reflections in a resonator.

First we look at the attenuation of an *absorbing* medium in which a plane wave of monochromatic radiation is travelling as illustrated in Fig. 15.5. The reduction in irradiance (power flow across unit area) as the wave travels from position z to $z + dz$ for a uniform medium is proportional to the magnitude of the irradiance and the distance travelled:

$$\Delta I(z) = I(z + dz) - I(z) = -\alpha I(z) dz. \qquad (15.21)$$

Here α is the *absorption coefficient*. In differential form:

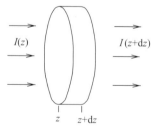

Fig. 15.5 Attenuation of a wave in a slab of material.

$$\frac{dI(z)}{dz} = -\alpha I. \tag{15.22}$$

On integration:

$$I(z) = I_0 \exp(-\alpha z), \tag{15.23}$$

where I_0 is the irradiance of the incident beam. This represents exponential attenuation.

If the number of stimulated emissions exceeds the number of absorptions, than rather than being attenuated the wave will grow in irradiance. The number of stimulated emissions depends on the energy density $\rho(\nu)$. The irradiance is the product of the energy density and the velocity, so that in free space

$$\rho(\nu) = \frac{I}{c}. \tag{15.24}$$

The change in irradiance dI of the wave in travelling a distance dz is now proportional to the difference between the number of stimulated emissions and absorptions:

$$dI = \left[N_2 B_{21} g(\nu) \frac{I}{c} - N_1 B_{12} g(\nu) \frac{I}{c} \right] h\nu \, dz. \tag{15.25}$$

Here we have introduced the normalized spectral function, or lineshape $g(\nu)$, for the transition, which describes the frequency spectrum of the spontaneously emitted radiation. A typical lineshape is shown in Fig. 15.6; this is a homogeneously broadened lineshape as described in Chapter 13. The normalization of this function is such that

$$\int_0^\infty g(\nu) d\nu = 1. \tag{15.26}$$

From Eq. (15.25), and the Einstein relations (15.18) and (15.19):

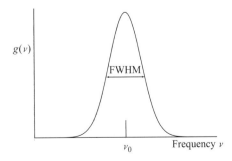

Fig. 15.6 A typical lineshape function $g(\nu)$.

$$\frac{dI}{dz} = \left(N_2 - \frac{g_2}{g_1}N_1\right)\frac{c^2 A_{21}}{8\pi\nu^2}g(\nu)I. \tag{15.27}$$

Integrating gives an exponential dependence on distance z:

$$I = I_0 \exp\{\gamma(\nu)z\}, \tag{15.28}$$

where I_0 is the irradiance at $z = 0$, and $\gamma(\nu)$ is the *gain coefficient*:

$$\gamma(\nu) = \left(N_2 - \frac{g_2}{g_1}N_1\right)\frac{c^2 A_{21}}{8\pi\nu^2}g(\nu). \tag{15.29}$$

If $N_2 > (g_2/g_1)N_1$, representing *population inversion*, then $\gamma(\nu) > 0$, and the irradiance grows exponentially with distance in the medium. The gain coefficient depends, as expected, on the transition probability A_{21}, and on the lineshape. Note, however, the frequency dependence (ν^{-2}); it is more difficult to make lasers for ultraviolet light than for infrared.

In comparing the suitability of different laser media it is convenient to specify a *stimulated emission cross-section parameter* σ_0, which is related to the gain coefficient $\gamma(\nu)$ by

$$\gamma(\nu) = \left(N_2 - \frac{g_2}{g_1}N_1\right)\sigma_0. \tag{15.30}$$

From Eq. (15.29):

$$\sigma_0 = \frac{c^2 A_{21}g(\nu)}{8\pi\nu^2}. \tag{15.31}$$

Since the lineshape $g(\nu)$ is normalized (Eq. (15.26)), the central height of the line $g(\nu_0)$ is inversely proportional to the linewidth[5] $\Delta\nu$:

$$g(\nu_0) = \frac{2}{\pi\Delta\nu}. \tag{15.32}$$

Then the cross-section parameter becomes

$$\sigma_0 = \frac{c^2 A_{21}}{4\pi^2\nu^2\Delta\nu}. \tag{15.33}$$

This shows that the stimulated emission cross-section for a homogeneously broadened transition is proportional to the ratio $A_{21}/\Delta\nu$ of the spontaneous transition rate to linewidth.

---

[5] The linewidth here is the full width at half-maximum, or FWHM, and is the result for homogeneous broadening, as described in Chapter 13. Note that the simple approximation $g(\nu_0) \sim 1/\Delta\nu$ applies to both homogeneous and inhomogeneous broadening.

15.5   POPULATION INVERSION

The population inversion condition $N_2 > (g_2/g_1)N_1$ derived in section 15.4 is a necessary condition for the gain coefficient to be positive. The two cases of thermal equilibrium and population inversion are shown in Fig. 15.7.

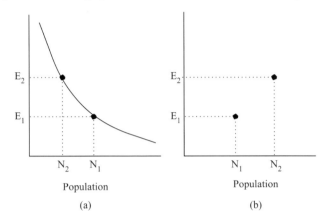

Fig. 15.7   Population inversion. The normal Boltzmann distribution (a) of population in two energy levels is shown inverted in (b).

To create population inversion energy is required to be put selectively into the laser medium such that the population of level 2 is increased over level 1 to form a non-equilibrium distribution. Excitation of the laser medium by pumping may be achieved in several ways. In gases at normal pressures the absorption lines have a narrow bandwidth and pumping is usually by electron collisions in an electrical discharge. Solid state crystals and glasses doped with an active ion have broader absorption lines than gases and are usually excited optically by the absorption of energy from a lamp or from another laser. In semiconductor lasers (Chapter 16) the relevant energy levels correspond to the conduction and valence bands, which are comparatively very broad. Here pumping is achieved by forward-biasing of a p–n junction in the semiconductor.

Lasers may conveniently be divided into three- and four-level systems depending on the number of levels active in their operation. This is illustrated in Fig. 15.8 which shows the inverted population at the laser transition.

15.6   THRESHOLD GAIN COEFFICIENT

Laser oscillation is initiated in a system with population inversion by the spontaneous emission of a photon along the axis of the laser. For the laser to sustain oscillation the gain in the laser medium must be greater than the losses in the cavity. The losses arise from transmission at the cavity mirrors (in order to provide the laser output, a typical transmission is 5% for continuous laser

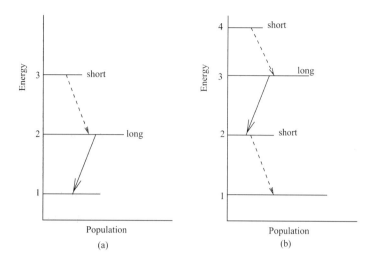

Fig. 15.8   Three- and four-level laser schemes. 'Long' and 'short' refer to the lifetimes of the levels.

operation). Other losses arise from absorption and scattering by the mirrors and in the laser medium, and diffraction out of the cavity. The threshold for laser oscillation will occur when the gain is equal to the losses. To calculate this threshold gain we combine all the sources of loss into one lumped loss coefficient k. At threshold the net round trip gain G will be unity.

Consider a cavity made up of mirrors M_1 and M_2 with reflectivities R_1 and R_2 and spaced by a distance L. A beam of irradiance I_0 starting at M_1 on reaching M_2 has become $I_1 = I_0 \exp\{(\gamma - k)L\}$, where γ and k are the gain and loss coefficients, respectively. On reflection from M_2 and travelling in return through the medium and undergoing reflection at M_1, the irradiance becomes $I_2 = I_0 R_1 R_2 \exp\{2(\gamma - k)L\}$. The round trip gain G is defined as I_2/I_0. Then,

$$G = I_2/I_0 = R_1 R_2 \exp\{2(\gamma - k)L\}. \tag{15.34}$$

The threshold condition for laser oscillation is $G = 1$, giving

$$R_1 R_2 \exp\{2(\gamma_{\text{th}} - k)L\} = 1, \tag{15.35}$$

where γ_{th} is the threshold gain coefficient, at which the laser will begin to oscillate.

From Eq. (15.35) we find

$$\gamma_{\text{th}} = k + \frac{1}{2L} \ln\left(\frac{1}{R_1 R_2}\right). \tag{15.36}$$

The first term is the loss within the cavity, and the second term is the loss due to the mirror transmission (or absorption), i.e. including that leading to the useful laser output.

Continuously operating lasers are called CW lasers, standing for continuous wave. In a CW laser the gain stabilizes at the threshold value, since if the gain were greater than or less than unity the irradiance would increase or decrease. The level at which the irradiance stabilizes depends on the pump power.

15.7   LASER RESONATORS

Most lasers require a long path through the active medium to obtain sufficient overall gain for oscillation. This is achieved by multiple reflections in an optical resonator, often referred to as a *resonant cavity*.

An optical resonator both increases laser action and defines the frequency at which it occurs. Optical feedback is provided by the optical resonator which retains photons inside the cavity, reflecting them back and forth through the laser medium. The simplest basic optical resonator is a pair of shaped mirrors at each end of the laser medium as in a Fabry–Perot interferometer. There are various configurations using plane and curved mirrors used in optical Fabry–Perot resonators; some of these are shown in Fig. 15.9. Not all configurations of mirror curvatures and spacings will give stable operation. Usually one of the mirrors is arranged to be as near as possible 100% reflecting at the laser wavelength; the other mirror (the output mirror) has a finite transmission, so that light will be transmitted out of the optical cavity to provide the laser output.

The optical Fabry–Perot resonator made up of two plane parallel mirrors is similar to the Fabry–Perot etalon or interferometer described in Chapter 9. The resonance condition for waves at normal incidence, along the axis of a cavity with optical length L, as for standing waves, is

$$m\frac{\lambda}{2} = L, \tag{15.37}$$

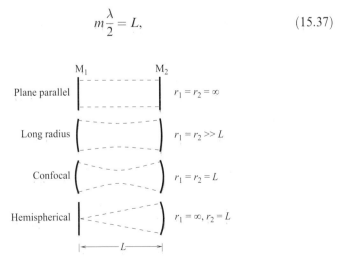

Fig. 15.9   Common laser resonator configurations. r_1 and r_2 are the radii of curvature of mirrors M_1 and M_2.

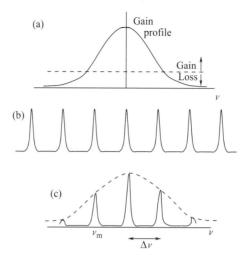

Fig. 15.10 Gain profile and resonant frequencies in a cavity laser: (a) gain profile of the laser transition; (b) allowed resonances of the Fabry–Perot cavity; (c) oscillating laser frequencies.

where m is an integer. Then the resonant frequency ν_m for each longitudinal mode of the cavity is

$$\nu_m = m\frac{c}{2L}.$$
(15.38)

This equation is important in defining the resonant frequencies at which the laser will oscillate, as they will if they fall within the gain profile of the laser transition, as illustrated in Fig. 15.10. The possible oscillating frequencies are termed the *longitudinal* modes of the laser; they are spaced by

$$\Delta\nu = \frac{c}{2L}.$$
(15.39)

Each of these frequencies may, however, be broken into a more narrowly spaced set; these are due to *transverse* modes, in which the field pattern may have different structures transverse to the beam direction. A transverse mode is an electric and magnetic field configuration at some position in the laser cavity which, on propagating one round trip in the cavity, returns to that position with the same pattern; some of these field patterns are shown in Fig. 15.11.

The laser output is at one or more frequencies from this set of modes. When only one longitudinal and transverse mode is selected, the bandwidth of the laser light is almost unbelievably small. For comparison, light from a single line of a low-pressure gas discharge lamp has a spectral width of about 1000 MHz. Non-pulsed laser light, in contrast, typically has a bandwidth of less than 1 MHz and may, with careful design, have a bandwidth of less than 10 Hz.

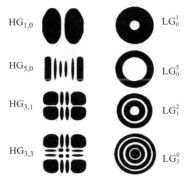

Fig. 15.11 Distribution of irradiance for various transverse modes: (a) Hermite–Gaussian (HG), where the subscript refers to the number of nodes in the x and y directions; (b) Laguerre–Gaussian (LG), where the subscript refers to the azimuthal phase and radial nodes, respectively.

As can be seen in Fig. 15.11, the transverse modes can have polar (or circular) symmetry or Cartesian (rectangular) symmetry; these are known respectively as Laguerre–Gaussian modes and Hermite–Gaussian modes. Although most lasers are constructed with circular symmetry, the modes with Cartesian symmetry are most common; this arises when some element in the laser cavity imposes a preferred direction on the transverse electric and magnetic field vectors. The lowest order mode is labelled TEM_{00} (not shown in Fig. 15.11). This is the fundamental mode with the largest scale pattern across the laser beam. The zero subscript indicates that there are no nodes in the x and y directions, transverse to the direction of the laser beam.

The cavity mirrors are, of course, required to reflect at the laser wavelength in order to make the cavity resonant. Typically one mirror has a reflectivity as close to 100% as possible and one is arranged to have a carefully selected transmission, chosen to produce the optimum laser output power; this necessarily means that the transmission must be less than the overall laser gain. For efficient operation the surface flatness of the mirrors is required to be within a small fraction of the laser wavelength (usually $\sim \lambda/20$).

15.8   BEAM IRRADIANCE AND DIVERGENCE

The beam of light leaving the laser is coherent both in relation to its narrow spectral linewidth and its spatial coherence over its emitted wavefront. As it leaves the laser, the beam will spread into a narrow angle by diffraction, the width depending on the field distribution across the beam. This can be viewed as the beam width acting as its own diffraction aperture. The simplest mode (the TEM_{00} mode), which has the narrowest beam, has a Gaussian radial

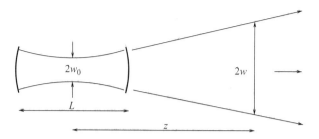

Fig. 15.12   The beamwidth w_0 at the waist and at a distance z from the waist.

dependence of irradiance $I(r)$ with peak irradiance at the centre, so that at radial distance r from the axis

$$I(r) = I_0 \exp\left(\frac{-2r^2}{w^2}\right). \tag{15.40}$$

The radial width parameter w is referred to as a *spot size*. (For $r = w$ the amplitude is $1/e$ of the amplitude on axis, but for convenience this is often referred to as the edge of the beam.) The spot size is smallest within the laser cavity, where there is a *beam waist*. Here the width w_0 (Fig. 15.12) is related to the length L of the resonator and the wavelength λ as[6]

$$w_0 = \left(\frac{\lambda L}{2\pi}\right)^{1/2}. \tag{15.41}$$

This applies for both the cavity with two plane mirrors and the symmetric confocal cavity. As we discuss below, the cavity mirrors must be significantly larger than this spot size to avoid diffraction loss.

The laser beam spreads by diffraction (Fig. 15.12) both inside and outside the resonator. Analysis of the Gaussian beam solutions of the paraxial wave equation leads to the width $w(z)$ of the beam at distance z:

$$w(z) = w_0\left[1 + \left(\frac{\lambda z}{\pi w_0^2}\right)^2\right]^{1/2}, \tag{15.42}$$

which approximates to

$$w(z) \simeq \frac{\lambda z}{\pi w_0} \quad \text{for} \quad z \gg \frac{\pi w_0^2}{\lambda}. \tag{15.43}$$

[6] See, for example, P. W. Miloni and J. H. Eberly, *Lasers*. Wiley, 1988.

Note that the larger the beam waist the smaller the angle of spread of the beam. As expected from Fraunhofer diffraction, the angular width is of order λ/w_0; for the TEM$_{00}$ mode, which has a Gaussian spatial profile, the half-angle θ of the divergence cone for the propagating beam is

$$\theta = \frac{\lambda}{\pi w_0}. \qquad (15.44)$$

For example, a helium–neon laser with $\lambda = 632.8$ nm operating with a symmetric confocal resonator of length $L = 30$ cm has

$$\text{minimum spot size } w_0 = \left(\frac{\lambda L}{2\pi}\right)^{1/2} = 0.17 \text{ mm}, \qquad (15.45)$$

$$\text{divergence angle } \theta \simeq \frac{\lambda}{\pi w_0} \simeq 1.2 \text{ mrad} = 0.66°. \qquad (15.46)$$

Note that the beamwidth w_0 is determined by the length and not the width of the laser. There is however a need for the resonator mirrors to be sufficiently wide so that the beam is not lost by diffraction at each reflection. Assuming initially that the beam fills mirror M$_1$, the diffraction half-angle at mirror M$_1$ is $\sim \lambda/d_1$, where d_1 is the diameter of the mirror M$_1$ and also of the beam at M$_1$. The angle subtended at M$_2$ by M$_1$ is d_2/L, where d_2 is the diameter of mirror M$_2$. For low loss $d_2/L > \lambda/d_1$, or

$$\frac{d_1 d_2}{\lambda L} \geq 1. \qquad (15.47)$$

This is known as the Fresnel condition. For a symmetrical arrangement where $d_1 = d_2 = d$ the condition is $d^2/\lambda L \geq 1$; the quantity $d^2/\lambda L$ is known as the *Fresnel number* of the optical arrangement. (Note the close relationship to the Rayleigh criterion (Chapter 8), which defines the boundary between Fraunhofer and Fresnel diffraction.)

15.9 EXAMPLES OF IMPORTANT LASER SYSTEMS

Gas lasers

Gas lasers may be divided into atomic, ionic and molecular, depending on the active amplifying species in the gas. Generally gas lasers are excited by an electrical discharge in which excitation of the gas atoms or molecules is by collision with energetic electrons. Optical excitation of a gas is usually inappropriate since the absorption lines of gases are very narrow (in contrast to solids).

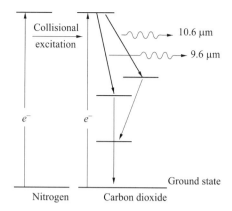

Fig. 15.13   Vibrational energy levels in the CO_2 laser.

The helium–neon laser described briefly in section 16.2 was the first gas laser to be operated (in 1960), and was the first continuously operating laser. It is still one of the most common lasers, operating on the 632.8 nm wavelength, and is used in many applications requiring a relatively low power, visible, continuous and stable beam.

The CO_2 gas laser provides large power outputs at the infrared wavelength of 10.6 μm. The laser action involves four vibrational energy levels, as in the scheme of Fig. 15.13. The broad highest level is closely equal to an excited level in nitrogen, which is an essential added gas component. The upper level of the CO_2 molecule is populated from this state by collisions with nitrogen molecules. The excitation of the nitrogen molecules is by electron collisions, and the electrons are produced in an electric discharge within the laser tube. The gas also contains helium, which assists the depletion of the lower levels by collisional de-excitation and stabilizes the plasma temperature. Large continuous power outputs, up to some tens of kilowatts, are obtainable; pulsed operation can give Joule pulse energies in microsecond pulses. As the gas densities are usually comparatively low, high-powered CO_2 lasers must be relatively large to contain a sufficient number of molecules.

Solid state lasers

A solid state laser, such as the ruby laser, may be in the simple form of a transparent rod with mirrors formed directly on the ends. The gain medium contains paramagnetic ions in a host crystalline solid or glass. The active ions may be substituted into the crystal lattice or may be doped as an impurity into the glass host. There are many combinations of dopant ion and host materials which provide a wide range of laser wavelengths. The doped solids exhibit broad absorption bands which makes them amenable to optical excitation from continuous or pulsed lamps or from semiconductor diode lasers.

A laser in which the active ion is distributed throughout a long silicon fibre is shown in Plate 15.1.

The dopant ion should fit readily into the crystal host by matching the size and valency of the element for which it is substituting. The optical quality of the doped medium is needed to be high so that there is low loss for the amplifying beam. Refractive index variations, scattering centres and absorption can contribute to loss processes. Suitable host media are garnets (complex oxides), sapphire (Al_2O_3), aluminates and fluorides. The glass hosts are able to be fabricated easily, in large sizes and with high optical quality. The energy levels are broader in glasses than in crystals making them more suitable for pumping by flashlamps. The lower thermal conductivity of glasses compared with crystals renders them susceptible to thermal distortion and induced birefringence. The increased linewidth leads to a reduced stimulated emission cross-section (section 15.4) such that the pumping threshold is higher. The crystal host is usual for CW operation and the glass host for pulsed operation, particularly at high pulse energy.

The paramagnetic dopant ions are usually from the transition metals and lanthanide rare earths. The Nd:YAG laser operating at 1064 nm is one of the most used solid state lasers; here neodymium ions Nd^{3+} provide the laser action, and yttrium aluminium garnet (YAG) is the usual crystal host. The crystal has a relatively high thermal conductivity which enables it to distribute heat efficiently following optical pumping. The laser can operate either pulsed or continuously.

Semiconductor lasers, which are dealt with in the following chapter, are derived from light-emitting diodes. They are distinct from the solid state lasers such as the ruby laser in their pumping and photon generating processes, deriving their energy from the electrical excitation of electrons within the semiconductor.

15.10   FURTHER READING

C. C. Davis, *Lasers and Electro-Optics*. Cambridge University Press, 1996.

P. W. Miloni and J. H. Eberly, *Lasers*. Wiley, 1988.

A. Miller and D. M. Finlayson, *Laser Sources and Applications*. Inst. of Physics Publishing, 1996.

A. E. Siegman, *Lasers*. Oxford University Press, 1986.

O. Svelto, *Principles of Lasers*. Plenum Press, 1998.

A. Yariv, *Optical Electronics in Modern Communications*, 5th edn. Oxford University Press, 1997.

NUMERICAL EXAMPLES 15

15.1 Calculate the rate at which stimulated emission is occurring in (i) a 1-W laser at wavelength 600 nm; (ii) a 1-mW maser at a frequency of 3000 MHz.

15.2 Calculate the total energy available in a single pulse and the peak power from (i) a λ 694 nm solid state laser in the form of a rod 10 mm diameter and 0.1 m long containing 3×10^{19} active ions cm^{-3} and operating at an efficiency of 1% and a pulse duration of 1 ms; (ii) a λ 10.6 μm CO_2 gas laser, 30 mm diameter and 2 m long containing gas with 6×10^{19} molecules cm^{-3}, operating at an efficiency of 10% with a pulse duration of 10 μs.

15.3 Calculate the longitudinal mode separation in the cavity of a $\lambda = 633$ nm helium–neon laser with mirrors separated by 0.3 m. How many of these modes could oscillate if the width of the gain curve is 2×10^9 Hz?

PROBLEMS 15

15.1 For a system in thermal equilibrium calculate the temperature at which the rates of spontaneous and stimulated emission are equal for a wavelength of 10 μm.

15.2 Show for a blackbody that the energy density per unit frequency interval ρ is related to the brightness B as

$$\rho = \frac{4\pi B}{c}.$$

15.3 The density of modes (per unit volume per unit frequency interval) for blackbody radiation is

$$\rho(\nu) = \frac{8\pi\nu^2}{c^3}.$$

Show that the rate of spontaneous transitions is equal to the rate of stimulated transitions if there is just one photon excited in each cavity mode.

15.4 Compare the Doppler broadened linewidth of the helium–neon laser with that of the argon ion laser given the following data:

	He–Ne	Ar$^+$
Atomic mass	20(Ne)	40
Wavelength(nm)	633	488
Gas temperature (K)	400	5000

15.5 In the helium–neon laser operating at 633 nm the Einstein A coefficient of the upper laser state is 3×10^6s^{-1}. The upper state has a degeneracy of 3 and a population of 10^{16}m^{-3} and the lower state a degeneracy of 5 and a population of 10^{15}m^{-3}. The mirror reflectivities are 1.0 and 0.95, the losses are 3% per round trip, and the gas temperature is 400 K (as in Problem 15.4). Calculate the gain coefficient required for the gain medium to achieve laser operation.

15.6 A helium–neon laser operating at 633 nm on the TEM$_{00}$ mode has an output power of 1 mW and a minimum spot size of 0.3 mm. Find

(i)   the beam divergence angle;
(ii) the laser brightness;
(iii) the temperature of a blackbody with the same brightness. (For a blackbody with energy density ρ and brightness B, $\rho = 4\pi B/c$).

15.7 Calculate (i) the threshold gain coefficient and (ii) the population inversion for a ruby laser operating at 694.3 nm. The spontaneous lifetime of the upper laser level is 3 ms, the linewidth of the transition is 150 GHz and the ruby crystal refractive index is 1.78. The laser cavity has reflectivities 1.0 and 0.96; the length is 5 cm and the losses are negligible.

15.8 Calculate the fraction of the beam power and the average photon flux within the beam waist for a 1-W argon ion laser operating at 515 nm.

15.9 A carbon dioxide laser operating on the 10.6 μm transition has a gas pressure of 1 atm and gas temperature of 400 K.
   (i) By estimating the contribution to the linewidth from Doppler and pressure broadening determine if the transition is broadened by homogeneous or inhomogeneous effects.
   (ii) At what pressure are the two contributions equal?

15.10 Determine the number of longitudinal modes in an argon ion laser of length 80 cm if the laser wavelength is 515 nm, the gas temperature is 5000 K and the loss coefficient is one-third the peak gain value.
   Calculate the maximum length of the laser cavity for only one longitudinal mode to oscillate.

15.11 Explain why the frequency spacing of modes for a laser in the form of a ring is twice that for a standing wave cavity of the same length.

16

Semiconductors and semiconductor lasers

How bright these glorious spirits shine!

Isaac Watts 1674–1748.

Semiconductors play a vital role in optics both as sources and as detectors of light. The light-emitting diode (LED) and laser diode are widely used, as are the various forms of photodiode detector.

The semiconductor laser in its many forms is the most numerous of all lasers. It has widespread applications, for example in optical fibre communication systems, barcode scanners, laser printers and the compact disc player. Semiconductor lasers, like the gas and solid-state lasers considered in the previous chapter, depend on stimulated emission to produce coherent light in a resonator. They have, however, different pumping and photon generating processes, which we now consider.

Semiconductor lasers have many valuable properties. They have high efficiencies (defined as laser power output/electrical power input) of typically 30–50%, which is higher than most other lasers. They are very small, typically with dimensions of less than 1 mm, and require only modest power supplies, operating typically at a few volts and currents of 10 mA to a few amps. Semiconductor lasers use direct electrical pumping; modulation at frequencies typically up to 20 GHz makes them very suitable for optical communications.

In this chapter we review the basic physics and radiative mechanisms of semiconductors and LEDs. We describe the structures and operation of practical forms of semiconductor lasers, including heterostructures and quantum well diodes.

16.1 SEMICONDUCTORS

An atom, of atomic number Z, consists of a positively charged nucleus, $+Ze$, surrounded by Z electrons of charge $-e$. The electron energies are quantized, and the electrons occupy the Z lowest energy states of the atom. When a large number N of such atoms are brought together to form a solid, the interactions between the atoms spread the energy levels into bands, each containing $2N$ energy states (the factor 2 results from the two-fold degeneracy of the atomic levels due to electron spin). Thermal excitation can raise electrons into higher energy states; however, if Z is even, as in silicon, the topmost occupied band of states is full at low temperatures, and the electron can only be excited into an empty energy state by surmounting a gap between the full and empty bands. The last fully occupied band is the *valence band* and the first empty band is the *conduction* band. The bands are separated by the *band gap* E_g; for silicon the band gap is 1.1 eV.

Silicon is called a semiconductor because of the relative ease of exciting electrons from the valence band to the conduction band. For a much larger band gap, the solid would be an insulator. If the band gap does not exist (i.e. the lowest energy of the upper band is less than the highest energy of the lower band), or if Z is odd giving a half-filled upper band, then the solid is a conductor, i.e. a metal. The schematic energy bands and their occupancy are shown in Fig. 16.1 for a metal, a semiconductor, and an insulator.

At absolute zero temperature there are no electrons in the conduction band of the semiconductor and the material is a perfect insulator. Thermal excitation raises the energies of a small number of electrons into the conduction band. This leaves a corresponding number of unoccupied energy states in the valence band; both the electrons and the vacancies are important in the behaviour of a semiconductor.

When the energy of an electron takes it into the conduction band, the unoccupied state, or *hole*, in the valence band allows some movement among the remaining electrons. This movement of the valence electrons is best understood by regarding the hole as a positively charged particle which has its own mobility and mass and which can contribute to the conductivity of the semiconductor. If an external electric field is applied, then the electron and the hole move in opposite directions, the electron moving faster than the hole. The thermally excited electrons in the conduction band and the holes in the valence

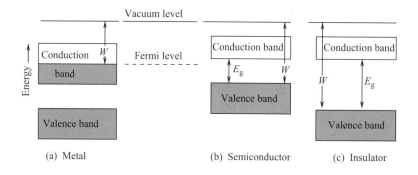

Fig. 16.1   Electron energy levels in (a) a metal; (b) a semiconductor; and (c) an insulator.

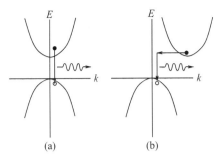

Fig. 16.2 Electron energy – wavevector diagram for (a) direct bandgap and (b) indirect bandgap semiconductors.

band are *carriers* and provide conduction in the semiconductor. Electrons may also be excited into the conduction band by absorbing the energy of a photon; this is the basis of a *photoconductor*, in which photons with sufficient energy to excite electrons directly from the valence band into the conduction band are detected by an increase in conductivity.

In the band model of the semiconductor, electrons of energy E are assumed to move through the periodic crystal lattice with wavevector k. The allowed energies of the electrons are related to the wavevector as $E = \hbar^2 k^2 / 2m^*$, where m^* is the effective mass of the electron. This parabolic E–k relation has important consequences for electron transitions in the semiconductor. Semiconductors may be divided into *direct bandgap* or *indirect bandgap* types depending on the relationship of E to k. For the direct bandgap case, shown in Fig. 16.2(a), the maximum of energy in the valence band and the energy minimum of the conduction band occur at the same value of wavevector. In the indirect bandgap case (Fig. 16.2(b)) the valence band energy maximum and conduction band minimum occur at different values of wavevector.

In a direct gap semiconductor, for example GaAs, an electron in the conduction band can make a transition to an empty level in the valence band and emit a photon of energy approximately equal to the bandgap energy E_g; no change of wavevector is involved. For the indirect bandgap case, for example Si or Ge, the transition of the electron from the conduction band to the valence band must involve a change in wavevector, which contravenes the selection rule for an allowed electron transition. The transition can occur only if there is also an interaction with a quantized unit of mechanical oscillation, or *phonon*,[1] within the crystal. For this reason only the direct bandgap materials are efficient light emitters.

The energy band structure of the semiconductor may be modified by introducing impurity atoms into the crystal lattice; this is known as *doping*.

---

[1] Phonons are discussed in detail in *Solid State Physics* by Hook and Hall. Their important property at issue here is that they have a larger wavevector for a small energy compared with photons; hence a transition for a phonon in Fig. 16.2 is almost a horizontal line.

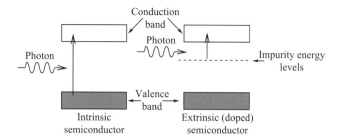

Fig. 16.3   Electron energy levels in a doped semiconductor.

The crystal lattice may be able to accept a replacement of 1% or more of its atoms, but doping usually extends only to the order of one in 10^3; a lightly doped semiconductor may have only one impurity atom in 10^6 of the host atoms. Silicon atoms are in group IV of the Periodic Table and so have four outer electrons. A donor impurity from group V with five outer electrons has one more electron than required for covalent bonding with neighbouring silicon atoms. This additional electron is much more easily lost to the conduction band, with an excitation energy of only 0.1 eV. Such an impurity is a *donor*. On the other hand an *acceptor* impurity, such as boron (group III) has three outer electrons and so contributes a hole to the valence band by allowing an electron from the valence band to be localized at the boron atom. A silicon semiconductor with a group V donor is known as n-type, and with a group III donor as a p-type.

The new dopant induced impurity energy levels are full, and are situated within the bandgap (Fig. 16.3). Semiconductors with added donor or acceptor dopants are known as *extrinsic*, in contrast to those which contain no dopants, which are known as *intrinsic*.

16.2   SEMICONDUCTOR DIODES

A semiconductor diode is a junction between the two types of doped semiconductor, n-type in which the dopant produces extra electrons, and p-type in which there are extra holes. When the p–n junction is formed, electrons and holes diffuse across the junction forming a contact region which is depleted of charge carriers, known as the *depletion layer*. The depletion layer has a high resistance, and there is a large electric field across it due to the dipole layer of positively charged donors in the n region and negatively charged acceptors in the p region. In photodiode detectors (Chapter 21) photons are absorbed and generate electrons and holes within the depletion layer of the junction; the intrinsic electric field between the n- and p-type regions then transports these free charge carriers to give a current in an external circuit.

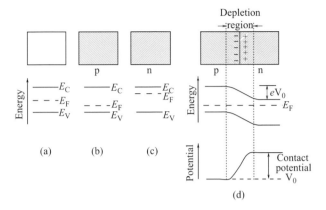

Fig. 16.4 The development of a contact potential across a p–n junction, showing energy levels: (a) an intrinsic semiconductor has no impurities and the Fermi level is halfway between the bottom of the conduction band E_c and the top of the valence band E_v; (b) p-type material has added acceptor impurities (energies – not shown – just above the valence band) that accept electrons from the valence band, becoming negatively charged and leaving positively charged holes in the valence band; (c) n-type material has added donor impurities (energies – not shown – just below the conduction band) that give electrons to the conduction band, becoming positively charged; (d) within the depletion layer of the p–n junction the negative charge of the acceptors in the p-region and the positive charge of the donors in the n-region create a dipole layer.

The development of the contact potential across the junction may be followed in the diagrams of Fig. 16.4. The valence and conduction bands of the pure semiconductor are shown at (a), with the Fermi energy level between.[2] Thermal excitation is sufficient for the energies of a small number of electrons to reach the conducting band, and the Fermi level is midway between the valence and conduction bands. The effect of doping is seen at (b), for p-type, and at (c), for the n-type; the Fermi levels are lowered and raised as shown. When the two types of semiconductor are joined, electrons and holes can flow across the junction until the Fermi levels are equalized. The energy levels adjust to give the same Fermi level, and the contact potential V_0 is developed.

This potential difference develops in the depletion layer at the junction. The n-type becomes positively charged, and the p-type negatively charged. An externally applied potential making the p-type more positive is a *forward bias*, which increases the flow of electrons from the n region and holes from the p region; the diode then has a low resistance. With *reverse bias* there is only a small reverse current, and the resistance is high. Figure 16.5 shows the voltage–current characteristic of a typical semiconductor diode.

---

[2] The Fermi level is that energy value for which the probability of a state being occupied is 1/2.

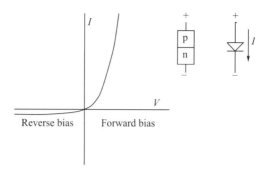

Fig. 16.5   The voltage–current characteristic of a typical semiconductor diode.

16.3   LIGHT-EMITTING DIODES AND SEMICONDUCTOR LASERS

The simplest light-emitting semiconductor diode is a p–n diode in a material such as gallium arsenide (GaAs), illustrated in Fig. 16.6. The active region of the diode is at the junction between layers of p- and n-doped GaAs. The n-doped side of the junction contains mobile electrons in the conduction band, and the p-doped side contains mobile holes in the valence band. Typical dimensions are sub-millimetre, as shown in the diagram. In a laser diode the parallel end faces of the crystal are cleaved and polished to form a laser resonant cavity. The surfaces are often not given reflective coatings, since the reflection coefficients can be large enough due to the high refractive index (for GaAs, $n = 3.6$, giving a reflectivity of 32%). A simplified diagram of the energy band structure is shown in Fig. 16.7. When a current passes in the forward direction, electrons from the n-doped side of the junction are injected at high density into the p-region of the junction, and holes from the p region into the n region. The electrons and holes recombine to emit photons; this mechanism of *radiative recombination* is the basis of the light-emitting diode (LED) and the semiconductor laser.

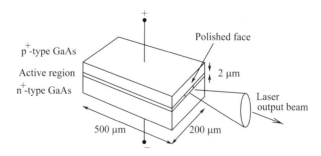

Fig. 16.6   Schematic illustration of a semiconductor homojunction diode laser.

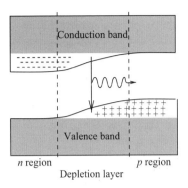

Fig. 16.7 Radiative recombination. An electron undergoes a transition from the conduction band to the valence band, providing a photon.

For the semiconductor diodes used in LEDs and lasers the p and n regions are heavily doped ($\sim 0.1\%$) to give a large population inversion; the n-type material is more heavily doped than the p-type material and may be denoted by n^+. When the junction is formed, the movement of electrons and holes causes the n region to be depleted of majority electron carriers and the p region to be depleted of majority hole carriers. The contact potential V_0 creates a barrier to further electrons moving from the n to p region or to further holes moving from the p to n region.

When the junction is forward biased (by giving the p region a positive potential V with respect to the n region), carriers are injected, the band energies are modified and the junction potential is reduced to $(V_0 - V)$. Filled electron states in the conduction band have energies above those of hole (empty electron) states in the valence band as shown in Fig. 16.7. This is equivalent to a *population inversion* and is the basis of gain and laser action in the diode laser.

Current flows in the p–n junction by injection of minority carriers – electrons into the p region and holes into the n region. To maintain electrical neutrality in the n and p regions the concentrations of mobile electrons in the n region and holes in the p region rise to balance the injected excess carriers. The injected excess carriers are removed by recombination of electrons and holes; an electron with energy in the conduction band falls into an empty electron state of lower energy in the valence band. This either produces an emitted photon or the energy is lost non-radiatively.

The light output power from a semiconductor laser as the diode current increases as shown in Fig. 16.8. Up to a threshold current light is emitted by spontaneous emission, and the diode acts as an LED. Above the threshold current laser action starts, and stimulated emission begins to dominate over spontaneous emission. Beyond the threshold the efficiency of conversion of electrical energy into light increases rapidly. A fully efficient laser would produce one photon for each injected electron. The formal definition of efficiency

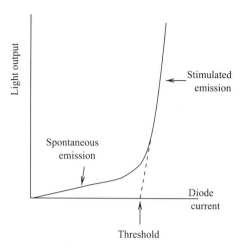

Fig. 16.8   Light output from a semiconductor laser diode.

η is as follows: for a laser with drive current I and a threshold current I_{th}, the output power of the laser at wavelength λ is

$$P = \eta \frac{hc}{e\lambda}(I - I_{th}), \qquad (16.1)$$

where η is < 1 and accounts for cavity losses.

The beam from a semiconductor laser such as that in Fig. 16.6 is emitted in the plane of the junction; here the path through the lasing material is greatest. The laser cavity is formed by polishing the ends of the diode. The beam is usually elliptical in cross-section, since the emitting area is rectangular. As expected from diffraction theory, small dimensions produce large divergence angles.

The emission photon energy of the semiconductor laser is close to the bandgap energy E_g. In GaAs $E_g = 1.42$ eV and the wavelength of the GaAs diode laser is about 904 nm; the typical linewidth is 20 nm, due to the energy distribution of electrons and holes in the conduction and valence bands, respectively. The addition of Al to the active layer forms GaAlAs which is also a direct bandgap material, and increases the bandgap. GaAlAs lasers are able to generate wavelengths from 750 to 850 nm; the most usual wavelength of 780 nm is used in the CD player, laser printers and other common applications. More efficient diode lasers employ layers of different materials at the junction: they are known as *heterojunction* lasers and have largely replaced the *homojunction* lasers we have described so far. The heterojunction is formed between two different semiconductors with different bandgap energies; typical materials are GaAs and AlGaAs. One form of double heterojunction is shown in Fig. 16.9; here the active region is a thin layer of GaAs which is sandwiched by the p and n regions of AlGaAs. In the heterostructure the crystal lattice periodicities should be closely matched to avoid interface dislocations; this is

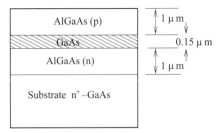

Fig. 16.9 Diagram of a double heterostructure laser in which the active GaAs layer (shown hatched) is sandwiched between the *p* and *n* regions of AlGaAs.

achieved in GaAs/AlGaAs where the lattice periods are respectively 564 pm and 566 pm.

The heterojunction has three significant advantages over the homojunction. First, it traps the charge carriers in a region where recombination is more probable. Second, an active region of higher refractive index can be formed which acts as a light guide, concentrating the light and increasing the efficiency of stimulated emission. Third, the laser emission is only weakly absorbed in the adjacent regions so that the losses are minimized. These three effects result in a much reduced threshold current density compared with the homojunction, allowing continuous wave operation at room temperature.

An important extension of the double heterostructure is made by reducing the thickness of the central active region so that it becomes comparable with the electron or hole de Broglie wavelength given by $\lambda = h/p$, where p is the electron or hole momentum. Since in the double heterostructure electrons and holes are confined to the central region, where the bandgap E_g is smaller than that for the cladding region, the electrons and holes are confined within a potential well. The energy levels for electrons and holes in the potential well are quantized to values dependent on the well dimensions. These structures are referred to as *quantum wells*. The quantum well lasers based on these types of structures have increased gain and reduced current threshold and have become the predominant structure for semiconductor lasers.

A semiconductor quantum well laser operating on a fundamentally different principle is the *quantum cascade laser*. It works on electron transitions between discrete levels in the conduction band which arise from the quantization of electron motion perpendicular to the plane of the active layer. This mechanism involves only the electrons in the conduction band and is to be contrasted with the normal semiconductor laser based on electron–hole recombination. The quantum cascade lasers produce radiation over a range of long infrared wavelengths.

16.4   SEMICONDUCTOR LASER CAVITIES

In the semiconductor laser the optical resonator is often made by directly using the cleaved ends of the crystal as mirrors. A more efficient alternative is

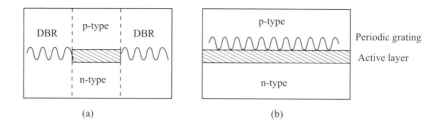

(a)   (b)

Fig. 16.10 Distributed Bragg reflectors in resonant laser cavities: (a) as mirrors at either end of a cavity; (b) distributed throughout the length of the cavity, in the DFB laser.

to use reflection from a periodic variation of refractive index within a layer coated on the ends of the active region, forming a multilayer mirror known as a *Bragg reflector*. Reflection at this periodic structure is wavelength sensitive; components reflected at each step are in phase if the periodic spacing Λ satisfies the Bragg condition $2n_{\mathrm{eff}}\Lambda = m\lambda$, where n_{eff} is the effective refractive index and m is an integer. A suitable choice of Λ selects a single mode of oscillation for the laser. An example is shown in Fig. 16.10(a), where a distributed Bragg reflector is positioned at each end of the active region. A similar effect is obtained if the periodic variation in refractive index extends throughout the active gain region, when the resonant modes of the cavity are restricted to wavelengths satisfying the Bragg condition; this is called a *distributed feedback (DFB) laser*. The selection of a wavelength that is available in the DFB lasers is important in the use of multiple wavelengths in optical communications.

Semiconductor lasers that generate laser light along the plane of the diode junction are called *edge emitting lasers*. An important alternative type is the *vertical cavity surface emitting laser* (VCSEL), shown in Fig. 16.11. In this structure, since the length of the active region is very small, high reflection coefficients are required to form the laser cavity. This is achieved by cladding with Bragg reflectors on both sides. VCSEL lasers operate with a high efficiency and a low current threshold. They can be packed into two-dimensional arrays in which the individual lasers can be individually addressed.

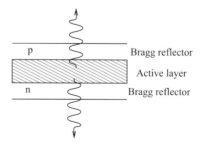

Fig. 16.11   Diagram of a vertical cavity surface emitting laser (VCSEL).

16.5   WAVELENGTHS AND TUNING OF SEMICONDUCTOR LASERS

The direct gap semiconductor GaAs is typical of the large group known as III–V lasers based on a combination of elements from the third group of the Periodic Table (for example Al, Ga and In) and from the fifth group (N, P, As and Sb). The ternary alloys AlGaAs and InGaAs and quaternary alloys such as InGaAsP also fall in this group. These produce emission wavelengths over the range 610–1600 nm which covers the range of optical communications.

As we remarked in Chapter 15, laser action becomes more difficult to induce at shorter wavelengths: covering the whole visible spectrum requires special attention to lasers producing blue light. Semiconductor lasers based on nitride compounds, for example InGaN, provide continuous emission in blue light, at 410 nm. Lasers based on the combination of elements from the second group (Cd, Zn) with those from the sixth group (S, Se), called II–VI lasers, give wavelengths in the blue-green region due to the larger bandgap compared with the III–V compounds. Wavelengths in the mid infrared (4–30 μm) are produced by the Pb salts of IV–VI compounds (S, Se and Te); however, these sources must be operated at low temperatures, and it may be preferable to use the quantum cascade laser described above.

The wavelength of oscillation of a semiconductor laser must lie within the comparatively broad band determined by the electronic band structure, within which there can be a number of cavity resonances (Fig. 16.12). The resonances are often well separated, since there is only a small spacing between the polished faces of the crystal forming the Fabry–Perot resonator. The separation between resonances is the *free spectral range* (FSR), which we have already encountered in the performance of spectrometers (Chapter 12). It is possible to tune the cavity resonance by mechanical pressure to change its length, by heating to change the refractive index, or by varying the junction current; the available range of laser action of a single mode is however limited to the free spectral range, since the laser oscillation will otherwise jump to an adjacent mode.

The free spectral range depends on the dispersion due to the refractive index of the semiconductor crystal as well as its length. For the Nth longitudinal mode in a resonator with length d and refractive index n:

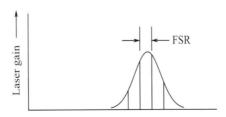

Fig. 16.12   The tuning range of a semiconductor laser. The free spectral range limits the range achievable by tuning the cavity resonances.

$$N = \frac{2nd}{\lambda}. \tag{16.2}$$

The spacing $\delta\lambda$ between resonant modes, where $\delta N = -1$, is found by differentiation:

$$\delta N = -1 = \frac{2d}{\lambda}\frac{dn}{d\lambda}\delta\lambda - \frac{2nd}{\lambda^2}\delta\lambda, \tag{16.3}$$

$$\delta\lambda = \frac{\lambda^2}{2nd}\left(1 - \frac{\lambda}{n}\frac{dn}{d\lambda}\right)^{-1}, \tag{16.4}$$

or in terms of frequency the tunable bandwidth is

$$\delta\nu = \frac{c}{2nd}\left(1 + \frac{\nu}{n}\frac{dn}{d\nu}\right)^{-1}. \tag{16.5}$$

For example, an infrared diode at wavelength $1\,\mu m$ ($\nu = 30\,000$ GHz) may have a resonator with $d = 0.5\,mm$, $n = 2.5$ and dispersion $(\nu/n)dn/d\nu = 1.5$. The free spectral range is then 24 GHz. Note the comparatively large contribution of the dispersion in the refractive index.

16.6   ELECTROLUMINESCENCE IN ORGANIC SEMICONDUCTORS

Organic materials are usually encountered as electrical insulators, and not as sources of light. Fireflies show us that there are other possibilities: they use the chemical excitation of organic molecules to produce light in a process with very high overall efficiency. *Electroluminescence*, which is the emission of light following electrical excitation, is achieved in certain solid organic materials by applying a high electric field to a thin layer of organic polymer. Electrons and holes are injected as in semiconductor diodes; recombination then leads to photon emission with the photon energy determined by the energy of a bandgap.

Electroluminescent organic materials offer considerable practical advantages in the production of matrix displays, which may be used, for example, in computer displays, instrument panels and motorway indicators. A matrix display might require many millions of individually controlled light-emitting elements, often using three colours. The development of such complex devices is a major task; we are concerned here only with the basic physical processes of a single element of such an array.

An organic polymer which is sufficiently conducting for electrical excitation may still require a field of order 10^7 V m^{-1} for efficient operation. Thin films only $100\,\mu m$ thick are therefore used, requiring only 2–3 V between electrodes,

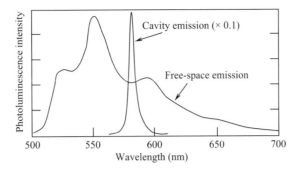

Fig. 16.13   Photoluminescence spectra of a thin film of PPV: (a) free-space emission; (b) within a resonant cavity, showing laser action. (Friend *et al.*, *Nature*, **397**, 121, 1999.)

one of which is a thin transparent conductor such as an alloy of magnesium and silver. The conducting polymer is usually a so-called π-conjugated polymer, in which there is a long chain of alternate double and single carbon–carbon bonds. The conductivity depends on electrons which are 'delocalized' along this backbone. Electrons emitted by one electrode, and holes emitted by the other recombine forming an excited complex known as an 'exciton': if this does not decay and lose its energy by interaction with adjacent molecules it will emit a photon at a wavelength determined by its molecular structure.[3] Efficient electroluminescence was demonstrated in 1990 by Richard Friend in Cambridge University, using the π-conjugated polymer poly(p-phenylene vinylene), known as PPV. Based on this material, a luminous efficiency of over 20 lumens per watt was achieved a few years later.

Figure 16.13 shows the luminescence spectrum of PPV when optically excited. Optical excitation is achieved without metal electrodes, so that the natural wide bandwidth of luminescence is observed. If the thin film is contained within a resonant cavity, which may be formed by electrodes or by tuned dielectric mirrors, then laser action may produce a sharp spectral line as shown.

16.7   FURTHER READING

P. K. Chen, *Fibre Optics and Optoelectronics*, 2nd edn. Prentice Hall, 1990.
J. R. Hook and H. E. Hall, *Solid State Physics*, 2nd edn. Wiley, 1991.

---

[3] The resonance of conjugated polymers at optical wavelengths is significant in their occurrence as opsins in vision and as chlorophyll in photosynthesis – see Chapter 22.

17

Laser light

How far that little candle throws his beams! / So shines a good deed in a naughty world.
Shakespeare, *Merchant of Venice*.

Stimulated emission, which is at the heart of laser action, produces a multiplicity of photons, identical in amplitude, phase and direction. This *coherence* in laser light contrasts sharply with the chaotic nature of light from spontaneous emission, and gives laser light its extraordinary properties of narrow spectral linewidth (i.e. temporal coherence) and spatial coherence. Some of these properties are familiar; a laser beam can be pencil-sharp over a large distance, and the *speckles* in a spot of laser light on a surface distinguish it at a glance from incoherent light. Coherence allows laser light to be focused to a spot only a few wavelengths across, with an intensity that is useful in heating and cutting many different materials. The highest power flux is obtained in lasers operating with short pulses, and techniques exist for producing pulses only a single cycle or a few femtoseconds (10^{-15}s) long. High intensity also means high electric fields; dielectrics may behave non-linearly at such high fields, producing effects such as the generation of shorter wavelength laser light at a harmonic of the laser frequency.

In this chapter we examine the temporal and spatial coherence properties of laser light, including its directionality and brightness, and the theoretical and attainable limits on linewidth, focusing, and pulse width. We also consider the effects of the extremely high electric fields in short laser pulses, including the nonlinear behaviour of dielectrics and the generation of harmonics.

17.1 LASER LINEWIDTH

In a laser cavity the frequency of the oscillation is determined by a resonance in the cavity rather than by a natural resonance in an atom or ion. The process of stimulated emission usually leads to laser light with a considerably narrower linewidth than that of the spontaneously emitted radiation, and the laser beam consequently has a very pure colour. The beam will, however, contain a small proportion of spontaneously emitted photons, which add to the beam with random phase. It is this addition of incoherent photons that ultimately limits the coherence of the laser light. We start with a simple situation when only the ratio between stimulated and spontaneous emission limits the coherence, which occurs when there is a large population inversion. The theoretically attainable limit of the bandwidth then depends only on the laser power and the width of the cavity resonance.

Consider a laser oscillating in a single mode, populated by $\bar{n}$ photons. Nearly all these photons have been produced by stimulated emission, and are completely coherent. There is also a population of photons generated by spontaneous emission from the excited atoms in the laser, and a small number of these must be in the same mode as the coherent population. The population of these incoherent photons relative to the coherent photons limits the coherence of the laser beam. We therefore need the relative rate of stimulated and spontaneous emission in a single mode. This ratio is simply the mean population $\bar{n}$ in the mode;[1] recalling that A_{21} and B_{21} represent incoherent and coherent photons, respectively, it follows that there is on average only one incoherent photon and $\bar{n} - 1$ coherent photons (or $\bar{n}$ to a very good approximation). The single incoherent photon contributes on average $1/\bar{n}$ of the power, and thus $1/\sqrt{\bar{n}}$ of the electric field. The effect of the vector addition of this incoherent field component is shown in Fig. 17.1; successive phasors of the incoherent photons follow a random walk. The photons in the population last for a certain time in the cavity before they are either emitted from the cavity or absorbed; this time τ_{cav} is the *decay time* for the cavity. The random walk of the incoherent photons builds up after many such changes. A single step of this random walk can modify the phase of the main field by the ratio of amplitudes $\Delta\phi \simeq \bar{n}^{-1/2}$ radians. The phase change builds up randomly, and on average reaches 1 radian after $(\Delta\phi)^{-2}$ changes, i.e. after a time $\tau_{\mathrm{cav}}\bar{n}$. This *phase diffusion time* determines the frequency width $\Delta\nu_{\mathrm{L}}$, from the bandwidth theorem (section 4.12):

----

[1] The ratio of the transition rates is

$$\frac{B_{21}\rho(T,\nu)}{A_{21}} = \frac{\rho(T,\nu)c^3}{8\pi h\nu^3}, \tag{17.1}$$

where $\rho(T,\nu)$ is the energy density and A_{21} and B_{21} are the Einstein coefficients (see section 15.3). The classical density of modes g(ν) in a blackbody cavity is

$$\mathrm{g}(\nu) = \frac{8\pi\nu^2}{c^3} \tag{17.2}$$

(see, for example, F. Mandl, *Statistical Physics*, 2nd edn. Wiley, 1988). The required ratio is

$$\frac{\rho(T,\nu)/h\nu}{\mathrm{g}(\nu)},$$

which is the number of photons per unit volume and unit frequency interval, divided by the number of modes per unit volume and unit frequency; the ratio of rates is therefore the average number of photons per mode.

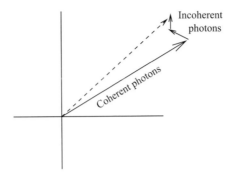

Fig. 17.1 Phasor diagram showing the small addition of an incoherent spontaneously emitted photon to coherent laser light. Further additions occur at random phase at intervals of τ_{cav}, the decay time of the laser cavity. The size of the incoherent components is greatly exaggerated in this diagram.

$$\Delta\nu_L \simeq \frac{1}{2\pi\bar{n}\tau_{cav}}. \tag{17.3}$$

The number of photons in the beam at any time is related to the power P and the cavity decay time by

$$\bar{n}h\nu = P\tau_{cav}, \tag{17.4}$$

giving the laser frequency width as

$$\Delta\nu_L = \frac{h\nu}{2\pi\tau_{cav}^2 P}. \tag{17.5}$$

The decay time of the cavity resonance τ_{cav} is related to the width of the resonance $\Delta\nu_c$ by

$$\tau_c = \frac{1}{2\pi\Delta\nu_{cav}}. \tag{17.6}$$

The laser linewidth may therefore be related to the width of the cavity resonance by

$$\Delta\nu_L = \frac{2\pi h\nu}{P}(\Delta\nu_{cav})^2. \tag{17.7}$$

Note that this theoretical minimum linewidth decreases as the power increases.[2]

---

[2] A more rigorous calculation takes into account the numbers N_2 and N_1 of species in the upper and lower laser levels and the population inversion $\Delta N(N_2 - (g_2/g_1)N_1)$, where g_2 and g_1 are the degeneracies of the levels. Then

$$\Delta\nu_L = \frac{2\pi h\nu(\Delta\nu_c)^2}{P}\frac{N_2}{\Delta N}.$$

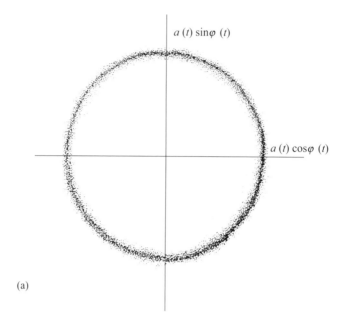

(a)

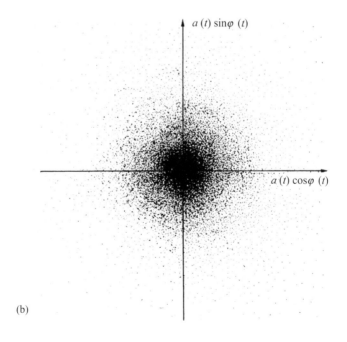

(b)

Fig. 17.2 Probability distributions of amplitude and phase for (a) laser light and (b) ordinary light. These are phasor diagrams showing how the amplitude and phase vary with time.

The theoretical limit is hard to attain in practice in gas lasers. A typical calculation from Eq. (17.7) shows that an ordinary helium–neon laser should give a line width of order 10^{-2} Hz, while in practice a width of a few kilohertz is commonly observed. The difference is due to thermal and mechanical instabilities which cause changes in the cavity length.

Theoretical linewidths are much larger in semiconductor lasers, because the small cavity length (typically 300 μm) leads to a very short decay time (5 ps) and a large cavity resonance linewidth (typically 10^{10} Hz). The linewidth $\Delta\nu_L$ of a semiconductor laser is usually around 10^6–10^7 Hz.

The further progress of the phasor diagram of Fig. 17.1 is shown in Fig. 17.2, where the essential difference between ordinary incoherent light and laser light is shown in the probability distribution of amplitude and phase. In ordinary light the arrival of photons follows Gaussian statistics, the probability distribution is a peak at zero, and all phases are equally probable. For laser light the distribution of amplitude is Poissonian about a mean; all phases are again possible, but on a long time-scale during which the phasor slowly wanders round the circle.

17.2   SPATIAL COHERENCE

The term *spatial coherence* usually refers to coherence transverse to the laser beam. As discussed in Chapter 14, the *transverse coherence length* L_t is related to the beam divergence angle θ_t by

$$L_t \sim \frac{\lambda}{\theta_t}. \tag{17.8}$$

The spatial coherence of the beam can be measured using the Young's double slit interferometer described in Chapter 7.

The angular width of the beam as it emerges from the laser is limited by the width w_0 of the waist, as shown in section 15.8. For the lowest order fundamental mode, the TEM$_{00}$ mode, the radial dependence of irradiance is the Gaussian function

$$I(r) = I_0 \exp\left(-\frac{2r^2}{w_0^2}\right), \tag{17.9}$$

where I_0 is the irradiance on the axis. The beam size increases with distance z as

$$w(z) = w_0 \left\{1 + \left(\frac{\lambda z}{\pi w_0^2}\right)^2\right\}^{1/2}. \tag{17.10}$$

At large distances from the beam waist w_0, when $\lambda z/\pi w_0^2 \gg 1$, the beam size becomes $w \simeq \lambda z/\pi w_0$ and the radius of curvature of the beam is $R \simeq z$. Thus the wave is approximately spherical with an origin at the beam waist. The divergence angle of the beam is then

$$\theta = \frac{w}{z} = \frac{\lambda}{\pi w_0}.$$ (17.11)

Thus the Gaussian TEM_{00} laser beam has a divergence of about one-half that of a plane wave beam diffracted by an aperture of the same size.

The beam remains almost parallel for some distance from the laser. In Eq. (17.10) the width is almost constant for distances $z \leq \frac{1}{2}z_0$, where $z_0 = \pi w_0^2/\lambda$ defines the Rayleigh range, i.e. the distance over which a laser beam is effectively collimated. For example, a red light beam from a laser with 1 mm aperture remains parallel for about 5 m. A longer but wider parallel beam can be achieved by using a beam expander, which is a telescope system used in reverse (Fig. 17.3). This effectively gives a larger coherent wavefront than the laser aperture alone. A survey theodolite with a 25 mm aperture would have a parallel beam over a distance of 3 km. Over longer distances the beam expander achieves a smaller angular spread than the laser alone. Given an optical system accurate to a fraction of a wavelength, and in the absence of atmospheric turbulence, a very narrow beam can be generated. A telescope with 1 m diameter aperture can transmit a laser beam with a divergence less than 1/2 arcsec across; this would illuminate a spot only 1 km across on the moon.

The focal spot

A laser beam may be focused to very small focal spot, not much more than a wavelength across, giving extremely high power densities. Since diffraction

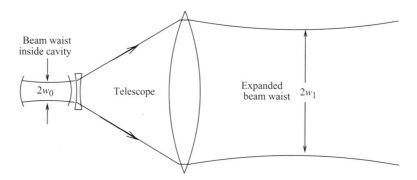

Fig. 17.3 A beam expanding telescope, used to increase the diameter of the coherent beam and reduce its angular divergence. The beam waist is increased by the ratio of the focal lengths of the two telescope lenses. In this diagram the beam is outlined at the $1/e^2$ irradiance level; the curvatures of the emergent beam are greatly exaggerated.

from a circular aperture with diameter D uniformly illuminated by a plane wave gives a beam with angular width

$$\theta = \frac{1.22\lambda}{D}, \tag{17.12}$$

the spot produced by a lens with focal length f is $(1.22\lambda/D)\,f$ across. The focal ratio, or F-number, is f/D, so that the spot size is $1.22\,\lambda F$. Even allowing for some lens aberration, spots of a few wavelengths diameter can easily be achieved, giving power densities high enough for cutting and welding metals. For example, CO_2 lasers operating at 10.6 µm wavelength with a power of 500 W can be focused to a spot 50 µm across, giving a power density of $100\,\text{kW mm}^{-2}$.

Laser speckle

A spot of laser light on a rough surface may easily be recognized by its grainy appearance, with a pattern of individual speckles which changes as the eye moves. This is a manifestation of the spatial coherence of laser light, both across the spot and within the depth of the rough surface profile. Light is scattered from each point of the surface with a phase depending on the height. At any point in front of the surface these sources combine coherently, but the combined amplitude depends on the random addition of their phases, which varies from point to point. Small involuntary movements of the eye change the relative phases and cause the pattern to shimmer.

Speckle observed directly by the eye has a definite pattern size S_E, which depends on the wavelength and the distance from the illuminated surface but not on the detail of the rough surface itself. The perceived luminance at any point is the combination of contributions from the smallest resolved area, which depends on the angular resolution of the eye and the distance from the surface. This is the scale size of the speckle pattern; the sum of the contributions changes outside this zone. If the eye is at distance D from the surface the scale size of the speckles is approximately

$$S_E \simeq D\frac{\lambda}{d}, \tag{17.13}$$

where d is the diameter of the pupil, and λ/d is the angular resolution. This is easily tested by moving to different distances D and by squinting to reduce d. Remarkably the pattern does not disappear when the eye is defocused, as when one's spectacles are removed.

When speckle is observed by the eye it is due to an interference pattern formed on the retina of the eye. A point on the retina ideally receives light from a single point of the patch of laser light, but actually receives light from a finite area, about the size of a Fresnel zone; contributions from scattering sources over this area are combined with random phases. There is, however, an interference pattern in the whole space in front of the scattering surface, as may be found simply by holding a piece of paper or exposing a photographic plate at a fixed position. A point on the plate is now receiving contributions from the whole of the illuminated surface and the luminance depends on the random addition of these contributions. Their relative phases change significantly at an adjacent point on the plate when the path difference between the two edges of the illuminated spot changes by λ. If the spot diameter is s, this requires an angular movement of λ/s; at a plate distance D this gives a speckle scale S_P on the plate, where

$$S_P = D\frac{\lambda}{s}. \tag{17.14}$$

Note the similarity of these two equations, and that the scale is proportional to D for both, as may easily be tested experimentally.

17.3   TEMPORAL COHERENCE AND COHERENCE LENGTH

The temporal coherence of the laser output is directly related to the spectral bandwidth. A spread of frequencies in a laser output having a bandwidth $\Delta\nu_L$ leads to a changing phase relation between the components in the spread, changing randomly the amplitude and phase of their sum (see Chapter 14). The temporal coherence is characterized by the *coherence time* τ_c, during which the frequency components maintain a fixed phase relation. Assuming a Gaussian line profile, τ_c is

$$\tau_c = \frac{1}{\Delta\nu_L}. \tag{17.15}$$

The *coherence length* l_c is the distance $c\Delta\tau$ travelled in the coherence time, so that $l_c = c\tau_c = c/\Delta\nu_L$. Very large coherence lengths are often encountered; even for a comparatively large linewidth of 1 MHz the coherence length is 300 m. Measurement of such narrow linewidths requires interferometers with correspondingly long optical paths.

The line emission from a low-pressure discharge lamp has a comparatively small coherence length; for example, the 546.1 nm mercury emission line would have a linewidth of about 2.5×10^{-2} nm and a coherence length of about 1 cm. In contrast, the stable single mode helium–neon laser with a bandwidth of 1 kHz gives a coherence length of 300 km. Even the normal helium–neon laser, which usually operates on several modes simultaneously and consequently has a larger bandwidth, gives a coherence length of about 50 cm.

The quantities τ_c and l_c can be measured using a Michelson interferometer or a Mach–Zehnder interferometer as described in Chapters 11 and 13. The open path Michelson interferometer is suitable for laser outputs with linewidths less than 1 GHz, since these correspond to path length differences less than 30 cm. A laser source with a linewidth of 1 MHz observed with a Michelson interferometer would require a path difference between the two interferometer arms of about 300 m. Such long path differences can be accommodated using a long optical fibre in one of the arms (section 11.7).

Lasers can be used for measuring distances of several kilometres or more in terms of the wavelength of the laser light; it must be emphasized, however, that the wavelength is determined by the resonant cavity and is not a fundamental physical parameter, so that only comparative measurements can be made. A change in distance of a fraction of a wavelength can be detected; for example, the gravitational wave detector (Chapter 11) is an interferometer measuring periodic changes in a path length of several kilometres, with the aim of detecting a proportional change as small as one part in 10^{20} (as may easily be seen, this represents a very small fraction of a wavelength).

17.4   LASER PULSE DURATION

Commonly encountered lasers such as the helium–neon laser and semiconductor lasers produce a continuous beam of light, although for communications purposes a semiconductor diode laser may be switched electrically at high rates. Other lasers are designed to produce individual very short pulses, often at extremely high intensity. Phenomenally high luminances can be achieved in these pulsed lasers; the mean power may however be kept to a manageable level by using a low pulse repetition rate. We describe here two techniques for producing these ultra-short pulses.

Q-switching

Intense short duration pulses may be obtained by the technique of *Q-switching*. In normal laser operation the resonant cavity has low losses: it is a resonator with a high quality factor Q. Laser action can be inhibited by reducing the Q of the cavity, for example by rotating one of the mirrors out of alignment. If the excitation (pumping) of the laser is maintained during the time the cavity is in a low-Q state, then the population inversion can build up to a very high value, as shown in Fig. 17.4. Restoring the original high Q by realigning the mirrors then gives an intense burst of laser light. The build-up of the pulse is very rapid, since the gain is very much larger than the threshold value. Laser action then removes the excitation in a very short pulse undergoing only a few passes through the laser medium.

The duration of the Q-switched pulse is approximately the same as the cavity lifetime τ_{cav} described in section 17.1. This depends on the length L of the resonator, the refractive index n, and the intensity reflection coefficient R of the mirrors:

$$\tau_{\text{cav}} = \frac{nL}{c(1 - R)}.$$ (17.16)

Q-switching can be achieved in several ways, most simply by rotating one of the mirrors, typically at about 10 000 rpm, giving a pulse for each rotation. Switching can also be achieved by opening an electro-optic modulator (an optical shutter such as a Kerr or Pockels cell, as in section 6.7) in the laser cavity.

As an example, the solid state Nd–YAG laser can be pulsed electrically to produce a pulse of about 1 ms duration, typically with a peak power of order 1–10 kW. When Q-switched the pulse duration can be reduced to 10–100 ns, with a peak power of several megawatts.

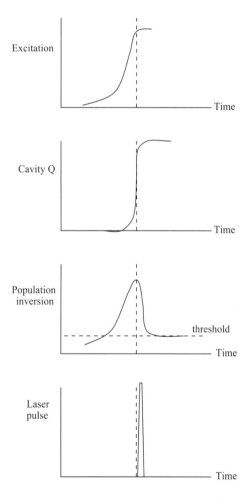

Fig. 17.4 Q-switching, showing (a) the growth of excitation; (b) the step increase of Q in the laser cavity; (c) the growing population inversion; and (d) the short laser pulse.

Mode-locking

As we saw in Chapter 4, a short pulse must contain components over a wide bandwidth. A laser oscillation has an inherently narrow bandwidth, but it may be able to oscillate in several modes with frequencies spaced within the bandwidth of the resonator and the linewidth of the lasing medium. If these modes can be excited simultaneously, with a suitable relation between their phases, then the effect of a broad bandwidth is obtained in a regular train of pulses which may individually be as narrow as 1 ps (10^{-12} s).

Consider a cavity resonator with two mirrors a distance L apart. The separation of the N modes in angular frequency is $\Delta\omega = \pi c/L$. If they are all excited simultaneously so that the different modes maintain the same relative phase, i.e. are *mode-locked*, and with equal amplitude, then their electric fields add as

$$E(t) = E_0 \sum_{n=1}^{N} \exp(i\omega_n t), \tag{17.17}$$

where $\omega_n = \omega + n\Delta\omega$. The sum of this series is already familiar in the context of diffraction gratings:

$$E(t) = E_0 \exp(i\omega t) \frac{\sin(N\pi ct/2L)}{\sin(\pi ct/2L)}. \tag{17.18}$$

The output irradiance is

$$I(t) = E_0^2 \frac{\sin^2(N\pi ct/2L)}{\sin^2(\pi ct/2L)}. \tag{17.19}$$

Such coherent oscillation in the set of N modes, known as mode-locking, may be achieved by the rapid opening of an electro-optic shutter in the laser cavity. The output is a train of pulses uniformly spaced in time at the period $t = 2L/c$, which is the round trip transit time of a pulse in the laser cavity. The duration of each pulse is approximately $(\Delta\nu)^{-1}$, where $\Delta\nu$ is the bandwidth of the set of longitudinal modes in the cavity. The maximum irradiance is $N^2 E_0^2$; note that without coherence between the modes the maximum would have been only NE_0^2.

Mode-locking can also be achieved by placing a cell containing a *saturable dye* in the laser cavity. The dye absorbs over a wide bandwidth, but at high intensities the absorption is reduced because a large proportion of the dye molecules are in the excited state. If the laser is oscillating with several modes covering the wide range $\Delta\nu$, then the pulse shape will contain structure with width $(\Delta\nu)^{-1}$, within a complex shape whose length is determined by a single mode (Fig. 17.5(a)). The highest peak will be amplified more than the lower

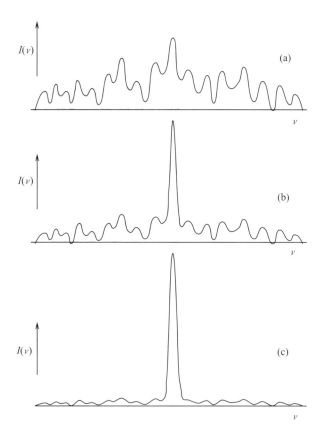

Fig. 17.5 Mode-locking with a saturable dye: (a) oscillation at several modes simultaneously produces a complex pulse; (b) the highest peak is amplified selectively; (c) after many passes through the dye cell the peak becomes a single narrow pulse.

peaks at each pass through the cell (Fig. 17.5(b)); repeated passages through the dye cell eventually amplify this peak into a single sharp pulse as seen in Fig. 17.5(c).

The pulse of a mode-locked laser can be compressed in time by factors of 20 or more to generate pulses as short as 5 femtoseconds (fs). This is achieved by a technique that induces a linear change in frequency along the pulse (known as a frequency *chirp*), followed by propagation through an optical system with dispersion in group velocity. The propagation delay for the back of the pulse is thus made less than for the front, and the pulse is correspondingly compressed. This technique is also used to make short radio pulses for high resolution radar systems.

Ultra-short laser pulses find many applications. Very fast processes in atoms, molecules and materials can be excited and probed. The short pulses, suitably amplified, are used as the pump source for X-ray lasers and to study high-temperature and high-density plasmas.

Plate 15.1 A high power fibre laser glowing with visible light. The core of the silica fibre, which is several metres long and has resonator mirrors at each end, is doped with thulium ions which lase in the infra-red at 2 μm. The pump is a semiconductor diode laser array, wavelength 790 nm, focused on one end. The laser power is 5 watts; the visible light emitted from the side of the fibre is from a wavelength up-conversion process. (Stuart Jackson, University of Manchester)

Plate 22.1 Double rainbow over Flagstaff, Arizona, photographed by W. Livingston (*Color and Light in Nature*, Cambridge University Press, 1995). Note the darker sky outside the primary bow, and the fainter supernumerary bow inside the primary

Plate 22.2 Aurora borealis photographed by Paul Neiman (*Color and Light in Nature*, Cambridge University Press, 1995)

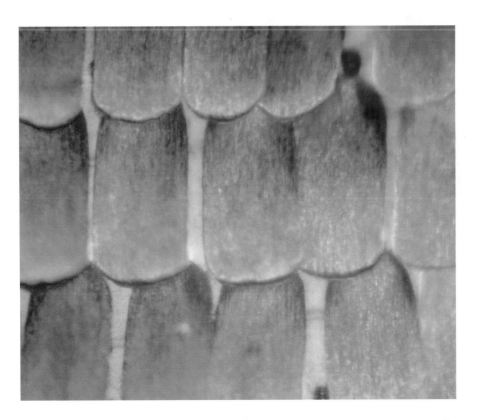

c

d   e

Plate 22.3 Iridescent colour in a butterfly wing: (a) The Morpho butterfly; (b) Wing scales ×300. The colour is due to diffraction and selective reflection in the structure of the wing scales, shown by scanning electron microscopy (SEM) and transmission electron microscopy (TEM); (c) the surface has a diffracting array of slots (SEM × 650); (d) the slots are deep and contain serrations (SEM × 10000); (e) a cross-section of the slots shows a pattern of grooves at quarter-wave intervals, which act in the same way as multilayer films (TEM × 25000) (Peter Vukusic, University of Exeter with G. Wakely for TEM)

17.5   BRIGHTNESS

The *brightness* even of low power lasers is often many orders of magnitude greater than the brightness of incoherent sources of light, because of the very high directionality of the laser beam. We recall that brightness B is defined as the power flow P per unit area A and per unit solid angle:[3]

$$B = \frac{P}{A \Delta \Omega} \, \mathrm{W \, m^{-2} sr^{-1}}. \tag{17.20}$$

We recall also that no optical system can increase the brightness of a light source (provided that object and image are in media with the same refractive index); for example, by focusing with a lens it is possible to create a smaller image of a source but at the expense of collecting less light from the original object.

As an example of a bright non-laser source, the brightness of the Sun is about $1.4 \times 10^6 \mathrm{W \, m^{-2}} \, \mathrm{sr^{-1}}$; this cannot be increased by focusing with a lens or a mirror. In contrast, even an ordinary low power, e.g. 1 mW, helium–neon laser operating at 632.8 nm has a brightness $B \sim 10^{12} \, \mathrm{W \, m^{-2} sr^{-1}}$, which is brighter than a million Suns.

An ultra-short pulse laser, such as a mode-locked 1.06 μm Nd–YAG laser producing 1 mJ pulses with a pulse duration of 50 ps, has a power of 20 MW during the pulse; this is equivalent to a brightness $B = 2 \times 10^{19} \, \mathrm{W \, m^{-2} sr^{-1}}$. High-power pulsed lasers followed by a train of amplifiers can achieve brightnesses approaching $10^{22} \, \mathrm{W \, m^{-2} sr^{-1}}$; furthermore, the coherent wavefront from a laser can be focused into a very small area.

17.6   FOCUSING LASER LIGHT

As shown in section 17.2, very high concentrations of power can be achieved by focusing laser light. The action of a lens with a short focal length f is illustrated in Fig. 17.6. Here the diameter d of the lens has been chosen to match the width of the wavefront of a beam at distance z from the waist, where the beam has expanded according to Eq. (17.10). Then

$$d = 2w_1 = 2 \frac{\lambda z}{\pi w_o} \tag{17.21}$$

where $2w_o$ is the width at the waist. The wavefront emerging from the lens converges to form a focal spot with width $2w_f$, limited by diffraction of the wavefront to

$$w_f = \frac{2f\lambda}{\pi d} = \frac{2}{\pi} \lambda F, \tag{17.22}$$

---

[3] Note that *spectral* brightness also includes 'per unit bandwidth'.

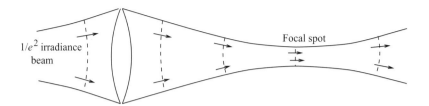

Fig. 17.6   Focusing a laser beam by a converging lens.

where F is the F-number of the lens. Provided that the lens diameter matches the width of the laser beam, the spot size is limited only by the F-number. A practical low value is $F = 1$, giving a smallest spot size approximately equal to the wavelength of the laser light.

A 1 mW helium–neon laser focused by a lens with $F = 1$ has a focal radius of $r_f = (2/\pi)(6.3 \times 10^{-7}) = 4 \times 10^{-7}$ m. The power per unit area at the focus is 2×10^9 W m^{-2}.

17.7   NONLINEAR OPTICS

From section 5.3 the root mean square (rms) electric field E in any electromagnetic radiation field is related to the irradiance I by $I = E^2/377$ W m^{-2} (in free space). The peak field E_{max} in a dielectric with refractive index n is therefore

$$E_{max} = 27.4 \left(\frac{I}{n}\right)^{1/2} \quad \text{V m}^{-1}. \tag{17.23}$$

A peak field reaching 10^{12} V m^{-1} is attainable in an ultra-short pulse from a high powered laser. This is greater than the typical internal field strength of a dielectric, or the field binding the electron to a proton in the hydrogen atom. The field of a point electric charge e at distance $r_0 = 0.1$ nm is

$$E = \frac{e^2}{4\pi\epsilon_0 r_0^2} \simeq 10^{11} \text{ V m}^{-1}. \tag{17.24}$$

A laser pulse can therefore completely disrupt a dielectric medium. Expensive optical components have been destroyed in a few picoseconds in this way!

At lower fields, in the range 10^7 to 10^9 V m^{-1}, the dielectric may respond non-linearly to the field and generate harmonics. We have previously treated the polarization of a dielectric as proportional to the electric field; we must now include further terms and write

$$P = \epsilon_0 \left(\chi_1 E + \chi_2 E^2 + \chi_3 E^3 + \cdots\right), \tag{17.25}$$

where χ_1 is the normal susceptibility of the dielectric, and χ_2, χ_3, etc. are higher order terms.[4] A light wave with a field $E = E_0 \cos \omega t$ induces a polarization

$$
\begin{aligned}
P &= \epsilon_0 \left(\chi_1 E_0 \cos \omega t + \chi_2 E_0^2 \cos^2 \omega t + \chi_1 E_0^3 \cos^3 \omega t + \cdots \right) \\
&= \epsilon_0 \left\{ \chi_1 E_0 \cos \omega t + \tfrac{1}{2}\chi_2 E_0^2 (2 \cos 2\omega t - 1) + \tfrac{1}{4}\chi_3 E_0^3 (3 \cos \omega t + \cos 3\omega t) + \cdots \right\}
\end{aligned}
$$

$$(17.26)$$

The polarization P is therefore oscillating at harmonics 2ω, 3ω, etc., and radiating waves at these higher frequencies. Frequency doubling, i.e. the generation of the second harmonic, is commonly achieved in non-isotropic materials; higher order harmonics may also be produced at higher field strengths in isotropic materials.

Frequency doubling is important as a way of producing laser light at short wavelengths. A practical problem is that the original laser light and its second harmonic must travel along the ray path through the dielectric with the same velocity; if they are different, the second harmonic light generated from different parts of the path will not add correctly in phase. Most dielectrics are sufficiently dispersive for this to be a serious limitation on the thickness of a harmonic generator. In some birefringent materials it can be arranged that the fundamental and second harmonic waves are polarized as ordinary and extra-ordinary waves (see Chapter 4), and a propagation direction can be chosen in which the two refractive indices are equal. A commonly used material for this purpose is potassium dihydrogen phosphate, known as KDP; the efficiency of frequency doubling can reach 30% with this material.

Two laser beams with different frequencies ω_1 and ω_2 propagating in a non-linear dielectric may induce polarization oscillating at the difference and sum frequencies $\omega_1 - \omega_2$ and $\omega_1 + \omega_2$ This is known as *mixing*. Again the efficiency of the process depends on matching refractive indices.

An interesting aspect of these processes is their interpretation in terms of photons. Two identical photons arriving nearly simultaneously at a molecule in a crystal lattice can emerge from the encounter as a single photon with twice the energy: this is frequency doubling. The probability of such close encounters depends on the flux of photons, since two must be found close to the same molecule for the interaction to occur; this is equivalent to the power-law dependence on the field strength in Eq. (17.24).

17.8 FURTHER READING

R. W. Boyd, *Non-linear Optics*. Academic Press, 1992.
R. Loudon, *The Quantum Theory of Light*. Oxford University Press, 1983.

---

[4] In this discussion we assume that P and E are in the same direction, although this is not generally true for anisotropic materials.

M. O. Scully and M. S. Zubairy, *Quantum Optics*. Cambridge University Press, 1997.
Y. R. Shen, *The Principles of Non-linear Optics*. Wiley, 1984.

NUMERICAL EXAMPLES 17

17.1 A 1 W laser beam is focused onto a spot 10 μm in diameter. Calculate the irradiance (see Appendix) and the mean electric field in the spot. What is the maximum temperature attainable in the spot?

17.2 A solid ruby rod laser 0.2 m long with refractive index 1.76 and coated end faces to form the resonator produces mode-locked pulses. What is the time interval between the pulses?

17.3 A collimated helium–neon laser beam, wavelength 632 nm, is required for surveying over a distance of 10 km. The beam will be expanded optically: what waist diameter will be needed?

PROBLEMS 17

17.1 From Eq. (17.7) calculate the minimum attainable linewidth $\Delta\nu$ obtainable from a 1 mW helium–neon ($\lambda = 633$ nm) laser if the cavity decay time is 10^{-7} s. Why is this theoretical limit never attained? If the laser length is 1 m, what change in length would give a frequency shift $\Delta\nu$ equal to the linewidth? If the coefficient of thermal expansion of the cavity is 10^{-6} K^{-1}, what temperature change would change the length by this amount?

17.2 Compare the coherence length of the following sources: (i) a heated filament lamp with a white light output over the wavelength range 400–700 nm; (ii) a stabilized CW Nd–YAG laser operating on a single mode with a line width of 20 kHz; (c) a helium–neon laser with a resonator length of 30 cm oscillating in three longitudinal modes.

18

Fibre optics

...beauty draws us with a single hair.

Pope, *The Rape of the Lock*.

If hairs be wires...

Shakespeare, *Sonnets*.

The transmission of light along a curved dielectric cylinder was the subject of a spectacular lecture demonstration[1] by John Tyndall in 1854. His *light pipe* was a stream of water emerging from a hole in the side of a tank which contained a bright light. The light followed the stream by total internal reflection at the surface of the water. Light pipes made of flexible bundles of glass fibres are now routinely used to illuminate internal organs in surgical operations in the *fibrescope* (or *endoscope*) which also transmits an image back to the surgeon. The overwhelmingly important use of glass fibres is, however, to transmit modulated light over large distances for communications.

Electrical cables and radio have largely been replaced by optical fibres in long-distance terrestrial communications. Hundreds of thousands of kilometres of fibre optic cables are now in use, carrying light modulated at high frequencies, providing the large communication bandwidths needed for television and data transmission. The techniques that made this possible are the subject of this chapter. These techniques involve the manufacture of glass with very low absorption of light, the development of light emitters and detectors which can handle high modulation rates, and the fabrication of very thin fibres which preserve the waveform of very short light pulses. An essential development has been the *cladding* of fibres with a glass of lower refractive index, which prevents the leakage of light from the surface.

Optical fibres are also useful in short communication links, especially where electrical connections are undesirable. They also offer remarkable opportunities in computer technology and in laboratory instrumentation such as interferometers and a variety of optical fibre sensors.

---

[1] J. Tyndall, *Proc. Roy. Inst.* **6**, 189, 1870.

We start by discussing the propagation of a light ray by internal reflection in a light pipe, and show how this approach may be developed into the concept of waves guided inside a dielectric slab or along a thin fibre. The cylindrical geometry of a fibre, and the boundary of a fibre, which in practice is a step or a gradient in refractive index, both need special consideration. Propagation in a light fibre may be *dispersive*, so that different wavelengths travel at different speeds; we show how the effects of dispersion are calculated and how they may be compensated for. We then briefly describe some of the many applications of fibres outside the field of communications.

18.1   THE LIGHT PIPE

The transmission of light along a glass rod depends on the total internal reflection of a ray reaching the surface at a glancing angle, i.e. at a high angle of incidence (section 1.3). A *light pipe* in the form of a glass rod can be used to conduct light round corners (Fig. 18.1), provided that the corners are not so sharp that the angle of incidence[2] $(90° − \theta)$ falls below the *critical angle*.

A bundle of thin glass fibres, usually coated with glass or plastic with lower refractive index, can transmit an optical image, as in the surgeon's endoscope. If the fibre ends are aligned in a plane, then the light distribution across them will be reproduced at the other end of the bundle. Rearranging the fibres along the bundle can be used to enlarge an image, or to compensate for distortions introduced in other parts of an optical system. A random rearrangement of the fibres within a bundle can be used to 'scramble' an image; the opposite rearrangement, using the bundle in the reverse direction, will then restore the image. Fibre optic bundles occur naturally both in the animate and the inanimate world. The retina of the human eye, and many other eyes, has an assembly of rods and cones which transmit light from the surface of the retina to the light-sensitive cells. In many insects the transmission is sensitive to polarization, providing the information which insects use for navigation. Natural inanimate fibre optics are found in the crystalline material known as ulexite, which is a fibrous form of borax.

Fig. 18.1   A light pipe, showing total internal reflection of a light ray.

---

[2] In fibre optics it is conventional to designate the angle between a ray or wavenormal and the fibre axis as θ, which is the complement of the angle used in the analyses of refraction by Snell and by Fresnel; see Chapters 1 and 3.

To deal with the propagation of light by the thin glass fibres used in communications a different approach is necessary; now the diameter of the fibre is comparable with the wavelength of light, and we must consider the light as a wave which is guided by the fibre. The configuration of the wave inside the fibre is constrained by conditions at the boundary of the fibre; we must also consider the extension of the wave field outside the boundary, into the cladding of the core fibre.

18.2   GUIDED WAVES

We shall develop the theory of the propagation of light along a cylindrical fibre in three stages. The requirement is to find wave configurations within the fibre which are solutions of Maxwell's equations, and which conform to the *boundary conditions*, i.e. the physical conditions at the surface of the core of the fibre. In Chapter 5 we applied Maxwell's equations to propagation in free space; we now consider a wave confined to a parallel-sided slab of the dielectric. Two major differences emerge: there can be components of both **E** and **B** fields in the direction of propagation, and only a limited number of wave patterns between the faces of the slab, known as modes, can propagate between the faces of the slab. The allowable mode patterns depend on the thickness of the slab and on the boundary conditions.

The effect of the boundary conditions is easiest to understand if the faces of the slab are perfectly conducting metal slabs; this is close to the practical case of waveguides for centimetric and millimetric radio waves. The second stage is to consider a slab guide bounded by a step in the dielectric constant; the boundary conditions are then more complicated and there is a component of the wave outside the surface of the slab. These two stages allow us to understand the fundamental characteristics of guided waves, and in particular their field patterns and their velocities. The geometry then needs to be adapted to the more complex mathematics of cylindrical rather than rectangular symmetry.

Maxwell's equations in free space (Eqs. (5.6) and (5.7)) are

$$\text{div } \mathbf{E} = 0, \qquad\qquad \text{div } \mathbf{B} = 0, \qquad\qquad (18.1)$$

$$\text{curl } \mathbf{B} = \epsilon_0\mu_0 \frac{\partial \mathbf{E}}{\partial t}, \quad \text{curl } \mathbf{E} = -\frac{\partial \mathbf{B}}{\partial t}. \qquad (18.2)$$

As we saw in Chapter 5, these lead to the wave equations

$$\nabla^2 \mathbf{E} = \epsilon_0\mu_0 \frac{\partial^2 \mathbf{E}}{\partial t^2}, \qquad\qquad (18.3)$$

$$\nabla^2 \mathbf{B} = \epsilon_0\mu_0 \frac{\partial^2 \mathbf{B}}{\partial t^2}. \qquad\qquad (18.4)$$

The **E** field may be expressed in Cartesian components:

$$\mathbf{E} = \hat{\mathbf{l}}E_x + \hat{\mathbf{m}}E_y + \hat{\mathbf{n}}E_z, \tag{18.5}$$

where $\hat{\mathbf{l}}$, $\hat{\mathbf{m}}$ and $\hat{\mathbf{n}}$ are unit vectors in the x, y and z directions.

The separate field components each obey the wave equation, so that the y-component obeys

$$\nabla^2 E_y = \epsilon_0\mu_0 \frac{\partial^2 E_y}{\partial t^2}. \tag{18.6}$$

For a plane wave in the z-direction E_y does not vary in directions x or y, and Eq. (18.6) reduces to

$$\frac{\partial^2 E_y}{\partial z^2} = \epsilon_0\mu_0 \frac{\partial^2 E_y}{\partial t^2}, \tag{18.7}$$

which represents waves of any form $E_y = E_0 f(z - vt)$, where the velocity $v = (\epsilon_0\mu_0)^{-1/2}$. In free space the corresponding **B** field is in the x-direction; both fields are transverse to the direction of propagation.

The same wave will propagate along the slab guide shown in Fig. 18.2, since it conforms to the boundary conditions at the conducting walls, where the tangential component of the electric field and the normal component of the magnetic field are required to be zero at the walls. Hence the electric field must be perpendicular to the walls. The wave velocity $v = \omega/k$ is the velocity of light in the medium between the plates. This mode is referred to as the transverse electric and magnetic, or TEM mode. We now find other modes that will propagate in the slab guide. As in Eq. (18.7) the wave travels in the z-direction, but the **E** field is now constant only in the x-direction. Setting $\partial/\partial x = 0$, Eqs. (18.2) reduce to

$$\frac{\partial E_z}{\partial y} - \frac{\partial E_y}{\partial z} = -\frac{\partial B_x}{\partial t}, \tag{18.8}$$

$$\frac{\partial E_x}{\partial z} = -\frac{\partial B_y}{\partial t}, \tag{18.9}$$

$$\frac{\partial E_x}{\partial y} = \frac{\partial B_z}{\partial t}, \tag{18.10}$$

$$\frac{\partial B_z}{\partial y} - \frac{\partial B_y}{\partial z} = \epsilon_0\mu_0 \frac{\partial E_x}{\partial t}, \tag{18.11}$$

$$\frac{\partial B_x}{\partial z} = \epsilon_0\mu_0 \frac{\partial E_y}{\partial t}, \tag{18.12}$$

$$\frac{\partial B_x}{\partial y} = -\epsilon_0\mu_0 \frac{\partial E_z}{\partial t}. \tag{18.13}$$

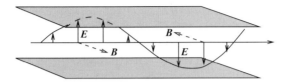

Fig. 18.2   A waveguide formed by two conducting plates, showing the simplest propagating mode (the TEM mode).

Two sets of solutions emerge from this array. Equations (18.9), (18.10) and (18.11) contain only E_x together with B_y and B_z; these form solutions in which the **E** field has no components in the direction of propagation, but the **B** field does. In contrast, Eqs. (18.8), (18.12) and (18.13) contain only B_x together with E_y and E_z; these form solutions in which the **B** field has no components in the direction of propagation, but the **E** field does. These two sets of solutions represent *transverse electric* (TE) and *transverse magnetic* (TM) modes, respectively. We now need a description of the field patterns in the individual modes.

Each mode has a simple field pattern that varies sinusoidally across the guide. This can conveniently be regarded as the combination of two crossing plane waves with wave vectors $\pm\mathbf{k}$, if $\mathbf{k}$ is correctly chosen. Consider first a pair of waves with the electric vector in the x-direction, and with vectors $\pm\mathbf{k}$ making angles $\pm\theta$ with the z-direction:

$$\mathbf{E_1} = \hat{\mathbf{l}}\, E_0 \exp i(\omega t - kz\cos\theta + ky\sin\theta), \tag{18.14}$$

$$\mathbf{E_2} = -\hat{\mathbf{l}}\, E_0 \exp i(\omega t - kz\cos\theta - ky\sin\theta). \tag{18.15}$$

The sum of these is

$$\mathbf{E} = \hat{\mathbf{l}}\, 2i\sin(ky\sin\theta) E_0 \exp i(\omega t - kz\cos\theta). \tag{18.16}$$

The boundary condition is that $E_x = 0$ at both plates, at $y = 0$ and $y = b$. This is achieved if the angle θ is chosen to give

$$kb\sin\theta = n\pi, \tag{18.17}$$

where n is an integer. There may be several pairs of waves with different values of θ which satisfy this criteria, provided that

$$n \le \frac{kb}{\pi}. \tag{18.18}$$

Each pair constitutes an allowable wave pattern, or *mode*, which can propagate independently along the guide in the z-direction. The propagation constant k_g along the guide is $k\cos\theta$; substituting for θ from Eq. (18.17) we have

$$k_g = \left(k^2 - \frac{n^2\pi^2}{b^2}\right)^{1/2}. \tag{18.19}$$

These modes are *transverse electric*, or TE modes; note that there are components of the magnetic field in the direction of propagation. In the *transverse magnetic*, or TM modes, the magnetic field is wholly transverse and there is a component of the electric field in the z-direction. The modes are designated TE_n and TM_n according to their mode number n. The wave velocity v_p of each mode is

$$v_p = \frac{\omega}{k_g} \qquad (18.20)$$

where the subscript p indicates the *phase* velocity in contrast to the *group* velocity (see Chapter 4). From Eq. (18.19) and recalling $c = \omega/k$:

$$v_p = c\left(1 - \frac{n^2\pi^2}{k^2b^2}\right)^{-1/2}. \qquad (18.21)$$

We see that the phase velocity is greater than the free space velocity c, and that it depends on the wave number k. The group velocity v_g is given by

$$v_g = \frac{d\omega}{dk_g}, \qquad (18.22)$$

which, with $k = 2\pi/\lambda$ and from differentiating Eq. (18.19), is

$$v_g = c\frac{k_g}{k}. \qquad (18.23)$$

The group velocity, as might be expected, is always less than c; the product $v_p v_g = c^2$.

The complete analysis of metallic waveguides must also involve boundaries in the x-direction, to form a rectangular waveguide.

18.3   THE SLAB DIELECTRIC GUIDE

A wave may be guided along a dielectric slab, such as a sheet of glass, provided that it is bounded by a material of smaller refractive index. The analysis is similar to that for the guide with conducting plates, but there are different boundary conditions to consider. The wave amplitude does not fall to zero at the boundary, and there is a component of the field beyond the boundary. We follow the same procedure of analysing pairs of crossing waves, each allowable pair constituting a propagating mode. It is convenient, however, to consider the pair of waves as a ray which is reflected to and fro between the boundaries of the slab, as in Fig. 18.3.

There must be total internal reflection at the boundary. From Snell's law (Chapter 1) this means that the angle of incidence $(90° - \theta)$ must be larger

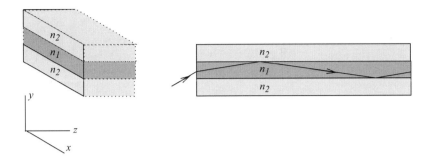

Fig. 18.3 Dielectric slab waveguide, showing total internal reflection at the interface between refractive indices n_1 and n_2.

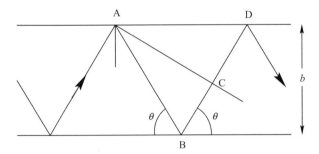

Fig. 18.4 The path difference between reflected rays in a dielectric guide.

than the critical angle, so that the ray angle must be closer to the axis than θ_{crit} given by

$$\cos \theta_{\text{crit}} = \frac{n_2}{n_1}, \qquad (18.24)$$

where n_1 and n_2 are the refractive indices inside and outside the slab. The pair of crossing waves that constitute a mode is now represented as a single ray which is reflected to and fro across the guide, as in Fig. 18.4. After the two reflections shown in Fig. 18.4 the ray CD must have the same phase as the incident ray AB, so that it constitutes the single wavefront of Eq. (18.14). The twice reflected ray has travelled an extra distance,[3] and in contrast to reflection at the conducting plate there is also a phase change $\phi(\theta)$ at each reflection. The extra path AB + BC for the reflected ray is found from

---

[3] In Fig. 18.4 the angle θ is shown larger than usual to help visualize the geometry. Note the similarity to the analysis of the Fabry–Perot plate in Chapter 9.

$$b = AB \sin \theta, \qquad (18.25)$$

$$BC = AB \cos(\pi - 2\theta), \qquad (18.26)$$

$$AB + BC = AB(1 - \cos 2\theta), \qquad (18.27)$$

$$= 2b \sin \theta. \qquad (18.28)$$

The rays arriving at A and C must be in phase, as they lie on the same wavefront. This gives a phase condition, including $2\phi(\theta)$ for the two reflections:

$$2b \sin \theta + \lambda_1 \frac{2\phi(\theta)}{2\pi} = N\lambda_1. \qquad (18.29)$$

With the additional term $\phi(\theta)$, this is similar to Eq. (18.17), except that λ_1 now refers to the wavelength λ_0/n_1 in the dielectric. N is again a mode number; there is a set of ray directions that can propagate, and each has its own group velocity. Equation (18.29) is a general condition for a propagating mode. Its solution is best approached by numerical methods; note that $\phi(\theta)$ depends on the polarization of the wave as well as the angle of incidence.

18.4   EVANESCENT FIELDS IN FIBRE OPTICS

The electric field does not fall to zero at the boundary of the dielectric slab, although the components of the propagating wave are totally internally reflected and, following Eq. (18.24), there is no refracted ray propagating away from it. The wave amplitude must therefore fall to zero in the y-direction. The wave outside the slab is an *evanescent wave*[4] (Fig. 18.5); we show that the amplitude decays exponentially with distance y.

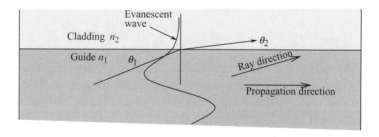

Fig. 18.5   Cross-section of the field pattern in a multi-mode dielectric guide, showing the penetration of an evanescent wave into the cladding.

Consider a refracted wave transmitted across the boundary when the angle of incidence $(90° - \theta_1)$ is less than the critical angle; θ_1 is the angle that the wave normal, i.e. the incident ray, makes with the surface of the slab (Fig. 18.6). As in

---

[4] *Evanescent* is fleeting or vanishing, from evanesce: to fade away.

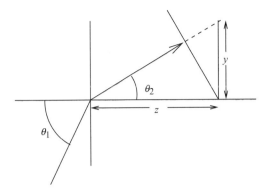

Fig. 18.6 Geometry of the refracted wave.

section 5.4, the amplitude of the refracted wave is E_t, and it propagates at an angle θ_2 to the surface as a wave E_2 with the form

$$E_2 = E_t \exp ik_2(z \cos \theta_2 - y \sin \theta_2). \tag{18.30}$$

Since from Snell's law

$$\sin \theta_2 = \sqrt{1 - \frac{n_1^2}{n_2^2} \cos^2 \theta_1}, \tag{18.31}$$

we can write Eq. (18.30) in terms of the angle of incidence as

$$E_2 = E_t \exp ik_2 \left(\frac{n_1}{n_2} z \cos \theta_1 - y \sqrt{1 - \frac{n_1^2}{n_2^2} \cos^2 \theta_1} \right). \tag{18.32}$$

For a ray beyond the critical angle the square root term becomes imaginary, and we can write

$$\sqrt{1 - \frac{n_1^2}{n_2^2} \cos^2 \theta_1} = \pm i\alpha. \tag{18.33}$$

The wave outside the boundary is now seen to be the evanescent wave[5]

$$E_T = E_t(\exp -\alpha k_2 y) \left[\exp \left(ik_2 \frac{n_1}{n_2} z \cos \theta_1 \right) \right]. \tag{18.34}$$

The wave propagates along the boundary, matching the guided wave inside the boundary, while the amplitude decays exponentially in the y-direction. The exponential decay constant αk_2 is the inverse of the *penetration depth* ξ, given by

---

[5] The solution with an exponentially *growing* wave is clearly unphysical.

$$\xi^{-1} = k_2 \left(\frac{n_1^2}{n_2^2} \cos^2 \theta_1 - 1 \right)^{1/2}. \tag{18.35}$$

The evanescent wave can penetrate a significant distance into the cladding of an optical fibre, which must be thick enough for the amplitude to fade away almost to zero. For an angle close to the critical angle in silica, and for a wavelength of $1.3\,\mu\mathrm{m}$, the value of ξ is about $10\,\mu\mathrm{m}$.

A considerable fraction of the wave energy travelling along an optical fibre is transmitted in the evanescent wave. The cladding must therefore be a glass with as high an optical quality as that of the fibre core, so as to avoid transmission losses.

18.5   CYLINDRICAL FIBRES AND WAVEGUIDES

The main principles of a slab or rectangular waveguide may be applied to a cylindrical dielectric guide without modification except for the detailed pattern of the propagating field. As before, we require solutions of Maxwell's equations in the form of the propagating fields:

$$\mathbf{E} = \mathbf{E_0}(x, y) \exp(-i\beta z), \tag{18.36}$$

$$\mathbf{B} = \mathbf{B_0}(x, y) \exp(-i\beta z). \tag{18.37}$$

The analysis requires the wave equations for $\mathbf{E}$ and $\mathbf{B}$ to be written in a cylindrical coordinate system. Solutions that satisfy the boundary conditions resemble those for the slab guide, showing discrete modes that will propagate for any given free space wavelength, provided it is less than a critical wavelength. The field patterns for each mode are in the form of Bessel functions rather than the sine and cosine functions in the rectangular guide.

Figure 18.7 shows some of the field patterns in the simplest case of a cylindrical waveguide; this is the metallic guide with conducting walls used in microwave transmission. In each mode there must be a small component of the electric or magnetic field along the axis, and the modes are often distinguished in terms of the field component which is wholly transverse as TE or TM modes. In a rectangular guide each mode is designated according to the number of cycles across the guide; the simplest mode in the rectangular waveguide is designated TE_{01}. In the cylindrical guide the modes are designated according to the order of the Bessel function describing the field pattern, as shown in Fig. 18.7. As already described for the dielectric slab, there is an evanescent wave in the cladding around the fibre. The detailed field configuration depends on the form of the interface between the fibre and the cladding; solutions can be obtained for a step change in refractive index, but in practice there are advantages in a more gradual transition of refractive index, involving a more complicated analysis.

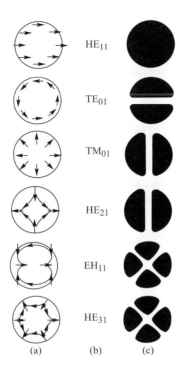

Fig. 18.7 (a) Electric field patterns in the three lowest order modes propagating in a circular cross-section waveguide. (b) The modes are designated in terms of TE_{lm} and TM_{lm} transverse modes for meridional rays and hybrid modes HE_{lm} and EH_{lm} for skew rays. (c) mode intensity distribution. (From J. M. Senior, *Optical Fibre Communications*. Prentice-Hall, 1992.)

We have already seen how the boundary conditions of dielectric slab guides affect the field patterns as compared with those of waveguides with conducting walls. The differences are more important in optical fibres, since they are usually clad with a dielectric with refractive index n_2 which is only slightly below n_1, the refractive index of the guide itself. The advantage is that rays at a large angle to the direction of propagation are not then internally reflected, and only low-order modes are propagated; in the limit there is only one propagating mode. Such a *single-mode fibre* is particularly important in long-distance communications, where the difference in group velocity between modes is a severe disadvantage. Single-mode fibres must have a core diameter less than a few wavelengths, and a small change in refractive index from core to cladding.

Exact analysis of the maximum diameter for a single-mode fibre is tedious, but an approximate analysis for a slab dielectric is easily done in terms of a ray in the slab at the limiting angle θ_{lim} for internal reflection. Then $n_1 \cos \theta_{\text{crit}} = n_2$. If there are N half wavelengths in the wave pattern across a slab with thickness b, then $b \sin \theta_{\text{crit}} = N\lambda/2$, where λ is the guide wavelength λ_0/n_1, giving the number of possible modes as

$$N = \frac{2b}{\lambda_0} \left(n_1^2 - n_2^2 \right)^{1/2}. \tag{18.38}$$

In practice the two refractive indices differ by only a small amount and may be written as n and $n + \Delta n$, so that $\left(n_1^2 - n_2^2 \right)^{1/2} \simeq (2n\Delta n)^{1/2}$. The maximum thickness $b_{\max}$ of a slab that carries only a single mode is

$$b_{\max} = \frac{\lambda_0}{2(2n\Delta n)^{1/2}}. \tag{18.39}$$

A similar but more complicated analysis for cylindrical fibre guides yields the useful parameter known as the *V number*, which for a step index guide with radius a is

$$V = \frac{2\pi a}{\lambda} \left(n_1^2 - n_2^2 \right)^{1/2}. \tag{18.40}$$

The analysis shows that if this V number is less than 2.4, then only a single mode can propagate. This requires very thin fibres: for example, if $n_1 - n_2$ is 0.2% of n_1, then the maximum radius for a single-mode fibre is only 8 wavelengths. For large values of V the total number of modes (including both polarizations) is approximately $V^2/2$.

18.6   NUMERICAL APERTURE

Although the rays in an optical fibre are at a small angle to the axis, they spread to a wider angle as they emerge at the end of the fibre. This is important in matching light detectors to fibres, where for efficient light collection the angle accepted by a detector should be approximately the same as the emergent light cone of the guide; similarly, in injecting light into a fibre the light cone from the source should match the acceptance angle of the fibre.

For a fibre with refractive index n_1 and cladding with refractive index n_2, the largest angle θ_{crit} within the fibre is shown in Fig. 18.8, where the refracted ray is along the surface. Applying Snell's law to a ray crossing the axis of the cylinder, the angle of incidence $\theta_i = 90° - \theta_{\text{crit}}$, giving

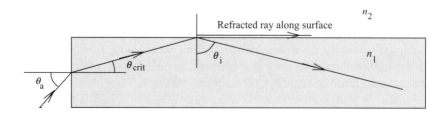

Fig. 18.8   The acceptance angle for light entering a dielectric guide.

$$\sin \theta_i = \cos \theta_{\text{crit}} = \frac{n_2}{n_1}. \tag{18.41}$$

A ray at this limiting angle enters a plane face at the end of the slab at angle θ_a to the normal, as shown in Fig. 18.8. Then if the refractive index outside the fibre is n_0,

$$n_0 \sin \theta_a = n_1 \sin \theta_{\text{crit}} = n_1 \left\{ 1 - \left(\frac{n_2}{n_1} \right)^2 \right\}^{1/2}. \tag{18.42}$$

All rays inside the acceptance angle θ_a will propagate within the slab.

The limited acceptance cone is usually expressed in terms of the *numerical aperture* (NA), which characterizes a cone of rays in any optical instrument, defined as

$$\boxed{\text{NA} = n_0 \sin \theta_a.} \tag{18.43}$$

In this case

$$\text{NA} = \left(n_1^2 - n_2^2 \right)^{1/2}. \tag{18.44}$$

A useful simplification can be made when the relative refractive index difference $\Delta = (n_1 - n_2)/n_1$ is small:

$$\text{NA} \simeq n_1 (2\Delta)^{1/2}. \tag{18.45}$$

Typically the fractional step Δ is around 1%, giving $\text{NA} \sim 0.2$ and an acceptance cone with half angle around $10°$.

18.7 MATERIALS FOR OPTICAL FIBRES

The most important requirement for the glass in an optical fibre is a low transmission loss.

Transmission loss is usually measured in decibels per kilometre (dB km^{-1}), as in communications engineering.[6] A slab of ordinary silica glass usually has a loss much greater than $100 \, \text{dB km}^{-1}$; this is due to absorption by impurities, particularly metallic ions such as iron, chromium and copper. Pure silica glass has remarkably low losses, below $1 \, \text{dB km}^{-1}$ at infrared wavelengths between 0.8 and $1.8 \, \mu\text{m}$. Beyond those wavelengths the losses increase sharply

---

[6] Loss in decibels is $10 \log_{10}$ (ratio of input to output powers). It is useful to remember that $10 \, \text{dB}$ is a factor of ten, $3 \, \text{dB}$ is close to a factor of two, and $1 \, \text{dB}$ loss is approximately 20%.

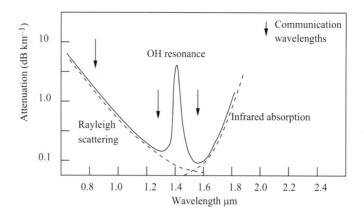

Fig. 18.9 Transmission loss as a function of wavelength in high quality silica glass.

(Fig. 18.9). In the visible region the principal losses are due to Rayleigh scattering from inhomogeneities frozen into the glass; this gives a loss increasing as λ^{-4}. There are also losses in the ultraviolet region due to electronic transitions. The increase in absorption at longer infrared wavelengths is the residual effect of vibrational states of the lattice and absorption bands such as that at $9.2\,\mu m$, due to a resonance in Si–O bonds.

Within the window between 0.8 and $1.8\,\mu m$ there is an appreciable rise in attenuation centred on $1.38\,\mu m$. This is related to water dissolved in the glass; the resonance actually occurs in the hydroxl ion (OH^-) at $2.7\,\mu m$ with a second

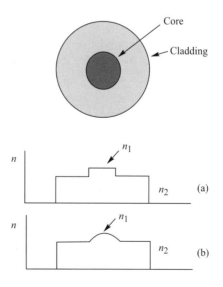

Fig. 18.10 Refractive index profiles of step and graded refractive index fibres.

harmonic at 1.38 μm. However, the practical situation is that there are two low absorption bands, at 1.3 and 1.55 μm, the longer wavelength band having attenuation loss of down to 0.2 dB km^{-1}.

Losses in optical fibres may also be due to geometric imperfections introduced in the manufacturing process, and from sharp bends which the guided waves may not be able to follow.

The simple cladding of a fibre with a different material results in the stepped refractive index profile of Fig. 18.10(a). There is however an important advantage in fibres manufactured with a gradient of refractive index, decreasing from the axis to join the lower refractive index of the cladding, as shown in Fig. 18.10(b). Such a *graded index* fibre can be made by allowing the cladding to diffuse into the fibre, but the manufacturing techniques that we describe below allow a more precise control of the refractive index profile. The advantage of a graded index fibre, as described below, is that the difference in velocity of the allowable modes is minimized.

18.8   DISPERSION IN OPTICAL FIBRES

Fibre optics communication systems usually use pulses of light. A typical train of pulses might be transmitted as in Fig. 18.11(a), and after travelling for a large distance might appear as in Fig. 18.11(b). In this figure the amount of pulse smearing is close to the limit which would still allow the signal to be decoded. Pulse smearing is inevitable in multi-mode fibres, although its effect can be reduced in graded index fibres. There are, however, important effects in the single-mode fibres used for long-distance communications, due to the spread in travel times over the wavelength band of the pulsed light. This may not matter if a narrow wavelength band is used, as in a laser light source, but if several adjacent spectral channels are used, or a wideband LED source, it is important to minimize the differences in travel time. We first examine the spread in travel time in a multi-mode fibre.

A simple way of appreciating this effect is illustrated in Fig. 18.12, where the two rays represent two different modes; as we saw in the analysis of a slab guide, the higher the order of the mode the larger the inclination of the equivalent rays to the axis. Here light pulses travel along the axis at velocity c/n_1,

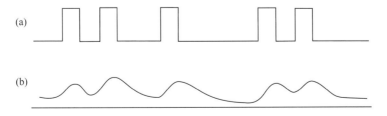

Fig. 18.11   The effect of dispersion in travel time on a train of pulses.

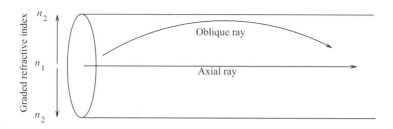

Fig. 18.12   Two rays representing two modes in a graded index fibre.

while an oblique ray at angle θ to the axis only progresses at the projected velocity $c\cos\theta/n_1$. For a step-index fibre, Eq. (18.43) allows the difference τ in travel time between rays on-axis and rays at the maximum allowed angle, over a length L, to be expressed in terms of the numerical aperture NA as

$$\tau = \frac{Ln_1\Delta}{c} \simeq \frac{L(NA)^2}{2n_1c}. \tag{18.46}$$

We now see the advantage of the graded index fibre. The path of the more oblique ray, although longer, is mainly in glass with a lower refractive index; the increased speed compensates for the extra path length.

Although multi-mode dispersion is reduced in graded index fibres, in practice it confines the use of multi-mode fibres to comparatively short or narrow-band communication links. Long-distance communications must use single-mode fibres, where dispersion effects are smaller.

There are two distinct causes for wavelength dispersion in travel time in a single-mode fibre; these are respectively the intrinsic dispersion of the glass (*material dispersion*) and *waveguide dispersion*, which is inherent in the waveguide geometry. The travel time for a pulse in the fibre is determined by the group velocity

$$v_g = \frac{d\omega}{dk}. \tag{18.47}$$

Consider first the effect of material dispersion, due to the variation of refractive index n with free space wavelength λ. The group velocity v_g is c/n_g, where n_g is the *group refractive index* given as a function of wavelength by differentiation as follows:

$$\omega = \frac{2\pi c}{\lambda} \qquad \frac{d\omega}{d\lambda} = -\frac{2\pi c}{\lambda^2}, \tag{18.48}$$

$$k = \frac{2\pi n}{\lambda} \qquad \frac{dk}{d\lambda} = 2\pi\left(-\frac{n}{\lambda^2} + \frac{1}{\lambda}\frac{dn}{d\lambda}\right), \tag{18.49}$$

$$\frac{d\omega}{dk} = \frac{d\omega}{d\lambda}\frac{d\lambda}{dk} = \frac{c}{n - \lambda(dn/d\lambda)}, \tag{18.50}$$

giving

$$n_g = n - \lambda \frac{dn}{d\lambda}. \tag{18.51}$$

The difference in travel time Δt for light at two wavelengths separated by $\Delta\lambda$, for a length L of fibre, is

$$\Delta t = \frac{L}{c}\frac{dn_g}{d\lambda}\Delta\lambda, \tag{18.52}$$

giving

$$\Delta t = -\frac{L}{c}\lambda\frac{d^2 n}{d\lambda^2}\Delta\lambda. \tag{18.53}$$

The dispersion in transmission delay $d\tau/d\lambda$ is shown in Fig. 18.13 for wavelengths near $1\,\mu m$ in silica glass. Fortunately the dispersion is very small at wavelengths close to the transmission band at $1.3\,\mu m$; this band has therefore been preferred for long-distance communications with a broad bandwidth. Techniques are, however, available for removing the effect of dispersion, and the low-loss band at $1.5\,\mu m$ is also now in general use for links with bandwidths of several gigahertz. The importance of dispersion delay may be illustrated by considering a fibre optic cable at $0.85\,\mu m$, where Fig. 18.13 gives a comparatively large delay of $98\,ps\,nm^{-1}km^{-1}$. An LED source at this wavelength might have a spectral width of $50\,nm$, so that the dispersion in delay would be $5\,ns\,km^{-1}$. In a communications link of $1000\,km$ this would spread a narrow pulse to a width of $5\,\mu s$, limiting the bandwidth to about $0.2\,MHz$.

The second cause of dispersion in a single-mode fibre is waveguide dispersion. This arises as a result of the dependence of the group velocity on the ratio between the core radius and the wavelength. An exact analysis is complex, since the effect depends on the refractive index profile. For a step-index fibre the

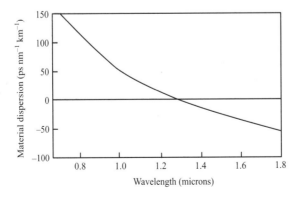

Fig. 18.13   The dispersion in group velocity as a function of wavelength.

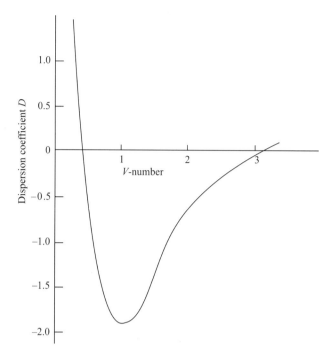

Fig. 18.14   The waveguide dispersion coefficient D as a function of V-number.

result is usually expressed as a delay Δt_w related to the parameter V, introduced in section 18.7 above, by[7]

$$\Delta t_w = \frac{L}{c}\left(\frac{\Delta\lambda}{\lambda}\right)(n_2 - n_1)DV \qquad (18.54)$$

where the dimensionless coefficient D is a function of the V-number, as shown in Fig. 18.14.

The importance of these various effects on the bandwidth of long-distance communications has prompted much analysis and experimentation. The results may be expressed as a product of bandwidth and fibre length. A typical step-index fibre bandwidth is less than 100 MHz km, due to multi-mode propagation. In a graded index fibre, where the effects of multi-mode propagation are reduced, the bandwidth/length product may be increased typically to 1 GHz km. The performance of single-mode fibres can achieve in excess of 3 GHz km.

A single fibre can be used to carry simultaneously several signal channels on different optical wavelengths, giving an increased overall signal bandwidth. This technique of *wavelength division multiplexing* requires optical filters at the transmitter and receiver.

[7] A. H. Cherin, *An Introduction to Optical Fibres*. McGraw-Hill, 1983, p. 103.

18.9 DISPERSION COMPENSATION

Comparison of Figs. 18.9 and 18.13 shows that dispersion is comparatively large at one of the wavelength bands with the lowest losses, at 1.55 μm. This band can nevertheless be exploited for long-distance broad bandwidth communication by the use of compensating devices, which introduce an equal and opposite dispersion. These devices use several simple principles which arise from the basic optics of earlier chapters in this book.

We recall first the principle of Bragg reflection in X-ray crystallography (Chapter 10), in which the reflection of X-rays from a regular crystal lattice occurred only at selected wavelengths depending on the spacing of planes in the atomic lattice. Selective reflection of light in an optical fibre can occur similarly if a periodic structure can be created along the length of the fibre as in Fig. 18.15(a). The periodic structure in this case is an artificially constructed cyclical variation of the refractive index, with a half wavelength period and extending for many wavelengths. Figure 18.15(b) shows a typical plot of reflection coefficient against wavelength for such a structure. The maximum reflection occurs when the small reflected waves from each peak in the refractive index add exactly in phase. As might be expected from considerations of coherence, as in Chapter 15, the selectivity of the reflection depends on the length of the structure.

Fibres with periodic structures can be used to isolate narrow wavelength bands in a communication system. They are known as *Bragg filters*.

The next stage is to vary the spacing linearly along the fibre, so that different wavelengths are reflected at different distances (Fig. 18.16(a)). This then becomes a dispersive element, in which the travel time in a return journey depends on wavelength (Fig. 18.16(b)). The magnitude of the dispersion-induced delay can

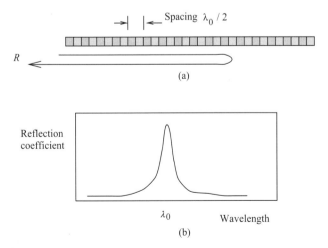

Fig. 18.15 Dispersion compensation: (a) Bragg reflection: a fibre with periodically varying refractive index; (b) the wavelength dependence of the reflection coefficient.

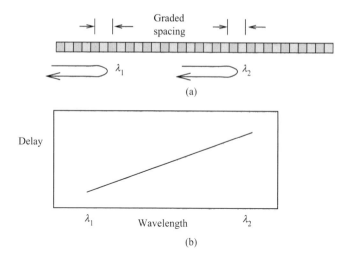

(a)

(b)

Fig. 18.16 Dispersion compensation: (a) Bragg reflector with graded periodicity; (b) delay as a function of wavelength.

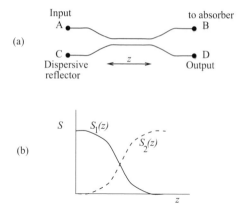

Fig. 18.17 (a) Directional coupler used to connect a dispersive reflector into a fibre optic communication system; (b) the exchange of light signal between the coupled light guides. The signal S_1 entering at port A is divided between the two fibres $S_1(z)$ and $S_2(z)$, as a sinusoidal function of the length z.

be made to match that of tens of kilometres of normal fibre in a device less than a metre overall.

The dispersive element can then be inserted into a fibre optic communication link by using a *directional coupler* shown diagrammatically in Fig. 18.17(a). The coupler consists of two light guides running close together so that their evanescent fields overlap. A signal in one of the two guides is then progressively transferred to the other as shown in Fig. 18.17(b). The length of the coupler is chosen so that half the signal from the input port A is transferred into the dispersive reflector at port C; the other half is lost in an absorber on port B. The returning signal from C, now compensated for dispersion, is then split again, half returning to the input fibre at A and half proceeding to the detector by the fourth port D.

Finally, we describe the method of imposing a periodic variation in the refractive index along the dispersive fibre element. A small but sufficient change in index (of order 1 in 10^4) can be induced in germanium doped silica by subjecting the glass to a flash of very intense ultraviolet light. The periodic structure is created by forming an interference pattern within the fibre from two coherent laser light beams incident at an angle to the fibre axis, as shown in Fig. 18.18(a). The wavelength ideally should be considerably shorter than the design wavelength of the dispersive element, so that ultraviolet light is needed; this may be obtained by harmonic generation from a longer wavelength laser, or it may be directly produced by a pulsed ultraviolet excimer laser. The regularly

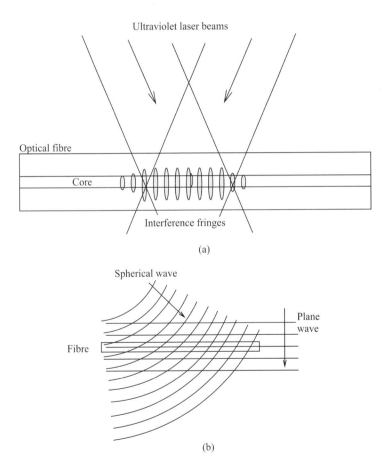

Fig. 18.18   Creating the periodic variation in the refractive index from an interference pattern (a) with uniform spacing (b). A curved wavefront provides a graded spacing.

spaced element in Fig. 18.18(a) is useful as a selective filter; by using a curved wavefront as in (b) the technique is extended to make the gradient in fringe spacing required for the dispersive element.

18.10   HOLE-ARRAY LIGHT GUIDE

The regularly spaced variation in refractive index along the length of a fibre, which is used in the Bragg filter of the previous section, may be extended to two or three dimensions to make a *photonic crystal lattice*. An example[8] with practical use in fibre optics is a regular hexagonal array of airholes along the length of an otherwise uniform silica glass fibre. If the spacing between the holes is comparable with the wavelength, the propagation of light waves within the lattice is subject to conditions similar to those of X-rays propagating in a crystal lattice, as described in Chapter 10. Such an array can be fabricated with a single missing hole, as seen in the micrograph of the cross-section (Fig. 18.19). The intact hexagonal region then acts as a light guide, in which light is trapped as it is in the core of a conventional fibre.

For light waves travelling in the direction of the fibre axis, the array of holes has the effect of lowering the refractive index. The central hexagon therefore acts as the core, and the surrounding array as the cladding, as in the conventional guide where the difference in refractive index is achieved by chemical doping. The behaviour of the hole-array fibre depends on the ratio of the diameter to the spacing of the holes, but if this ratio is less than about 0.2, then light will be propagated in only a single fundamental mode; this applies over a wide range of wavelengths. The distribution of intensity across the core is shown in the figure.

An important advantage of such a fibre is that a single-mode propagation can be achieved over a large cross-section of the core, so that the energy density can be very much lower than in the conventional fibre, avoiding the non-linear effects associated with the transmission of higher powers. Single-mode operation has been demonstrated in a fibre with core diameter up to 50 free-space wavelengths.

18.11   FABRICATION OF OPTICAL FIBRES

A crude glass fibre is easy to make by heating the centre of a glass rod and pulling the ends apart. A useful fibre, with a constant diameter and consisting of a core and cladding, needs a more sophisticated technique. Two methods are available: drawing from a pre-formed thick rod, which already contains the core and cladding, and drawing from a concentric double crucible in which the

---

[8] J. C. Knight, T. A. Birks, P. St. J. Russell and D. M. Atkin, *Optics Letters*, **21**, 1547, 1996.

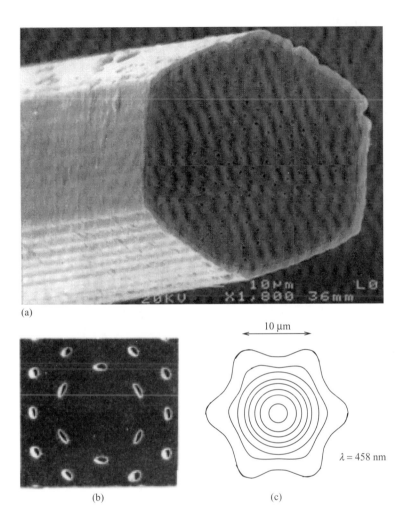

(a)

(b)   (c)

Fig. 18.19   (a) Scanning electron micrograph of a cleaved end face of a large mode area photonic crystal fibre. The fibre shown here has a core diameter of 22.5 μm and a relative air hole diameter $d/\Lambda = 0.11$, and is monomode at all wavelengths $\lambda > 458$ nm at least. (b) The central hole pattern. (c) Contour map of the near field intensity distribution for the guided mode in the fibre shown in (a) at a wavelength of $\lambda = 458$ nm. The contours are plotted at 10% intervals in the modal field intensity distribution. (Optoelectronics Group, Department of Physics, University of Bath).

two components are separately melted. The temperature (about 2000°C) at which pure silica has a workable viscosity is considerably greater than that for common glass (around 1000°C). In the double crucible method the properties of the core and cladding glasses must be reasonably well matched so that they can flow together and not be under stress when they cool. A controlled gradient of refractive index, which is essential in most applications, is obtained by drawing from a pre-form which already contains the graded index.

A 'pre-form' with graded index can be made by diffusing various dopants such as GeO_2 and P_2O_5 into silica glass. Both these dopants increase the refractive index; typically an addition of 10% gives an increase from 1.46 to 1.47. The dopant can be added by gas deposition to the inside of a tube of pure silica; the tube is collapsed later by melting. The 'hole-array' fibre described in the previous section is drawn from a pre-form which is assembled from thin hexagonal rods, which themselves have been drawn from larger diameter hollow tubes. A single solid rod is packed into the centre to form the core.

Drawing the fibre must be controlled so that the diameter is maintained within about ±2%. After the main drawing process an extra coating of some plastic material is added to protect the fibre. All these processes are adaptable to continuous operation, which typically runs for some days at a rate of up to $1 \, m \, s^{-1}$, i.e. more than 80 km per day.

18.12   FURTHER READING

A. Ghatak and K. Thyagarathan, *An Introduction to Fiber Optics*. Cambridge University Press, 1998.

G. Kaiser, *Optical Fibre Communications*, 2nd edn. McGraw-Hill, 1991.

J. M. Senior, *Optical Fibre Communications: Principles and Practice*, 2nd edn. Prentice Hall, 1992.

A. W. Snyder and J. D. Love, *Optical Waveguide Theory*. Chapman Hall, 1983.

NUMERICAL EXAMPLES 18

18.1 Light from the end of an optical fibre in air forms a patch of light 3 cm across on a screen 10 cm away. Find the numerical aperture. If the core refractive index is 1.5, find the fractional step in index between the core and the cladding.

18.2 A single-mode optical fibre has core diameter 4 μm and step in index $\delta n/n = 2\%$. From Eq. (18.40) find the minimum wavelength that will propagate in a single mode.

18.3 If the loss in a fibre is 0.5 dB km^{-1} and there is an added loss of 1 dB at joints that are 10 km apart, find the necessary interval between amplifiers when the transmitter power is 1.5 mW and the detector level is 2 μW.

PROBLEMS 18

18.1 From Eq. (18.46) find the dispersion τ in propagation time for a 5 km length of fibre with ungraded index $n = 1.5$ and $\Delta = 1\%$. What bit rate B_T could be used in this length if $B_T = 1/2\tau$?

18.2 Consider an idealized graded index fibre in which the index n varies linearly with distance x from the axis. Show that a ray follows an arc of a circle with radius $n(dn/dr)^{-1}$. Show that the optical path along the ray is independent to first order of the angle at which it crosses the axis.

(This is an extreme example of the reduction in geometrical dispersion obtained from a graded refractive index.)

18.3 The effect of material dispersion on pulse travel time in a fibre depends on the bandwidth of the light source. Find the dispersion (i) for an LED with bandwidth 20 nm and (ii) for a laser with bandwidth 1 nm for a 1 km length of glass fibre used at $\lambda = 0.85\,\mu$m where $\lambda^2(\mathrm{d}^2n/\mathrm{d}\lambda^2) = 0.025$.

19

Holography

But soft! what light through yonder window breaks?

Shakespeare, *Romeo and Juliet.*

If a scene is viewed through a window or any aperture large compared with the wavelength it is seen three-dimensional and is completely lifelike. The scene changes as we alter our viewpoint: we approach a window and look up through it to see an object in the sky for example. As the viewpoint alters, objects in the scene show *parallactic* displacements relative to each other; if we move from left to right nearby objects seem to move from right to left compared with distant ones. Another effect is that of being able to focus the eye on a particular object at a specific distance.

How different is this view through an aperture from a photograph of the same view! The photograph may give an impression of depth, but is only two-dimensional. No parallactic displacements of objects within it may be seen by a shift of viewpoint. The eye must be focused on the plane of the photograph to see it, and no eye focusing can make sharp any part of the photograph not originally brought into focus by the camera.

What is the information that has been lost in the photograph? According to the Huygens–Fresnel principle (and more precisely according to Kirchhoff's diffraction theory), the amplitude and phase at any point on the viewer's side of the screen can be deduced from the amplitude and phase in the aperture. The photograph, however, only records the square of the amplitude, and not the phase. It is the *complex amplitude* in the aperture that is the true paradigm that would enable us to reconstruct, completely lifelike and indistinguishable, the view through the window.

In this chapter we show how holography enables us to record the complex amplitude over an aperture, and so store all the information necessary to construct a three-dimensional image of the original scene behind the aperture. Hence the term holography, from the Greek word 'holos' meaning whole.

19.1   RECONSTRUCTING A PLANE WAVE

The wavefront passing through any aperture can be regarded as an assembly of elementary plane waves at various angles and with various amplitudes. We show how one of these plane waves may be recorded by combining it with a reference wave, and how it may be reconstructed. In Fig. 19.1(a) two coherent plane waves are incident on a photographic plate at equal angles α to the normal. The upper beam is to be recorded in the hologram, and the lower beam is a reference beam. As in Fig. 4.17, there is an interference pattern on the plate with a field irradiance $I = 2I_0[1 + \cos(\Delta \mathbf{k} \, \mathbf{r})]$, where $\Delta \mathbf{k} = \mathbf{k}_1 - \mathbf{k}_2$ for the two wavevectors $\mathbf{k}_1$ and $\mathbf{k}_2$. This can be developed to produce a grating with spacing d given by the familiar grating formula

$$d = \lambda/2 \sin \alpha. \tag{19.1}$$

The grating ideally has a sinusoidal distribution of amplitude; it is then referred to as a *sinusoidal diffraction grating*. Now we place the developed grating at its original position, illuminated by only one of the plane waves, which is the reference beam. Figure 19.1(b) shows the diffracted beams emerging from the grating, labelled according to the order M of diffraction at the grating; the undeviated beam is at $M = 0$. The beam at $M = 1$ is now a reconstruction of the second beam, travelling in the original direction. As a check on the direction, note that for a beam to emerge at angle θ

$$\sin \alpha - \sin \theta = M\lambda/d. \tag{19.2}$$

Since $d = \lambda/2 \sin \alpha$ the beam must be at angle $-\alpha = \theta$ as shown. The beam at $M = -1$ is an unwanted second image.

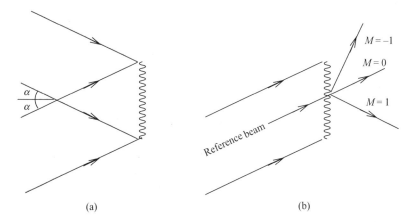

(a)   (b)

Fig. 19.1 Holographic reconstruction of a plane wave: (a) two crossing plane waves form fringes on a photographic plate; (b) diffraction at the sinusoidal fringe pattern illuminated by one of the beams. The first-order diffracted beam is the holographic image of the other beam.

Having reconstructed an elementary plane wave, we now see that any plane wave will produce a grating pattern on the photographic plate; the whole wavefront can therefore be recorded simultaneously and reproduced by illuminating the grating with the reference beam. The reference beam itself may take almost any form, provided that exactly the same beam is used in the reproduction; it need not even be a plane wave, as we now show in a more general analysis.

Let $E_0(x, y)$ be the complex amplitude in the plane of the photographic plate, and $E_R(x, y)$ the complex amplitude of the reference beam. Then the plate records the luminance $|E_0 + E_R|^2$. We assume the photographic process is linear, so that its developed opacity is k times this luminance. The light $T(x, y)$ transmitted through the grating when it is illuminated by E_R in the reconstruction is

$$T(x, y) = E_R\left\{1 - k|E_0 + E_R|^2\right\} \tag{19.3}$$

$$= E_R - kE_R E_0 E_0^* - kE_R^2 E_0^* - k|E_R|^2 E_0 - kE_R|E_R|^2. \tag{19.4}$$

The useful term in this transmitted complex amplitude is $k|E_R|^2 E_0$, which is the required complex amplitude E_0 multiplied by the luminance of the reference beam. The other four terms are best understood by considering the reference beam to be a plane wave with unit amplitude at an angle α to the normal (as in Fig. 19.1); this is $E_R = \exp(i\phi)$, where $\phi = (2\pi/\lambda)y \sin \alpha$. Then since $|E_R|^2 = 1$ we find

$$T(x, y) = \left[1 - k\left(1 + |E_0|^2\right)\right]\exp(i\phi) - kE_0^* \exp(2i\phi) - kE_0. \tag{19.5}$$

Here the last term is the required reconstructed beam. The first term is the zero-order beam in Fig. 19.1(b), in the direction of the reference beam. The second term, with twice the exponent, is at an angle α', where $\sin \alpha' = 2 \sin \alpha$; this is the unwanted beam at $M = -1$ in Fig. 19.1(b). The reconstructed rays can be interpreted as the Fourier transform of the sinusoidal grating, giving a zero-order and two first-order images.

The essence of the process is that the recording of the hologram enables both the amplitude and the phase of the object wavefront to be stored, even though the photographic plate only responds to irradiance.

19.2 HOLOGRAPHIC RECORDING

The light used for holography must be coherent over a large volume, which includes the object and the photographic plate. This requires laser light both for illuminating the object and for the reference beam. The laser provides a high degree of spatial and temporal coherence. A typical arrangement is shown in

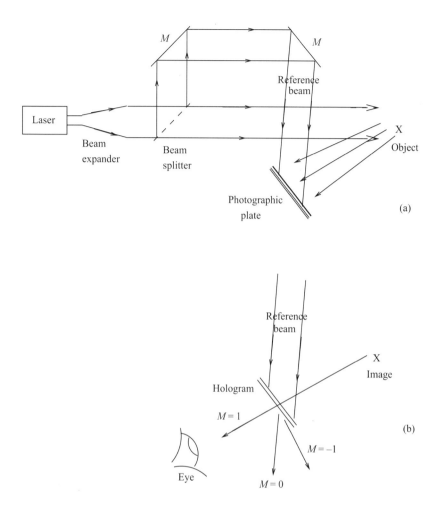

Fig. 19.2   Recording and reproducing a holographic image. The object is illuminated by laser light, and the beam splitter produces a reference beam from the same source. The developed hologram is illuminated from behind by the same laser beam.

Fig. 19.2. Here a divergent beam of laser light illuminates the object through a beam splitter; one beam provides light which is scattered back from the object to interfere with the other reference beam at the photographic plate. After development the plate is a hologram; when it is placed in the same position and illuminated by the reference beam in the same way it shows the image from behind the plate, as though the object is being seen through a window. This is three-dimensional photography, achieved without a camera lens!

The process may be understood in the simple example of Fig. 19.3. Here the photographic plate receives a coherent plane wave, which is the reference beam, and light from the same wave scattered from a point object O. The reference and scattered waves interfere to produce the holographic pattern on the plate.

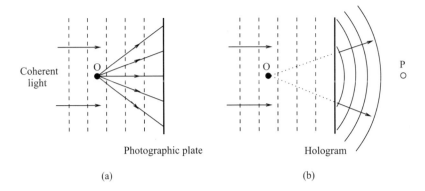

Coherent
light

Photographic plate   Hologram

(a)   (b)

Fig. 19.3 Holography using plane wave illumination. The wave diverging from the point object O interferes with the reference wave to form the hologram.

When the developed hologram is illuminated by the same plane wave, diffraction creates a diverging spherical wave which appears to originate in a point source at the position of O. (Light scattered back from the hologram will also form a diverging wave, which appears to diverge from a pseudoscopic image P.)

19.3 GABOR'S ORIGINAL METHOD

Dennis Gabor discovered the principle of holography[1] while searching for improvements in electron microscopy. His aim was to reconstruct an object from its diffraction pattern, as in X-ray diffraction from a crystal lattice. This was to be achieved by using a photographic diffraction pattern itself as a diffraction grating, so reversing the Fourier transformation of the original diffraction. As we noted in section 10.9, the problem was that phase information had been lost in the original diffraction process. Gabor's idea was to add a reference beam, which would be very difficult to achieve in X-rays or in electron microscopy, so he set out first to demonstrate the principle using light.

Gabor's first demonstration of holography was in 1948, before the invention of the laser. Only a very small-scale demonstration was possible because the coherence volume of ordinary monochromatic light sources is so small. His original system is shown in Fig. 19.4. A pinhole source of monochromatic light illuminates a small transparent three-dimensional object close to the pinhole. A photographic plate behind the object records the intensity of the diffracted light. From the viewpoint of geometrical optics the object would cast a shadow on the photographic plate; what actually happens is that at each point

---

[1] D. Gabor, A new microscopic principle, *Nature*, **161**, 777, 1948. Microscopy by reconstructed wavefronts, *Proc. Roy. Soc.* **A197**, 454, 1949.

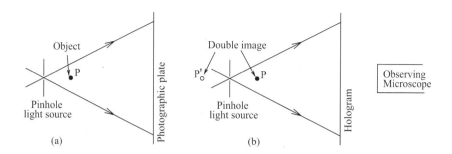

Fig. 19.4 Gabor's original system of making a hologram and reconstructing the image: (a) the hologram is recorded on a photographic plate; (b) the object is removed and the hologram is illuminated by the original plane. The image and its pseudoscopic partner can be seen by looking through the hologram.

of the plate interference occurs between the undisturbed light wave and the transmission diffraction field of the object. The undisturbed wave is then the reference beam. The developed hologram is illuminated through the same pinhole, using a lens or microscope to see the tiny object. The true three-dimensionality of the image can be demonstrated by racking the focal plane of the microscope through different layers of the holographic image.

Another interesting property is revealed by further movement of the microscope's focal plane. This is the existence of a pseudoscopic second image of the original object located behind the pinhole. This is inverted and has the property that each point on it is the same distance from the pinhole as the corresponding point on the first image. This is the second image discussed in the previous section.

19.4 ASPECT EFFECTS

When the reconstructed object is viewed through a hologram the edges of the hologram act rather like a window frame. Within the limits set by the frame, movement of viewpoint changes the aspect of the reconstructed scene, so that if it is three-dimensional, one can see more of the image to the right by moving the head to the left. Similarly, parallactic displacements may be observed.

A consideration of these aspect effects brings out another interesting property. From a given direction of observation light reaching the eye from the image only comes from a small portion of the hologram determined by the position of the eye and the angle subtended by the object, as shown in Fig. 19.5. Evidently if all the rest were removed except this piece, the object could still be seen, but only from that aspect. Thus if a hologram is broken into fragments, the reconstructed object can still be seen through each fragment, as seen from the appropriate aspect. This is rather like looking through a window that is

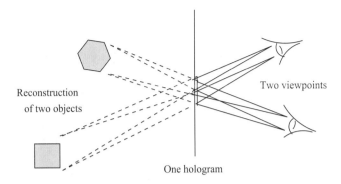

Fig. 19.5   To see a reconstructed object from one aspect, only a small portion of a hologram is used. The same portion viewed from another angle allows a different reconstruction (or a different part of the same one) to be seen.

completely obscured except for a small hole. The view is still to be seen, but only from the position allowed by the hole.

An amusing aspect of these properties is the recording of several quite different pictures on one hologram. Supposing a number of photographic transparencies are arranged so that during the recording process none obscures any other as seen from any point of the photographic plate. Then looking through the resultant hologram, or *any part of it*, the different pictures may all be seen. A very large amount of information can evidently be stored on a hologram; three-dimensional holographic memories for data storage and computing offer enormous possibilities of a vast capacity and rapid access (see section 19.10).

19.5   HOLOGRAPHIC INTERFEROMETRY

In the reconstruction of a holographic image the object is normally removed. If instead it is replaced in the same position, it will appear superposed on its image, so that light from any point will originate from the laser and reach the eye by two routes, directly and via the hologram. These will interfere, so revealing any movement of the object between the recording and the reconstruction. In this way holographic interferometry can measure displacements or distortions of objects within a small fraction of a wavelength. An example is shown in Fig. 19.6.

Dynamic effects, such as small but rapid vibrations of mechanical components, can be followed by an electronic TV camera rather than a photographic plate. The object beam is imaged onto the camera detector, together with a reference beam from the same laser source. These combine to form a speckle pattern (see Chapter 17) which can be scanned and recorded at 25 frames per second or even faster. Any small movement of the object is immediately

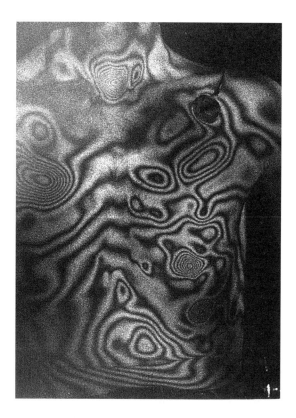

Fig. 19.6 Holographic interferometry of a human torso, showing surface movement due to the action of the beating heart. The movement in 70 ms is recorded by super-posing two holographic exposures (Hans Bjelkhagen, De Montford University).

obvious as a movement of the speckles. This technique is known as *electronic speckle pattern interferometry*.

19.6   PHASE HOLOGRAMS

The holograms discussed so far are recorded as negative or positive photo-graphic images, which are then used as amplitude transmission gratings. They have an inherent disadvantage of low efficiency in the brightness of the recon-structed image, since light must be lost in the grating. The amplitude hologram may, however, be converted into a phase hologram in which the hologram grating changes the phase instead of the amplitude of the light wave. This is achieved by storing the interference pattern as a corresponding distribution of refractive index changes within the recording film, and bleaching out the developed amplitude hologram. For silver halide photographic plates the silver metal can be converted into a transparent silver compound, with a suitable refractive index.

19.7   HOLOGRAPHY IN COLOUR

A disadvantage of the methods described above is that both the recording and reconstruction processes demand laser light and produce a monochromatic image. Reconstructions seen through such holograms are in bright red, or whatever monochromatic laser light is used; the three-dimensional and aspect effects have been gained at the expense of unreality of colour. If white light is used to illuminate such a hologram, then no reconstruction is seen at all, as an infinite number of overlapping and different sized images are produced by each wavelength present.

Curiously enough the modern key to making holograms that can be viewed in white light and produce naturally coloured images is to be found in the historic work of Lippmann in 1891 (section 4.6) and of Bragg in 1912 (section 10.9). It will be recalled that Lippmann demonstrated the real existence of standing waves close to a reflecting surface by showing that a thick photographic emulsion on top of a mirror was darkened in layers corresponding to the maxima of the interference pattern between the direct and reflected waves. Such a plate serves as a selective reflector of light of the wavelength in which it was made, for only at that wavelength do the reflections from the different layers in depth add constructively. Bragg's work on crystals shows how a three-dimensional structure can single out a particular direction or directions and send radiation to it selectively. This effect was the basis of Lippmann's process of colour photography.

The holograms we have considered so far have been essentially flat two-dimensional patterns. To make a three-dimensional hologram with sufficient structure in depth two main changes are necessary. First, a thick emulsion (up to a few millimetres, much thicker than the fringe spacing) is used for making the hologram; throughout its depth the film is transparent except where it has been blackened by an interference maximum during exposure. Secondly, the angle between the reference beam and the scattered light from the object is made large. In the most extreme case of this the angle is made almost 180°, so that the scattered light and the reference beam arrive at the photographic plate from opposite sides. In Eq. (19.1) the angle α becomes 90° and an interference structure of the order of $\lambda/2$ in depth is produced, with interference fringes parallel to the emulsion surface. The hologram is reconstructed in reflection rather than transmission.

Such holograms when illuminated by diffused white light from say a tungsten filament lamp will only transmit light of the right colour which is going in the appropriate direction. To produce full colour holograms it is necessary to illuminate the object with red, green and blue light from three separate lasers, three corresponding reference beams being used. When illuminated with an ordinary white light source this hologram produces a passably convincing three-dimensional coloured reconstruction.

19.8   HOLOGRAPHY OF MOVING OBJECTS

We now come to an interesting technical challenge in holography. The process of forming the hologram depends on the phase relationship between the reference beam and the scattered light from the object remaining constant within a few degrees during the exposure. Clearly the object must not move more than a fraction of a wavelength. Any larger movement will not just cause blurring of the reconstruction; there will be nothing to reconstruct. A consideration of a simple case is helpful. Suppose we were making a diffraction grating by allowing two coherent beams of light to meet at an angle and interfere at a photographic plate. Then obviously the movement of a wavelength or so of the source of one of the beams would move the interference fringes on the photographic plate so we would get no grating at all. The grating here is of course a simple case of a hologram.

Happily lasers that make the light for holographic work can produce incredibly short pulses. It is instructive to calculate how short an exposure is needed. If we take it that for human scenes we need to record objects moving at up to $10 \, \text{m s}^{-1}$ the exposure must be so short that movement of only $\lambda/10$ (say) happens in that time. If $\lambda = 5 \times 10^{-7} \, \text{m}$, then the exposure can last only for $5 \times 10^{-9} \, \text{s} = 5 \, \text{ns}$. Five nanoseconds is a short time by any standards: light itself only moves $1.5 \, \text{m}$ in that time. A laser pulse can however be as short as $1 \, \text{ns}$ (the shortest is about $4 \, \text{fs}$), and repetitive pulses are easily obtained. The three-dimensional cinematograph is no longer an impossible dream.

19.9   HOLOGRAPHIC OPTICAL ELEMENTS

Diffraction gratings made by the holographic technique of interfering two laser beams in a photographic emulsion may be a simple amplitude grating in a thin film of emulsion, or they may be three-dimensional, using a thick emulsion; they may also be phase changing rather than amplitude gratings. More generally, a hologram may be regarded as an optical component which will modify a light wavefront in ways that are usually associated with conventional components such as lenses, spatial filters, beam splitters and optical connections used in microelectronic systems.

The three-dimensional grating made by interfering two plane waves behaves like a crystal in X-ray diffraction. The interference pattern in the film is a regular lattice; transmission through or reflection from the hologram follows Bragg's law. If instead one of the beams is diverging, then the resulting grating will behave as a lens, since the reconstructed beam is a copy of the original beam. A plane laser beam will be focused to a spot. Movement of the hologram causes the spot to be scanned; this is the basis of the holographic scanner. The barcode scanner used in shops uses a mosaic of such holograms with different orientations formed on a circular disc, providing a multiple scan pattern when the disc is rotated and with each scan line focused to a different position in space.

19.10   HOLOGRAPHIC DATA STORAGE

A holographic image stored in a thick recording medium (a volume image) may be reconstructed by a reconstructing laser beam at the same angle as the reference beam used in the recording. At this angle the condition for Bragg reflection is satisfied, but at other angles no reconstruction takes place. This allows many holograms to be superposed in the same volume of recording medium, each able to be accessed by its own particular reference beam angle, or by its own wavelength at a particular angle. A very large amount of information can be stored in this way, which is the basis of holographic data storage and holographic memories.

A high-capacity holographic memory must be transparent through many wavelengths thickness of recording medium. This makes an amplitude grating unsuitable, and phase grating techniques must be used. Phase grating volume holograms are based on photorefractive crystals or polymers, in which the refractive index is altered by a pattern of space charge formed by photoexcited electrons. The process is reversible; the grating may be erased by illuminating the grating uniformly, so that the same material can be used as an optically rewritable memory. The potential performance is phenomenal: data may be stored at a density of 10^{11} cm^{-3} (100 Gbit cm^{-3}) and may be accessed in less than 100 μs, or transferred at a rate of 10^9 bits per second.

19.11   FURTHER READING

P. Hariharan, *Optical Holography*. Cambridge University Press, 1989.

R. Jones and C. Wykes, *Holographic and Speckle Interferometry*. Cambridge University Press, 1989.

H. M. Smith, *Principles of Holography*. Wiley, 1975.

C. M. Vest, *Holographic Interferometry*. Wiley, 1979.

20

Radiation, scattering and refraction

Fiat lux.

Genesis i: 3 (Vulgate).

Although the interaction of light with matter must ultimately be considered in terms of photons, there are many circumstances when a classical electromagnetic theory of radiation in terms of accelerated charges provides a practically complete description. Scattering and absorption of light by individual charges and atoms is also usefully considered by a classical electromagnetic approach.

In this chapter we consider examples of radiation processes in which an isolated electron is accelerated either in the electric field of a positive ion, as in a dilute ionized gas, or when its motion is deflected by a magnetic field, as in *synchrotron radiation*. Both processes are enhanced in an interesting way when the electron has a very high energy. We will also outline the theory of *Thomson scattering* by free electrons and *Rayleigh scattering* by individual atoms and molecules.

These radiation and scattering processes are all concerned with isolated particles behaving independently of one another. The emitted or scattered radiation from many particles adds with random phase, so that it is the power and not the amplitude that is proportional to the number of radiating or scattering events. In dense materials it is no longer appropriate to consider individual particles, and we consider instead the *polarization* of a dielectric and its effect on an electromagnetic wave.

20.1 RADIATION PROCESSES

We first discuss briefly the basic equation for the radiation from a single accelerated charge. In classical electrodynamic theory the electric field generated by a single charge has three components. The first is the electrostatic field, whose strength varies as the inverse square of the distance from the charge. A uniformly moving electron generates a second field component which is

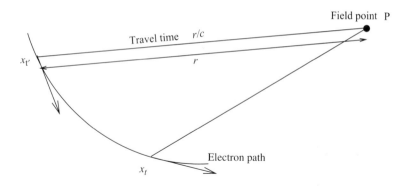

Fig. 20.1   Radiation from an electron. The radiated electric field at the field point P at a time t depends on the acceleration of the electron at time $t' = t - r/c$.

proportional to its velocity; this field also falls away as the square of the distance from the charge. The third component of the field is only present when the charge is accelerated. This is the radiation field, which falls away only as the first power of the distance.

The strength of the radiation field at a point some way from the accelerating charge depends on the component of the acceleration perpendicular to the line of sight. Since radiation reaches the field point after a finite travel time, the acceleration must be measured at the time the radiation leaves the charge rather than when it arrives (Fig. 20.1). The field is given by

$$E = -\frac{q}{\epsilon_0 c^2 r} \times (\text{acceleration}), \qquad (20.1)$$

where by (acceleration) we mean:

(i) the component of acceleration projected onto a plane perpendicular to the line of sight from the field point to the charge;
(ii) the acceleration is measured at time r/c earlier, where r is the distance from the charge and c is the phase velocity of the radiated wave.

The radiated electric field is therefore always transverse and perpendicular to the line of sight. The magnetic field B is also transverse; it is perpendicular to the electric field E.

20.2   THE HERTZIAN DIPOLE

A dipole is a pair of equal and opposite charges separated by some distance. The product of charge and distance is the *dipole moment*. A dipole whose moment oscillates sinusoidally, and whose dimensions are small compared with the corresponding radiated wavelength, is known as a Hertzian dipole. It radiates in the same way as a single oscillating charge, since the two opposite charges oscillate with opposite accelerations and their radiated fields must therefore add. We therefore consider only one charge q, moving linearly according to

$$x = x_0 \exp(i\omega t). \tag{20.2}$$

The component of acceleration as seen from the field point P at a distance r in free space, and in a direction θ from the axis of the dipole (Fig. 20.2), is then easily found from Eq. (20.1), giving a field:

$$E = -\frac{\omega^2 q x_0}{\epsilon_0 c^2 r} \sin\theta \exp i\omega(t - r/c). \tag{20.3}$$

The variation with θ is known as the *radiation pattern* or *polar diagram*. This is shown in Fig. 20.2(b); note that the radiation pattern of a radio antenna is often plotted as variation of power (E^2) rather than field (E). The term r/c in the oscillatory function is already familiar in a propagating wave as the phase term $\phi = \omega r/c$.

The total power W radiated by the dipole is found by integrating over a sphere the power flux $\epsilon_0 E^2 c$, giving

$$W = \frac{2}{3c^3} \omega^4 q^2 x_0^2. \tag{20.4}$$

The radiated power W depends on the square of the dipole moment (qx_0), and on the fourth power of the frequency.

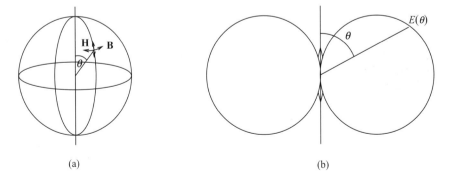

(a)

(b)

Fig. 20.2 Radiated field from a Hertzian dipole: (a) the field at P is aligned along the surface of the sphere, with strength proportional to $\sin\theta/r$ (b) the variation of field strength with θ can be represented as a 'polar diagram'.

20.3   FREE–FREE RADIATION

Light may be emitted by the electrons of an ionized gas either in discrete spectral
lines or as a continuum caused by their acceleration in the Coulomb field of
ionized nuclei. The radiation from a single electron in an ionized gas, accelerated
as it encounters a positive charge, is known as *free–free radiation*, to distinguish
it from processes involving quantized energy levels. In the encounter the electron
loses energy; the radiation is therefore also known as *bremsstrahlung* (German
for 'braking radiation'). (The acceleration of the nucleus need not be considered
as it is smaller than that of the electron by the ratio of their masses.)

   The acceleration of the electron is in the form of an impulse, shown in Fig.
20.3; it therefore has a broad spectrum, in contrast to the narrow bandwidth
radiation from an oscillating dipole. The closest encounters give the narrowest
pulses, containing the highest frequencies: a full calculation of the overall
spectrum requires integration over a range of impact parameters and electron
energies. An upper limit to the emitted frequency is set by the photon energy:
this cannot be greater than the original electron energy. This limit is encountered in an X-ray discharge tube, where electrons are accelerated by a high
voltage V. The minimum wavelength of the X-rays emitted when the electrons
reach the target anode is

$$\lambda_{\min} = \frac{hc}{eV} \tag{20.5}$$

and there is a continuum of radiation emitted at lower frequencies. The emis-
sivity of the gas is proportional to $n_e n_i Z^2$, where n_e and n_i are the electron and
ion densities and Ze is the nuclear charge.

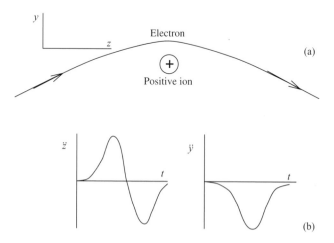

Fig. 20.3   Free–free, or *bremsstrahlung*, radiation from an electron accelerated in its
encounter with a charged nucleus. Two components of acceleration are shown; in the
direction of travel z the electron speeds up and slows down, while there is a single
impulse normal to z.

Free–free radiation is encountered over a wide range of the electromagnetic spectrum, including radio emission from interstellar clouds and the continuum emission from X-ray tubes; it is also important in high-energy laboratory physics, since it is responsible for the energy loss of high-energy electrons in solid matter.

20.4 SYNCHROTRON RADIATION

Electrons may be accelerated by a variety of forces, such as the electric fields from a radio transmitter or the electrostatic field near ions in an ionized gas. In synchrotrons high-energy electrons and protons are continuously accelerated radially by the magnetic field which confines them to a circular orbit. This acceleration, as for any other acceleration, produces radiation, and the power radiated is a serious limitation on the energy which can be given to electrons in a synchrotron. This *synchrotron radiation* mechanism is also responsible for light and radio waves from astronomical sources such as quasars and supernovae.

Consider an electron with large energy, γ times the rest energy, moving with velocity near c perpendicular to a magnetic field B (Fig. 20.4). The electron orbit is a circle, which is followed at a frequency ν_L/γ, where ν_L is the Larmor frequency $eB/2m_0$. Seen perpendicular to the orbit, the acceleration is towards the centre of a circle, and a circularly polarized wave is emitted along the magnetic field. The much more intense radiation in directions close to the plane of the orbit must however be calculated using the acceleration at the corrected time $(t - r/c)$; this has a large effect when the electron velocity is close to c. Instead of a sine wave the radiated field appears as in Fig. 20.4(b). The effect is to compress the time during which the electron moves towards the observer, and stretch the time during which it is receding (the relativistic Doppler effect: see section 4.7).

Each time the electron travels towards the observer a sharp peak is radiated. The electric field therefore consists of a regular series of pulses, separated by the orbital period of γ/ν_L. The width τ of the peaks decreases rapidly with increasing electron energy; it may be shown geometrically that

$$\tau \sim \frac{1}{\gamma^2 \nu_L}. \tag{20.6}$$

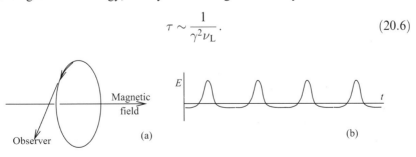

Fig. 20.4 Synchrotron radiation: (a) an electron gyrating in a magnetic field is viewed from a point in the plane of its orbit. The radiated electric field (b) has a sharp maximum each time the electron travels towards the observer.

The waveform of the radiation can now be Fourier analysed. It contains harmonics of the period γ/ν_L up to a frequency of the order of γ^{-1}, i.e. up to a frequency of approximately $\gamma^2\nu_L$.

The light from the Crab Nebula is generated by electrons with cosmic ray energies, probably up to 10^{12} eV, moving in a magnetic field of the order of 10^{-6} tesla (10^{-2} gauss). Radio astronomy has shown that the interstellar gas contains a magnetic field of only 10^{-10} tesla, but this is still sufficient to accelerate the cosmic ray electrons and produce measurable radio waves. By contrast an electron synchrotron machine might have a field of 1 tesla, and visible light would be emitted by electrons with energies of 10^9 eV. Synchrotron machines with energies up to 10^{10} eV provide powerful sources of X-rays for diffraction analysis of many different types of organic and inorganic materials.

20.5   CERENKOV RADIATION

A charge moving with high relativistic velocity in a dielectric may radiate without any acceleration transverse to its direction of motion, if its velocity is higher than the local phase velocity; this is *Cerenkov radiation*, which is important for example when a cosmic ray particle enters the Earth's atmosphere.

It is easy for a boat to travel faster than the waves it generates, and it is commonplace for an aircraft to travel faster than the speed of sound. In both cases a sharp wavefront travels out in a V-shape, at an angle whose sine is the ratio between the group velocity of the waves and the velocity of the vehicle. By analogy, it might be expected that electromagnetic waves can be generated by a charge moving faster than the wave velocity c/n. Light emitted from an electron moving with a relativistic velocity in a medium with $n > 1$ was first observed by Cerenkov, and is named after him.

We note first that the electron itself has no acceleration normal to its velocity, which can account for the radiation. It is in fact possible to analyse the radiation as generated by the electron by expressing the field in terms of the electron charge, its velocity and its acceleration, at a suitable retarded time $(t - nr/c)$, but it is simpler to regard the electron as exciting the dielectric medium as it progresses, and then to consider the dielectric as the source of the radiation (Fig. 20.5). The dielectric becomes polarized, the electron attracting positive charge and repelling negative, so that as the electron passes each part of the dielectric it acts like a small dipole giving only one impulsive oscillation. The radiation from all impulses along the track adds in phase where the Huygens' wavelets coincide.

What is the angle α between the Cerenkov ray and the electron track? Consider a single wavelength component of the impulsive radiation from a point in the medium. The track is now a one-dimensional array of radiators, phased progressively as

$$\phi(x) = \frac{2\pi n\upsilon x}{\lambda}\frac{}{c},$$ (20.7)

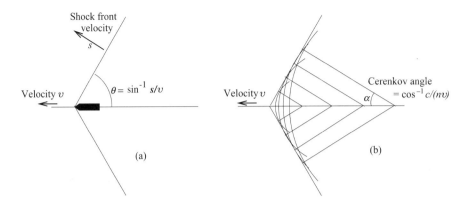

Fig. 20.5 Comparison of a shock wave in air with Cerenkov radiation from an electron: (a) a shock wave from a bullet, forming a cone with a semi-angle θ; (b) a Cerenkov wave from an electron in a medium with refractive index n. Huygens' wavelets form the wavefront.

where v is the velocity of the electron. By analogy with the diffraction grating (Chapter 10) we know that this radiates a cone at an angle $\alpha = \cos^{-1}(c/nv)$, provided that n/c is the *phase* velocity of the wave. There is a small paradox to resolve here, since the Huygens' wavelets of Fig. 20.5 obviously travel at the *group* velocity, leaving us uncertain about the true value of α. The solution is simply that a true impulse only continues without change of form if the medium is non-dispersive. The correct value for α when the medium is dispersive is obtained from the group velocity in the range of wavelengths that is being observed.

The high-energy particles of a cosmic ray shower generate a flash of Cerenkov light, which shines within a few degrees of the direction of the shower axis. Light generated by high-energy particles as they pass through liquids or gases can be used as a sensitive detector for cosmic ray showers, or for particles from accelerators. When the Cerenkov angle α can be measured it gives a direct and simple measurement of the velocity of high-energy particles.

20.6   RAYLEIGH SCATTERING

The scattering and absorption of electromagnetic waves in gases often involves quantum processes. In a neutral gas, however, and for low photon energies, it is often appropriate to treat the neutral atoms as classical oscillators, as in section 20.11; these oscillators respond to the electric field of the electromagnetic wave, and reradiate some of the incident energy at the same frequency but in directions away from the direction of the incident wave. The amplitude of the oscillation is found from the response of a resonator of mass m to an applied electric field with frequency ω:

$$\frac{d^2x}{dt^2} + \frac{1}{\tau}\frac{dx}{dt} + \omega_0^2 x = \frac{eE_0}{m}\exp(i\omega t), \tag{20.8}$$

where τ is the classical lifetime and ω_0 is the resonant frequency. The response is an oscillation with displacement $x(t)$:

$$x(t) = \frac{eE_0}{m}\frac{\cos(\omega t + \phi)}{m\left\{\left(\omega_0^2 - \omega^2\right)^2 + \left(\omega\Delta\omega_{1/2}\right)^2\right\}^{1/2}}, \tag{20.9}$$

where E_0 is the incident field. The phase difference ϕ is of no consequence; furthermore, at low frequencies (i.e. low photon energies) the terms in ω^4 and $\left(\omega\Delta\omega_{1/2}\right)^2$ can be neglected in comparison with ω_0^4. The result is that the amplitude of oscillation is independent of the wave frequency.

The radiated electric field from the electron is proportional to its acceleration, found by twice differentiating Eq. (20.9) with respect to time. This is proportional to ω^2, and the reradiated power is therefore proportional to ω^4, or to λ^{-4}. This is the *Rayleigh law of scattering*. Its most familiar application is to the scattering of visible sunlight in the Earth's atmosphere, where blue light is scattered more than the longer-wavelength red light.

The Rayleigh law can also apply to the scattering of light in transparent solids or liquids, where the scattering elements are irregularities in density or structure rather than individual atoms. An important example is in the transmission of light along glass fibres, as used in long-distance communications (Chapter 18). Here the lowest achievable attenuation is mainly limited by Rayleigh scattering by small irregularities in the glass. The Rayleigh law dictates the use of the longest infrared wavelengths for which light sources and detectors are available, but avoiding the hydroxyl ion resonance at $1.38\,\mu\text{m}$; in practice the lowest attenuation is obtained at about $1.55\,\mu\text{m}$.

20.7   RAMAN SCATTERING

The scattering of light by atoms and molecules can be described by the Rayleigh law only for light with low photon energy. When the photon energy is comparable with or greater than the resonance energies in the scatterers, there may be a quantum interchange of energy, so that a photon emerges from a collision with a different energy.

The response of the damped driven classical oscillator, as in Eq. (20.9), is maximum at resonance, when $\omega = \omega_0$ and the photon energy is close to a resonant energy. The scattering that follows from this response corresponds to the re-emission of photons with unchanged energy. This is not always the case: the atom or molecule may interact with a photon via a virtual state and emit a photon of different energy; this inelastic scattering is known as *Raman scattering*.

Two types of Raman scattering may be distinguished. Normally the emergent photon has a lower energy than the incident photon: this is the usual inelastic scattering. If, in contrast, the scatterer is already in an excited state before the interaction, a photon may be emitted with higher energy than the incident photon: this is referred to as super-elastic scattering. Analysis of both types must provide both for the conservation of energy and for the conservation of momentum, as for Compton scattering of free electrons.

20.8   THOMSON AND COMPTON SCATTERING BY ELECTRONS

The electrons of an ionized gas may scatter electromagnetic radiation in a simple classical process, analogous to Rayleigh scattering by neutral atoms. This is known as *Thomson scattering*, for which the individual electrons have the *Thomson cross-section*

$$\sigma_{\mathrm{T}} = \frac{8\pi r_0^2}{3} = 6.652 \times 10^{-29} \ \mathrm{m}^2, \tag{20.10}$$

where r_0 is the classical electron radius. In Thomson scattering the radiation is not absorbed, but reappears as radiation travelling in a different direction. There is no change of frequency; in quantum terms the photons collide elastically with the electrons. This applies when the photon energy $h\nu$ is much less than the rest mass of the electron $m_0 c^2$.

For higher photon energies the scattering can only be regarded as a quantum process in which a collision between an electron and a photon involves an exchange both of energy and momentum. This is the realm of *Compton scattering*, in which the photon emerging after the collision has a lower energy, i.e. a longer wavelength, the amount of the change depending on the geometry of the collision. The analysis, which depends on the conservation both of energy and of momentum, gives the Compton scattering formula for the increase in wavelength $\Delta\lambda$ as a function of θ, the angle through which the photon is deviated:

$$\Delta\lambda = \frac{h}{m_0 c}(1 - \cos\theta). \tag{20.11}$$

Compton scattering is observed in X-rays passing through a solid or a gas. The essential interaction is between a high-energy photon and an individual electron, whether or not that electron is bound to an atomic nucleus.

20.9   POLARIZATION IN DIELECTRICS

The processes so far described in this chapter are concerned with radiation from or scattering by individual isolated electrons or atoms. When they are close

together, as in a solid, they can no longer be considered as individuals. Instead
we consider the effect of a continuous dielectric on an electromagnetic wave in
terms of the *electric polarization* induced by the electric field of the wave. The
polarization represents a charge separation within the material, forming a
dipole moment that oscillates at the wave frequency: this reradiates a secondary
wave, which combines with the original wave. The combination may have a
different phase from that of the original wave: this means that the phase
velocity of the resultant wave has been determined by the polarization properties of the dielectric; that is, by its dielectric constant.

The velocity of electromagnetic waves is given generally by $c/(\epsilon\mu)^{1/2}$, where c
is the velocity in free space. For a pure dielectric, whose magnetic susceptibility
is unity, the refractive index therefore equals $\epsilon^{1/2}$, the square root of the
dielectric constant. The relation between ϵ and the polarization P is given by
simple electrostatic theory as

$$\epsilon E = E + \frac{P}{\epsilon_0}. \tag{20.12}$$

The polarization P is defined as the electric dipole moment per unit volume.
If there are N atoms per unit volume, each polarized by the field E to form a
dipole with moment αE, then

$$\epsilon = 1 + \frac{\alpha N}{\epsilon_0}. \tag{20.13}$$

We must allow for the possibility that the atomic polarization is not in phase
with the field E, which may happen when the atom behaves like a resonant
atomic oscillator. The refractive index $\epsilon^{1/2}$ is therefore complex, having real and
imaginary parts. If it is $n - i\kappa$, the effect is that the wave becomes

$$E = E_0 \exp i\{\omega t - (n - i\kappa)kx\} \tag{20.14}$$

in which k is the wave number in free space. The quadrature component κ then
represents absorption, since Eq. (20.14) becomes

$$E = E_0 \exp(-\kappa kx) \exp\{i(\omega t - nkx)\}. \tag{20.15}$$

The wave proceeds with phase velocity c/n, while the amplitude decays exponentially with distance. Equation (20.13) becomes

$$(n - i\kappa)^2 = n^2 - \kappa^2 - i2n\kappa = 1 + \frac{\alpha N}{\epsilon_0}, \tag{20.16}$$

and if κ is small enough, as in a dielectric where the attenuation is small,

$$n^2 \sim 1 + \frac{\alpha N}{\epsilon_0}. \tag{20.17}$$

A further approximation may be made if n is close to unity:

$$n \sim 1 + \frac{\alpha N}{2\epsilon_0}.$$ (20.18)

We now have the required relationships between the refractive index of a medium and the response of its charged particles to the oscillatory electric field. The response, represented by the polarizability α, will now be found for two particularly interesting cases.

20.10   FREE ELECTRONS

A particularly simple situation arises when the dielectric polarization is entirely due to electrons that are not bound to nuclei. This is the case for radio wave propagation through the ionosphere, and also for X-rays in metals, which behave like dielectrics for these short wavelengths. Neglecting the effect of collisions between electrons and ions in the ionosphere, and the effects of lattice irregularities in metals, the polarizability α for a single electron is found from the simple equation of motion

$$\frac{d^2x}{dt^2} = \frac{eE}{m} = \frac{eE_0}{m} \exp(i\omega t).$$ (20.19)

The electrons all follow an oscillatory displacement $x = x_0 \exp(i\omega t)$, and the polarizability α is the dipole moment per unit field:

$$\alpha = \frac{ex}{E} = -\frac{e^2}{m\omega^2}.$$ (20.20)

Substituting in Eq. (20.17) we find

$$n^2 = 1 - \frac{Ne^2}{\epsilon_0 m\omega^2}.$$ (20.21)

This gives the refractive index for any dielectric where the polarization is due to free electrons. It may be written as

$$n^2 = 1 - \frac{\nu_p^2}{\nu^2},$$ (20.22)

where ν_p is typical of the dielectric. For the low-density terrestrial ionosphere ν_p is at radio frequencies, reaching 10 MHz where the electron density is greatest (i.e. in the F-region), while for metals it lies in the ultraviolet. For frequencies above ν_p the refractive index is real and less than unity, and the phase velocity is therefore greater than the free space velocity c. It is easy to show that if Eq. (20.22) holds, then the product of the group velocity and phase velocity is c^2.

20.11   RESONANT ATOMS IN GASES

The second case for which we can easily evaluate the polarizability, and hence obtain the refractive index, concerns neutral atoms in gases. In a low-pressure gas the atoms may be assumed to act independently. Electrons are bound within the atoms at a series of energy levels; the atoms emit and absorb light with photon energies equal to the differences between energy levels. The classical analogy is a simpler picture, in which the atoms are considered as oscillators that resonate at the appropriate spectral line frequencies. These resonators respond to the electric field of a wave with a forced oscillation, whose amplitude depends on the resonant angular frequency ω_0, and the sharpness of resonance, characterized by a width $\Delta\omega_{1/2}$ corresponding to a lifetime $\tau = 2/\Delta\omega_{1/2}$. The approach of Eqs. (20.19) and (20.20) then leads to a polarizability

$$\alpha = \frac{e^2}{m} \frac{1}{\omega_0^2 - \omega^2 + i\omega\Delta\omega_{1/2}}. \tag{20.23}$$

Some awkward algebra is needed to relate α through Eq. (20.23) to the refractive index $n - i\kappa$, but if κ and $n - 1$ are both small, as is appropriate for a low-pressure gas, then a good approximation may be found:

$$n = 1 + \frac{Ne^2}{2\epsilon_0 m} \frac{\omega_0^2 - \omega^2}{\left(\omega_0^2 - \omega^2\right)^2 + \left(\omega\Delta\omega_{1/2}\right)^2}, \tag{20.24}$$

$$\kappa = \frac{Ne^2}{2\epsilon_0 m} \frac{\omega\Delta\omega_{1/2}}{\left(\omega_0^2 - \omega^2\right)^2 + \left(\omega\Delta\omega_{1/2}\right)^2}. \tag{20.25}$$

The form of the variation of n and κ with the wave frequency is shown in Fig. 20.6. Well away from resonance, the refractive index increases steadily with

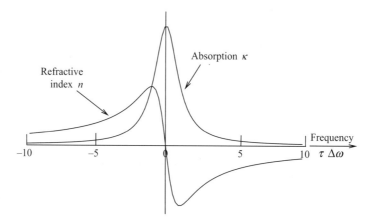

Fig. 20.6   The variation of absorption and refractive index in the region of a resonance. The width of the resonance is determined by the lifetime τ.

frequency, and the absorption is small. As the wave frequency approaches the atomic resonance, and the amplitude of the atomic oscillators increases, the absorption increases and the dispersion in the refractive index changes sign. This *anomalous dispersion* is found at every resonant frequency of the atoms in the gas. Anomalous dispersion is most marked in gases at low pressure, since it is only when the atoms behave independently that a sharp resonance is found. When resonators of the same frequency are coupled together the combined resonance curve is broadened; in quantum terms this means that the lifetime of an excited state is reduced, while in terms of refractive index there are less marked effects of anomalous dispersion.

In real gases the atoms will have many resonances, each of which has an associated anomalous dispersion. Molecular gases also have rotational and vibrational resonances, occurring generally at lower frequencies. A complete curve showing refractive index and absorption for a gas therefore shows a series of the dispersion curves of Fig. 20.6. Away from resonances, and where the absorption is negligible, the refractive index increases with wave frequency; this is seen in glass, where the refractive index is higher at the violet than at the red end of the spectrum. When the wave frequency is very much higher than the resonant frequencies, the refractive index becomes less than unity; this is the free-electron case discussed in the previous section.

The quadrature component κ of the complex refractive index is the *extinction coefficient*. The wave propagates with phase velocity c/n, and with an amplitude decreasing exponentially with distance. This attenuation is expressed by an *absorption coefficient* μ, such that the fraction of the intensity lost in a thin slab with thickness dx is μdx. An expansion of Eq. (20.15) shows that

$$\mu = \frac{2\omega\kappa}{c}.$$

(20.26)

The absorption may be regarded as the sum of the absorptions by all the individual atoms, each of which must then have a *cross-section for absorption* $\sigma = \mu/N$.

20.12   ANISOTROPIC REFRACTION

In most substances the refractive index is independent of the direction of propagation and independent of the polarization of the light. From the foregoing discussion this is obviously due to isotropy in the structure of the substance: the polarization is independent of the direction of the electric field. For some materials, however, this may not be so. In a crystal the unit cell may be anisotropic, so that the polarizability depends on the direction of the electric field in relation to the directions of the crystal axes. This leads to the birefringence already discussed in section 6.4. Again, a liquid may be composed of molecules with a permanent dipole moment, so that they may be aligned by an

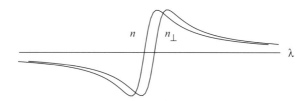

Fig. 20.7 Anisotropic dispersion. Strong magnetic or electric fields can change the resonant frequency of atomic oscillators so that the dispersion curve is displaced between two planes of polarization. The dielectric then shows magnetic or electric double refraction. The magnetic effect is known as the Voigt effect. Some substances show an induced birefringence well away from resonances. This is due to the reorientation of molecules in the field. These effects are known as the Kerr effect (electric) and the Cotton–Mouton effect (magnetic).

external steady electric field. In this way a dielectric liquid may be made birefringent, and used to modify the polarization of light passing through it. Induced birefringence (section 6.7) in a liquid is known as the *Kerr effect*; a related phenomenon in which the existing birefringence of a crystal is modified by an electric field is known as the *Pockels effect*.

Both these effects relate to the refractive indices of plane-polarized components of a light ray. The two components will also show anomalous dispersion for frequencies close to atomic resonances. Figure 20.7 shows a further effect in which the resonant frequency of an atomic oscillator is changed by a strong electric or magnetic field. The dispersion curve is then displaced in the wavelength between two planes of polarization, giving a large birefringence in the region of the resonance. This is the *Voigt effect*.

Optical activity, discussed in Chapter 6, is due to a difference in refractive index between the two hands of circular polarization.

20.13   FURTHER READING

H. C. van der Hulst, *Scattering of Light by Small Particles*. Dover, 1984.

21

The detection of light

Get the glass eyes; / And like a scurvy politician, seem to see those things thou dost not.
Shakespeare *King Lear*.

The range of photon energies, from the negligibly small quanta of long radio waves to the overwhelmingly large quanta of cosmic gamma rays, is reflected in the wide variety of methods of detecting radiation. At the extremes, there is no obvious connection between the measurement of the oscillating electric field of a radio wave and the measurement of the momentum interchange in a collision between a gamma-ray photon and a material particle. All methods lead to a measurement of the flow of energy in radiation, and when we consider the ultimate sensitivity of any measurement we specify it in terms of the smallest quantity of radiant energy that can be detected. At short wavelengths this smallest quantity is the energy in a single photon; at long wavelengths the sensitivity is limited by thermal effects in the detector and individual photons cannot readily be detected.

Light occupies a middle position in the electromagnetic spectrum, where most detectors are photonic. In this chapter we concentrate on photonic detectors and include only a brief section on thermal detectors. The photonic detector may involve photoemission, in which an electron is excited and detected outside the photoemitting surface, or where a photon excites an electron from the valence band to the conduction band of a semiconductor, so creating an electron–hole pair. Mobile electrons and holes within a semiconductor may be detected by applying an external voltage, as in photoconduction, or as a current in the internal electric field of a semiconductor photodiode.

21.1 PHOTOEMISSIVE DETECTORS

In all *photoemissive* detectors an incident photon frees an electron from a solid by ionization. If the electron has sufficient energy it may escape from a photoemissive surface; this is the *photoelectric effect*. The photoemitting material may be a metal or a semiconductor. The minimum photon energy for photoemission depends on the material; it is called the *work function W*. If the photon has sufficient energy, the electron may leave the surface with energy E given by

$$E = h\nu - W. \tag{21.1}$$

The minimum frequency ν for photoemission corresponds to a maximum wavelength λ_0 given by

$$\lambda_0 = \frac{1.24 \times 10^3}{W} \text{ nm}, \tag{21.2}$$

where W is in electron-volts. Each photoemissive material has its own work function; for example, for gold $W = 4.5$ eV, so that only photons of light with a wavelength shorter than 275 nm can release electrons from it. W is smaller for the alkali metals; for caesium $W = 2.1 eV$, giving a limiting wavelength of 590 nm, while for a widely used alloy of alkali metals NaKCsSb (the S20 cathode), the wavelength limit is 850 nm.

Even if an electron is given enough energy to overcome the work function, it may be trapped in the bulk of the photoemissive material. The proportion of emitted electrons to incident photons is the *quantum efficiency η*. Typically only about 10% of the electrons energized by photons reach the surface and are detected, corresponding to $\eta = 0.1$.

Figure 21.1(a) shows the wavelength dependence of quantum efficiency for a typical S20 photocathode, with $\eta = 0.2$. Figure 21.1(b) shows the *responsivity R*,

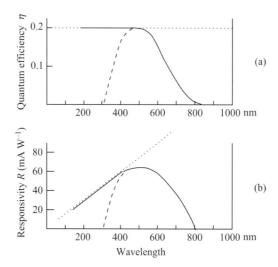

Fig. 21.1   (a) The quantum efficiency η and (b) the responsivity R of a typical photo-emitter, showing the fall in performance at lower quantum energies. The dotted line shows the performance expected with a quantum efficiency $\eta = 0.2$, independent of wavelength. The broken line indicates the effect of absorption at higher quantum energies.

which takes into account the wavelength dependence of the photon energy; it is
defined as the ratio of the photoelectric current A (amps) to the input signal
power I (watts). Semiconductors such as gallium arsenide (GaAs) or caesium
oxide (Cs_2O) are widely used as photoemitters; they have a lower surface
reflectivity than metals, and they can provide quantum efficiencies of up to 30%.

In a simple photoelectric detector (Fig. 21.2) the electrons from the emitting
surface (the *photocathode*) in a vacuum tube may be collected by an electrode
(the *anode*); the current is then proportional to the rate of incidence of photons.
At low light levels it is possible to detect the arrival of individual photons by
amplification in a *photomultiplier* tube (Fig. 21.3). Here the electrons emitted
from the photocathode are accelerated to a second metal surface, or *dynode*,

Anode   Photocathode

Fig. 21.2   Photoelectric tube. An electron emitted by the photocathode is collected by
the anode.

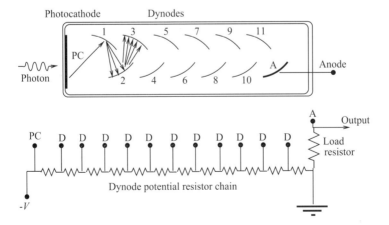

Fig. 21.3   Photomultiplier tube. An electron emitted by the photocathode PC is accelerated to D_1, the first of a series of dynodes. At each dynode each electron stimulates the
emission of several secondary electrons, and a large pulse of current is collected by the
anode A. The potentials of the dynodes are set by the resistance chain, and the output
voltage pulse is developed across the load resistor.

which emits several electrons each time a primary electron strikes it. An accelerating potential of typically 1–2 kV is required between the anode and the photocathode. A series of these dynode secondary emitting stages can be used, usually from 6 to 14, eventually multiplying the charge by a factor of 10^5 to 10^8. Photomultipliers may be used either to determine an average light intensity by recording the direct current, or to detect pulses of light, including low light levels corresponding to individually detected photons.

The usefulness of all photoelectric detectors at low light levels depends both on the proportion of photons that produce a detectable output, and on any random output, or *noise*, generated inside the detector. (We consider in section 21.5 below how to quantify the overall performance of a detector at low light levels.) In the photomultiplier most of the noise is generated by the random emission of electrons from the first photoemissive surface due to thermal excitation; there is however an added noise from the dynodes, mainly from the first. The background current from thermal emission is known as the *dark current*. The added noise from the photocathode depends on the temperature and the work function. At room temperature the thermal energy of electrons is of order 1/40 eV, so that for a metal such as gold with a high work function this random background is negligible. Photomultipliers for high photon energies may therefore be made to be virtually noise free.

A single primary photon falling on the photocathode produces an output pulse of appreciable length; this may be important since it limits the photon counting rate which can be achieved without overlapping pulses. The output pulse length from a photomultiplier is determined mainly by the spread in travel time of electrons between the dynodes, which typically restricts the time resolution to about 10 ns. A time resolution of less than 1 ns (10^{-9} s) is achieved in photomultipliers designed for a high counting rate.

Photomultipliers are widely used for their high sensitivity and time resolution, especially for operation at room temperature.

21.2 SEMICONDUCTOR DETECTORS

As we saw in Chapter 16, electrons in the valence band of a semiconductor can become mobile if given sufficient energy. In a photonic detector the energy to free the electron is derived from an incident photon. Given sufficient energy the electron freed by a photon may escape the solid entirely, in the process of photoemission. Given a lower energy (but greater than the energy gap), the electron may become mobile within the semiconductor. A mobile hole is also created within the valence band. Semiconductor detectors may detect individual electrons and holes, or a number of mobile electrons within the crystal lattice may be detected as an increase in the conductivity.

We recall from Chapter 16 that the electron energies within the crystal lattice of a pure semiconductor are almost all constrained to lie within the valence band, where they occupy almost all available energy levels. Above this band of

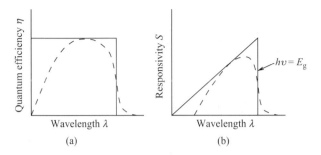

Fig. 21.4 Ideal (full line) and practical (broken line) photoconductor detector responses: (a) quantum efficiency η; (b) responsivity S.

energies is the bandgap, and above this gap again is the conduction band. The excitation of an electron into the conduction band leaves a hole. If an external electric field is applied, the electron and the hole move in opposite directions, the electron moving faster than the hole. This is the action of a *photoconductor*, in which the rate of arrival of photons with sufficient energy is measured by an increase in conductivity.

The *responsivity* of a photoconductor detector is the output current divided by the input photon power. Figure 21.4 shows how the quantum efficiency η and the responsivity S vary with wavelength, showing the departure from the behaviour of an ideal detector. Ideally η is unity and S rises linearly up to the cut-off wavelength. In practice the photon absorption falls off at shorter wavelengths, while the cut-off is broadened by thermal excitation.

Most photoconductors use *extrinsic* semiconductor material, in which electrons are excited from an impurity energy level into the conduction band. In *intrinsic* materials the excitation is from the valence band; the energy gap is greater, so that intrinsic materials are appropriate for higher photon energies. Figure 21.5(a) shows the spectral response of silicon, an intrinsic semiconductor. The response of CdS is close to that of the human eye, so that it is particularly suitable for use in exposure meters for cameras; while that of the alloy HgCdTe can be adjusted to give a peak between 1 and 30 μm depending on the ratio of HgTe to CdTe in the alloy. Figure 21.5(b) shows the response of three doped germanium extrinsic semiconductors, extending well into the infra-red spectrum. All such semiconductors with small bandgaps must be cooled to avoid excessive thermal excitation.

The current in the detector circuit of a photoconductor may be much larger than that caused by a single electron–hole pair for each absorbed photon. If the applied electric field sweeps a single electron across the semiconductor in a drift time τ_d, and the lifetime before recombination of the freed electron is τ_e, then each photon can result in a total current pulse of τ_e/τ_d electrons. This ratio, known as the *photoconductive gain*, may be as large as several hundred; it does not, of course, affect the quantum efficiency, which is the proportion of photons that are absorbed and produce mobile electron–hole pairs.

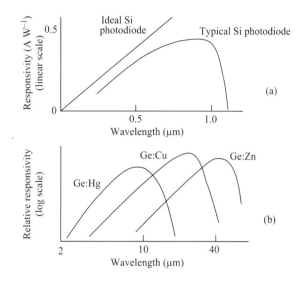

Fig. 21.5 The spectral response of typical photoconductive detectors: (a) intrinsic; (b) extrinsic semiconductors.

21.3 SEMICONDUCTOR JUNCTION PHOTODIODES

In the depletion layer of a junction photodiode (section 16.2) an electric field is already present within the semiconductor. Absorption of a photon in the depletion layer generates an electron–hole pair which is separated by the internal electric field, creating a change in current and potential across the photodiode.

A p–n photodiode is usually operated with a reverse bias voltage, as in Fig. 21.7(a); the increased current due to the generation of electrons and holes in the depletion region is shown as I_p in Fig. 21.6. The reverse bias increases the field

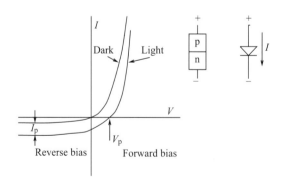

Fig. 21.6 Current–voltage relation for a p–n junction showing the change with incident light. The photodiode current at reversed bias is I_p, and the open circuit voltage is V_p.

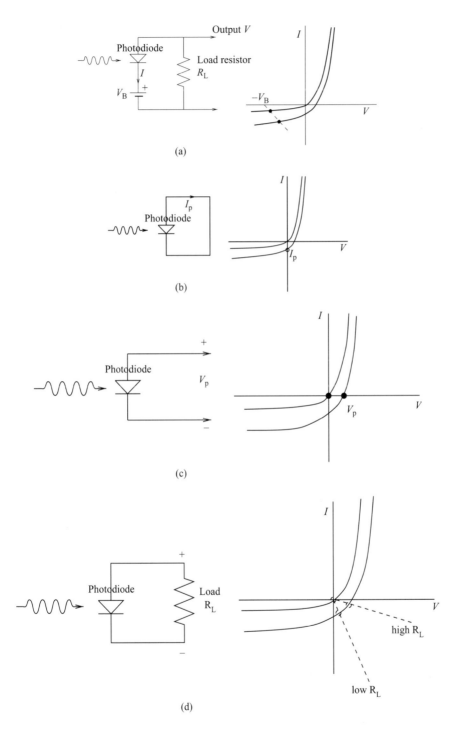

Fig. 21.7 A p–n junction diode used (a) with reverse bias; (b) short circuit; (c) open circuit, photovoltaic mode; (d) photovoltaic mode with load R_L, as in a solar cell.

in the depletion region, which has two advantages; it increases the width of the depletion region, which improves photon detection, and it shortens the transit time for electrons and holes. The output is in the form of a voltage developed across a load resistance R_L, following the load line shown in Fig. 21.7(a).

A photodiode can also be used in the photoconductive mode as a source of current, as in the short circuit mode of Fig. 21.7(b).

Operation at an open circuit, as in Fig. 21.7(c), produces the voltage V_p; this is known as the *photovoltaic mode*. A *solar cell* is required to deliver power into a load resistor R_L, as in Fig. 21.7(d). The highest power is produced for the maximum product IV. A solar cell can convert about 15% of incident radiation into electrical power.

The efficiency of a photodiode may be improved by including a layer of undoped, or *intrinsic*, material between the n- and p-type layers. Such a p–i–n diode has a larger effective depletion layer, giving it a higher quantum efficiency. A silicon p–i–n diode may have a quantum efficiency approaching unity at its maximum spectral response in the region of 0.8 μm. The p–i–n diode has a rapid response time (some tens of picoseconds) because of the larger depletion region, which has a reduced capacitance. The response of silicon diodes, however, does not extend beyond about 1.3 μm. Diodes with a response in the important wavelength range 1.3–1.6 μm in which fibre optic communication systems operate (Chapter 18), are often *heterojunctions*, using different materials on either side of the junction. An example is the InGaAsP/InP diode, which has a quantum efficiency approaching 0.75 over this infrared communication band.

The electric field across the depletion layer may be increased by operating a junction diode with a large reverse bias. The electrons and holes created by photons may then acquire sufficient energy to ionize more atoms, generating more electron–hole pairs which are in turn accelerated and may create further ionization. A diode designed to operate in the voltage region immediately below this 'avalanche' level is known as an *avalanche photodiode*, or APD. The current amplification in an APD may be greater than 100; the response time is not much affected by the avalanche process and may be less than 1 ns. APDs are widely used in optical communications.

The p–n boundary is usually between layers of the two materials in a thin slab or wafer; for example, a thin layer of n-type dopant may be deposited on a substrate of p-type material, and diffused into it in sufficient quantity to form the junction. A thin metal film, forming a transparent electrode, is then deposited on the surface.

Metal–semiconductor photodiodes, also known as *Schottky diodes*, are formed by depositing a thin transparent metallic film (usually gold) on a doped semiconductor, usually n-type. The structure and the energy bands are shown in Fig. 21.8. When a metal with work function W is in contact with an n-type semiconductor equilibrium is attained by charge transfer occurring until the Fermi levels attain the same value. The electrostatic potential of the semiconductor is raised in relation to the metal and a depletion layer forms in

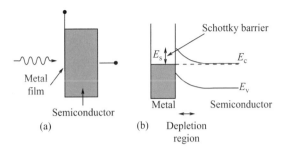

Fig. 21.8 (a) Structure and (b) energy band diagram for a metal–semiconductor photodiode, in which a metal is deposited on an n-type material. The energy barrier E_s, known as the Schottky barrier, determines the lowest photon energy at which the photodiode will operate.

the semiconductor near the junction. Photons of energy greater than the Schottky barrier height E_s are able to release electrons from the metal, which rapidly move into the semiconductor under the electrostatic field in the depletion region, generating a photocurrent. Schottky diodes have some advantages over p–n diodes: the depletion layer is close to the surface, which minimizes the loss of photons before they reach the active region of the diode, and they may also have a shorter response time, giving bandwidths in the region of 100 GHz. The metal film can be sufficiently thin to transmit blue and near ultraviolet light, giving the diode short wavelength sensitivity.

21.4 IMAGING DETECTORS

Photoelectric detectors have been described so far as single detectors. There are a number of ways in which a two-dimensional array of detectors may be assembled so that an image may be recorded, as in an electronic camera. An array of photodiodes on a silicon chip can be used as a very efficient detector of an image, with sufficient pixels for use as a television camera. It would be quite impractical to make a separate wired connection to each diode, so the outputs are scanned sequentially after exposure to light. An arrangement which is widely used is the *charge-coupled device* (CCD).

Three principles are involved in the CCD: photodetection, charge storage and charge transfer. The photodetectors may be photodiodes or a photo-metal-oxide–semiconductor (MOS) capacitor structure. In Fig. 21.9(a) a MOS photodetector is shown as an electrode insulated from a semiconducting silicon substrate by a thin film of metal oxide. The substrate is usually p-type, and the electrodes are biased positively. Under each electrode the positive carriers (holes) are repelled, leaving a depletion layer at a positive potential. Light transmitted to the metal electrode and oxide layer generates an electron–hole pair in the p-type silicon, and the pair separates under the electrostatic field.

The electrons move towards the metal electrode but are unable to penetrate the oxide layer and are trapped in the depletion layer, which acts as a potential well. The amount of charge accumulated is proportional to the integrated light flux. Photoelectrons accumulating in this potential well are stored with very little loss for periods up to some hours, allowing long integration times for faint-light photography, as for example in astronomical photography.

To provide read-out the accumulated charge is required to be transported to an output by the technique of charge transfer. An example of the charge transfer process in a line of MOS photodetectors is shown in Fig. 21.9(b) and (c). In this example the charge is accumulated under every third electrode, where a deeper potential well is created by applying a larger voltage. Charge can be transferred along a row of electrodes by the successive transfer of this bias voltage, as shown in Fig. 21.9(b). The action is often likened to a 'bucket brigade', in which buckets of water are passed from hand to hand to fill a tank. The last electrode on the line is the input to a transistor amplifier. The outputs of the individual rows can similarly be read sequentially. Efficient illumination of a CCD array requires the surface electrodes to be transparent: they are often made of a metallic compound such as PtSi, forming a Schottky barrier diode. Alternatively the diode array can be illuminated from the rear, when the silicon chip must be thinned to allow the photodiode action to occur close to the potential wells.

Cooled CCD cameras are efficient in infrared light, with low noise, high sensitivity, low dark current and wide dynamic range. CCDs are extensively used in digital cameras both for visible and infrared light; they provide an electronically stored image which can be read into a digital store and subsequently processed in a computer. They are very valuable in astronomy, where they provide images with up to a million pixels, operating with a high quantum efficiency (detective quantum efficiency (DQE) reaching over 50%) and with a linear response. Schottky barrier diodes can be used to extend the wavelength coverage to about $6\,\mu m$. Diodes operating at these long wavelengths must be cooled to reduce thermal emission.

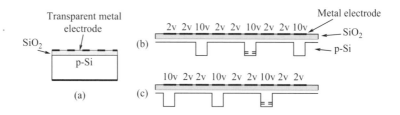

Fig. 21.9 The charge transfer system in a CCD, showing the deeper potential wells under the electrodes with higher voltages. One of the wells in (a) has accumulated electrons, which are shifted to the adjacent store in (b). A three-phase cycle of voltages moves the charges progressively along the line, as shown in (c).

21.5 NOISE IN PHOTODETECTORS

Ideally the output of a photodetector is simply proportional to the input optical power. In practice, there may be an output current with no optical input: this is known as a *dark current*. Both the desired output and the dark current have a random component, known as *noise*, which limits the accuracy of measurements at low light levels. We briefly describe the various sources of noise in photodetectors, and define the parameters that specify their overall performance when a signal is to be detected in the presence of noise.

The first source of noise is inherent in the photon nature of light itself. Each detected photon gives a pulse of output current, and the fluctuations in the photon stream become electrical noise, known as *shot noise*. If a photon stream is random, the number of photons m arriving at the detector in a given time interval varies about the mean rate n according to Poisson statistics, with a probability $P(m)$ given by

$$P(m) = \frac{e^{-n}n^m}{m!}. \tag{21.3}$$

For Poisson statistics the standard deviation, root-mean-square noise N, is the square root of the mean n, and since the mean n is also a measure of the signal strength S the *signal-to-noise ratio*

$$S/N = n^{1/2}. \tag{21.4}$$

The same argument applies to the photoelectron flux from a detector with quantum efficiency η, when the randomness in the photoelectron output is the statistical fluctuation of a photon stream with mean rate ηn, giving a signal-to-noise ratio

$$S/N = \eta^{1/2}n^{1/2}. \tag{21.5}$$

The degradation of the signal-to-noise ratio in a detector in comparison with that of the input photon stream is specified as the *detective quantum efficiency* (DQE), defined as

$$\text{DQE} = \frac{(S/N)_{\text{out}}^2}{(S/N)_{\text{in}}^2}. \tag{21.6}$$

If the degradation is entirely due to the quantum efficiency, then $\text{DQE} = \eta$.

In a communication system with a bandwidth B the noise relates to the number of detected photons arriving in a time interval $1/2B$. If the photon flux is Φ, then

$$S/N = \left(\frac{\eta\Phi}{2B}\right)^{1/2}. \tag{21.7}$$

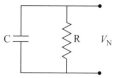

Fig. 21.10   Noise voltage V_N in a simple CR circuit.

We have noted that detectors sensitive to low-energy photons may have to be cooled to minimize the random thermal excitation of electrons. This is an example of *thermal noise*, also called *Johnson noise*, which arises from the random motion of charge carriers in resistive materials. This may be evaluated from the simple circuit in Fig. 21.10.

From the classical statistical thermodynamics of a system in equilibrium,[1] an average random thermal energy $kT/2$ is associated with each degree of freedom.[2] The energy stored in a capacitor is $CV^2/2$, so that the noise voltage V_N in a single degree of freedom is given by

$$\tfrac{1}{2}C\langle V_N^2\rangle = \tfrac{1}{2}kT. \tag{21.8}$$

The corresponding kinetic component is the Johnson noise current I_J given by

$$I_J^2 = V_N^2 R^{-2} = \frac{kT}{CR^2}. \tag{21.9}$$

The number of degrees of freedom is the bandwidth $\delta f = (4CR)^{-1}$ of the circuit,[3] giving the Johnson noise current

$$\langle I_J^2\rangle = \frac{4kT\delta f}{R}. \tag{21.10}$$

Thermal noise may arise in a photodetector either within the detector itself or in the electronic circuit which detects the output. Its importance depends on the photon flux; if the photon flux is large, the photon noise dominates, while at low photon flux the combined detector and circuit noise dominates.

The detector noise may be regarded in terms of an input signal giving the same output as the noise. In communication circuits the relevant input signal is a sinusoidally modulated light signal, which must be distinguished from the noise in a small signal bandwidth. The practical performance of a photodetector at low light levels is therefore specified by its *noise equivalent power* (NEP),

---

[1] See, for example, F. Mandl, *Statistical Physics*, 2nd edn. Wiley, 1988.
[2] It may be necessary at very low temperatures and very high signal frequencies to include the quantum factor $h\nu/[\exp(h\nu/kT) - 1]$.
[3] See, for example, I. S. Grant and W. R. Phillips, *Electromagnetism*, 2nd edn. Wiley, 1990.

which is the rms value of a sinusoidally modulated monochromatic power incident upon a detector which gives rise to an rms signal equal to the rms noise from the detector in a 1 Hz bandwidth. The inverse of NEP is termed the *detectivity*[4] *D*, which is a measure of the ability to distinguish a light signal from the detector noise.

21.6   IMAGE INTENSIFIERS

An image on an extended photoemissive surface may be detected by focusing the emitted electrons onto a phosphor screen, as in the *image intensifier* (Fig. 21.11). Here the electrons emitted from the photocathode surface in a vacuum tube are accelerated towards a phosphor screen which is at high potential. The focusing is achieved by an electromagnet, as shown in the figure, or by electrostatic field lenses. The brighter image obtained in this way is useful for night vision, especially if the photocathode has a spectral response extending into the infrared.

In television cameras an image stored on a photosensitive screen is scanned by an electron beam, providing a sequential electronic signal. In the *vidicon* camera (Fig. 21.12) the material of the screen is photoconductive, such as lead oxide (PbO). The current from the scanning electron beam is collected by a transparent conductor, such as stannic oxide, on the front of the screen; the current then depends on the conductivity of the PbO at the point scanned by the electron beam.

The *image orthicon*, shown diagramatically in Fig. 21.13, depends on photoemission. Electrons emitted from the photocathode are accelerated to a thin non-conducting plate P, which may consist of glass or magnesium oxide. The potential of P is about 300 V above that of the photocathode, so that secondary electrons are emitted on impact; these are collected by a fine wire mesh at a potential slightly above P. The plate P therefore accumulates a positive charge

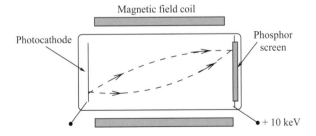

Fig. 21.11   An image intensifier using magnetic focusing.

---

[4] The performance of several types of detectors varies approximately as the square root of their area *A*, leading to the definition of a *figure of merit*:

$$D^* = \frac{A^{1/2}}{\text{NEP}}. \tag{21.11}$$

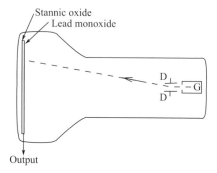

Fig. 21.12   The vidicon. The conductivity of the lead oxide layer is proportional to the number of photons that fall on it. An electron beam from the gun G scans over the layer, and a current is collected by the transparent conducting layer of stannic oxide.

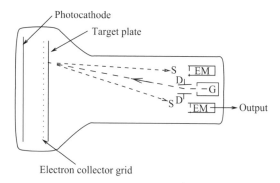

Fig. 21.13   The image orthicon. Photoelectrons from the photocathode are accelerated to the target plate, where they release a large number of electrons. These are collected by a fine wire grid, leaving an image on the target plate P in the form of a stored positive charge. An electron beam from a gun G is scanned by the deflector plates D. The number of scattered electrons S depends on the charge on P. The electrons are detected in an electron multiplier (EM) surrounding the electron gun.

density whose distribution represents the image on the photocathode. A beam of electrons now scans P, with the electron energies so arranged that the beam is scattered back more or less according to the potential of the part of P close to the beam. At the same time the beam neutralizes the charge on P. The scattered electrons are detected in the secondary electron multiplier surrounding the electron gun.

Another approach to an imaging detector is the *microchannel plate*, or channel plate multiplier, shown in Fig. 21.14. This is essentially a close packed array of photomultiplier tubes, each of which is a thin glass tube coated inside with photoemissive material. The array of tubes, each about 16 μm diameter, forms an insulating plate about 1 mm thick. As in the electronic image

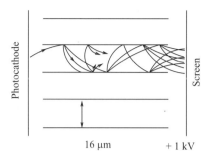

Fig. 21.14 Microchannel plate. An electron emitted from the photocathode enters a channel, strikes the wall and stimulates a shower of secondary electrons. Part of the shower development is shown. A large bunch of electrons reaches the fluorescent screen S.

intensifier this microchannel plate is placed between the photoemitting and phosphor plates.

The photomultiplier action is illustrated in the figure. A potential difference of around 1 kV between the surfaces of the plate provides a gradient of potential along each tube, and several stages of multiplication are effectively achieved. The phosphor plate may of course be replaced by a CCD or other array detector. The microchannel plate can then be used as a photon-counting imaging detector.

21.7 PHOTOGRAPHY

The photographic process has two important advantages over photoelectronic devices. First, it can store permanently an enormous amount of information. The linear resolution of a photographic plate is about 1 μm, while plates over 30 cm across are available for use in wide field cameras, such as the Schmidt telescope (Chapter 2). Second, it can be used to build up an image during very long exposure times; this is a process of *integration*.

A photographic emulsion on a glass plate or embedded in a plastic film contains individual grains of silver halide crystals, each about 0.1–1 μm in diameter. Each crystal can be developed by a reducing agent to produce a grain of silver. The reducing agent is not, however, powerful enough to develop grains unless they contain an imperfection in the form of a single silver ion which has already been converted into a silver atom. This conversion can occur when a photon is absorbed by the crystal. The photon acts by exciting an electron into the conduction energy band, where it is free to move until it is trapped by an imperfection in the crystal lattice; here a silver atom is produced. Groups of silver atoms within the crystal grain form a *latent image*. The latent image is converted into a visible image by the process of development. A chemical reducing agent is used to reduce the silver bromide to silver. Grains containing groups of silver atoms are more quickly reduced and form the visible image.

There is a natural relaxation process in the crystal, and a developable grain is only produced if more than one photon is absorbed within some minutes of time. Furthermore, red and green light has insufficient quantum energy for direct action on the crystal, and absorption must be arranged to occur in organic dyes which coat the grains. Photography is not therefore a process with a high DQE. The spatial resolution of the emulsion is limited by the size of the grains. More sensitive films have larger grains, and lower spatial resolution. The resolution is specified in terms of the closeness of a resolvable pattern of parallel lines: photographic films have a spatial resolution from 200 to 2000 lines per millimetre.

Unlike most electronic detectors the response of the photographic emulsion can be non-linear at both low and high light levels. Figure 21.15 shows a typical

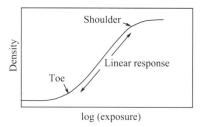

Fig. 21.15   The characteristic curve for the response of a photographic film.

relation between density on the developed plate and the logarithm of the light exposure.[5] The lower non-linear portion, referred to as the *toe*, is due to the requirement for a minimum rate of arrival of photons in an individual grain. The *shoulder* is the effect of saturation, where all grains have become developable. Between these is the straight-line region in which the response is more or less linear.

21.8 THERMAL DETECTORS

For the low photon energies of long infrared wavelengths ($100\,\mu$m to 1 mm), and occasionally also at shorter wavelengths, we may require a measurement of a steady flow of energy through its heating effect on an absorbing surface. The resultant temperature rise may be detected by a change of resistance, in a *bolometer*, or by a change in thermoelectric potential at a junction of two dissimilar metals, in a *thermocouple*.

---

[5] This *characteristic curve* is known as the Hurter and Driffield, or HD, curve, after its originators.

In both types of detector it is essential to use a sensitive element with a small thermal capacity, well insulated from its surroundings, so that its temperature can respond rapidly to the incident energy. The response time τ is the ratio of thermal capacity K to the rate of heat loss Q:

$$\tau = \frac{K}{Q}. \tag{21.12}$$

A response time of a few milliseconds can be achieved. A rapid response requires a small heat capacity K; however, for a very small detector thermal fluctuations may be important. The sensitivity limit may then be set by spontaneous temperature fluctuations with mean value ΔT given by

$$\Delta T = T\sqrt{\frac{k}{K}}, \tag{21.13}$$

where k is Boltzmann's constant.

The response of a bolometer element depends on the temperature coefficient α of resistance in the detector material, defined as

$$\alpha = \frac{1}{\rho}\frac{d\rho}{dT}, \tag{21.14}$$

where ρ is the resistivity of the material. Most metals at room temperature have α values of around $0.005\,\mathrm{K}^{-1}$; larger values are available in so-called *thermistor* semiconductors, usually oxides of manganese, nickel or cobalt. Here the temperature coefficient depends on the bandgap in the semiconductor; values of $0.06\,\mathrm{K}^{-1}$ are obtained at room temperature. Carbon provides a useful resistance element at low temperature; infrared astronomy from satellites uses such detectors cooled by liquid helium.

Thermocouples are generally less sensitive than bolometers, but they are more rugged and generally more convenient to use. They operate at room temperature, and their wide spectral response is useful for wide-band spectroscopy. A larger and more sensitive detector can be made by connecting an array of thermocouples in series; this arrangement is known as a *thermopile*.

Pyroelectric detectors provide a wide and flat spectral sensitivity over the range 1 to 100 μm with a detectivity comparable to that of a thermopile but with a rapid (sub-nanosecond) response time. They are used for infrared sensing in fire detection and security alarm systems. The pyroelectric detector is made from a ferroelectric crystal which has a permanent electric dipole moment. A change in the temperature of the crystal alters the overall dipole moment, inducing a charge on electrodes on the surfaces of the crystal.

21.9   FURTHER READING

G. C. Holst, *CCD Arrays, Cameras and Displays*. JCD Publishing, 1996.

R. J. Keyes, *Optical and Infrared Detectors*, 2nd edn. Springer-Verlag, 1980.

W. R. McCluney and R. McCluney, *Introduction to Radiometry and Photometry*. Arlech House, 1994.

G. H. Rieke, *Detection of Light from the Ultraviolet to the Submillimeter*. Cambridge University Press, 1994.

S. M. Sze, *Physics of Semiconductor Devices*. Wiley, 1981.

D. Wood, *Optoelectronic Semiconductor Devices*. Prentice-Hall, 1994.

NUMERICAL EXAMPLES 21

21.1 A photomultiplier detecting a weak source has a gain of 10^7 and a quantum efficiency $q = 0.1$. What photon rate will produce a current of 10^{-9} A? How long would it be necessary to integrate to detect a 1% change in the source?

21.2 An intrinsic semiconductor used as a photoemitter has an upper wavelength limit of 200 nm; when used as a photoconductor the upper limit is 1.8 μm. Find the widths E_g of the bandgap and E_c of the conduction band.

22

Optics and photonics in nature

In Nature's infinite book of secrecy/A little I can read.
Shakespeare, *Anthony and Cleopatra.*

A lover of Nature responds to her phenomena as naturally as he breathes and lives.
Marcel Minnaert, *The Nature of Light and Colour in the Open Air*,
Dover Publications, 1954.

In common with most academic physics, this textbook has presented an analysis of optics and photonics as a product of the human intellect. There are, however, many examples of applied optics in nature, which would be regarded as major technological achievements if they were the product of human design, but which are solely the products of evolution and natural selection. We have already referred to the outstanding example of the human eye, and to the use by insects of polarized light in navigation.

In this chapter we look in more detail at the techniques of vision that occur throughout the animal kingdom, and at the photonic detectors used both by animals and plants. We also describe examples of fibre optics and polarimeters, and examples of interference phenomena involved in the colouration of insects.

22.1 LIGHT AND COLOUR IN THE OPEN AIR

We start by following Minnaert (see the quotation above) in celebrating two of the many non-biological phenomena that occur in the natural world, the rainbow and the aurora. These glorious spectacles are examples of dispersive refraction and of spectral line emission, respectively.

The simple geometry of the rainbow (as set out in Problem 1.3) gives a single arc at an angular distance of 42° from the anti-solar point. Plate 22.1 shows two bows: the primary at 42° and a secondary at 52°. This secondary bow, which although fainter is often seen, is accounted for by two reflections inside the waterdrop instead of one.[1]

The colours of the bow are due to dispersion: the refractive index for red light is less than for violet, so that the primary bow is red on the outside, while in the secondary bow (in which the rays cross over) the sequence is reversed. The colours are not a pure spectrum; each angle has a cusp-like maximum of one wavelength with additional light from longer wavelengths. Note that the space between the bows is dark: this shows that each bow is tracing a limiting value of angular deviation, allowing light inside the primary and outside the secondary but not vice versa.

Our second example of atmospheric physics is the aurora (Plate 22.2). This is an airglow in which molecules are excited by energetic particles, mostly electrons, streaming from the Sun and channelled by the Earth's magnetic field into the two auroral zones round the north and south magnetic poles. Collisions with the atmospheric molecules lead to their excitation and dissociation, and light is emitted in a series of spectral lines due to de-excitation and recombination of several different molecular species. The colour of the aurora depends on the balance between these different lines. The spectacular shapes follow the paths of the solar particles, typically forming rays or sheets at heights of a few hundred kilometres.

The reader is recommended to pursue the study of these meteorological phenomena through the reading list at the end of this chapter.

22.2   THE DEVELOPMENT OF EYES

It has long been recognized that more than one route of evolution has led to the development of eyes. All vertebrate animals have similar eyes, which must share a common ancestry. Octopus eyes, however, despite their external appearance, cannot have followed the same route of evolution, since the retina has its network of nerve connections behind the surface of photodetectors, while in the vertebrate eye they are in front. Evolution could not involve such a topological inversion, and a separate path of development must have been followed. There is also a fundamental difference in eyes that work in air and under water; in the human eye, for example, most of the converging power is at

---

[1] The angular deviation D for k reflections is $D = 2(i - r) + k(\pi - 2r)$. The bow occurs when dD/di is zero, giving $dD/dr = k + 1$. Using Snell's law we can then find for the minimum deviation

$$\cos i = \left(\frac{n^2 - 1}{k^2 + 2k} \right)^{1/2}.$$

The bows with $k > 2$ would occur towards the Sun rather than away from it, and cannot be seen against the light scattered and reflected directly from the raindrops.

The theory of the rainbow was first given by Descartes in 1637. An interesting account is to be found in the classic *Physical Optics* by R. W. Wood, 3rd edn, Macmillan, 1934, including a discussion of the so-called *supernumerary* bows seen immediately inside the primary in Plate 22.1.

the air–cornea interface, with the lens acting as a focusing device with only one-third of the power of the cornea, while in fish the water interface is relatively weak and focusing is achieved in a spherical lens inside the eye. Insect eyes are more obviously of different types; many are multiple eyes, with several different types of focusing systems. It appears that many different and independent evolutionary routes have been followed to produce this diversity.

The basic element, from which all separate developments have evolved, is the photoreceptor. Many plants as well as animals have internal clocks which mark out a diurnal rhythm; this is kept in phase with day and night by simple photodetectors. Even the human eye has a special set of photodetectors dedicated to this task. The succession from a simple light detector to the wonderful optical systems now to be found in nature, follows a simple and logical route. The first step is the eye-cup, which is commonly found in lower phyla such as the flatworms. If the receptors are recessed in a cup or pit, some crude directivity is achieved; the animal can then react by moving towards or away from the light. If the pit is filled with transparent tissue, the directivity is improved, again with obvious advantage. If the surface of the transparent tissue is curved, the directivity is again improved, and a simple eye has been evolved. Beyond this there is an amazing diversity of optical systems, some of which we now describe.

22.3   CORNEAL FOCUSING

Eyes that depend on the focusing power of the front surface of a cornea are universal among the land vertebrates, and are also found in spiders. The sharp focusing which is so important in human eyes requires an optical system that is relatively free from chromatic and spherical aberrations. There is no known system in nature for correcting chromatic aberration; the only relief is the restricted wavelength range of the photoreceptors, and especially those that are involved with high acuity vision. Spherical aberration, however, is reduced in two ways: by shaping the surface of the cornea and by grading the refractive index inside the lens.

As we saw in Chapter 2, axial and peripheral rays refracted at a spherical lens surface focus at different distances. Camera lenses correct for this by using non-spherical surfaces, which are more difficult to make, but biological processes lead naturally to such a solution. In many types of eye, including human eyes, the front surface of the cornea is non-spherical, with flattened outer edges; the radius of curvature of the outer parts is typically twice that of the centre.

A second type of correction for spherical aberration is a graded refractive index in the lens itself. This is the main source of correction in fish eyes, but it plays a smaller part in the eyes of terrestrial animals. Together these corrections provide a precision of focusing which, on the axis, matches the diffraction limit set by the diameter of the eye pupil. In daylight the diameter of the pupil of the human eye is about 2 mm, and the image of a point source in the centre of the field of view is about 1 arcmin.

Fish and other aquatic animals cannot follow the same focusing system as eyes operating in air, since there can be only a small step in refractive index at the front surface of the eye. A simple lens, either at the surface or internally, would necessarily have too long a focal length, since the maximum refractive index in biological materials is around 1.56. Focusing is instead achieved in a spherical lens with graded refractive index; the *fish-eye lens* first described by Maxwell is an example of a perfectly stigmatic system (Chapter 2).

A spherical lens with a single ungraded refractive index (Fig. 22.1(a)) has serious spherical aberration. The fish-eye lens (Fig. 22.1(b)), with the continuous gradation of refractive index, bends rays throughout the lens and gives a short focal length as well as a perfect focus. The spherical geometry also gives a wide field of view over which sharp images are obtained. The ratio of focal length to radius is around 2.5, giving a compact and very efficient eye.

It is not surprising, in view of the excellent performance of the graded-index lens, that it is found in many unrelated marine animals. It is very surprising, however, to find a marine animal which uses instead a totally different lens system, with multiple elements resembling those of the cameras described in

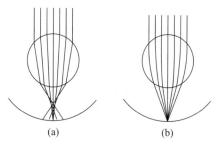

(a)   (b)

Fig. 22.1   Ray-tracing through a spherical lens (a) with a homogeneous refractive index and (b) with a graded refractive index.

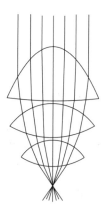

Fig. 22.2   The multi-element eye of *Pontella* (after M. F. Land, *Contemp. Phys.*, **29**, 435, 1988).

Chapter 3. The animal is the copepod crustacean *Pontella* described by Land (1984). The lens, which is sketched in Fig. 22.2, has three elements with refractive index 1.52. How such a complex system evolved, and with what advantage, is a mystery.

22.4   COMPOUND EYES

The fly's eye (Fig. 22.3) is a well-known example of the compound eyes that are found in more than half of the species of the animal kingdom. Instead of the single camera system, there are in the various types of compound eye many optical units packed together over a spherical surface. In the simplest types the units act independently, each with its own lens and photoreceptor, and each covering a limited field of view. These are known as *apposition* eyes (Fig. 22.4 (a)).

The field of view of each unit in this simple compound eye is typically about 1°, in contrast to the typical 1 arcmin resolution provided by the receptors of the human eye. Light from the whole of this field is concentrated into a single detector unit, a *rhabdom*, in which the light is transmitted through a light-guide to receptor cells. A rhabdom may taper down to a cylinder only a few wave

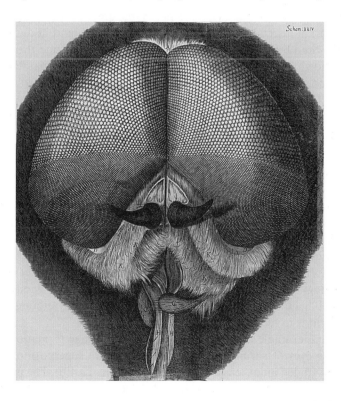

Fig. 22.3   The compound eyes of a male horsefly, from Hooke's *Micrographia*, 1665 (Royal Society Library).

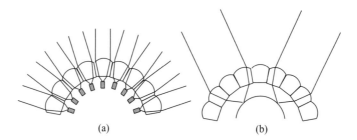

Fig. 22.4 Compound eyes. (a) The apposition eye, in which each lens acts independently; (b) the superposition eye, in which many facets contribute to each image point.

lengths across, so that propagation is determined by wave optics rather than simple geometry. Here nature has anticipated the development of fibre optics! Propagation down an asymmetrical rhabdom can be sensitive to polarization, providing insects such as bees with the polarimeter which they use for navigation by detecting scattered sunlight.

A more complex use of a multiple lens system is shown in Fig. 22.4(b). Here an individual receptor receives light from several lenses, giving a large increase in sensitivity; eyes of this *superposition* type are common in moths and other nocturnal insects, while the apposition eyes are found in diurnal insects such as flies and bees. The individual lenses are elongated, and their refractive index is graded. The retina is placed at a focal surface, well separated from the multiple lens.

22.5   REFLECTION OPTICS

Although there is no biological process which can make a metallic mirror, there are many examples in nature of reflecting surfaces made from steps in the refractive index. The reflectivity is often enhanced by thin coatings, as described later in this chapter, and is sufficient for reflection optics to be used in two rare and totally different types of eye. The first, which is found in some crustacea and molluscs, including scallops, is shown in Fig. 22.5(a). It uses a concave mirror to form an image on a retina facing away from the light source; the closest analogy is the Newtonian telescope (Chapter 3). An obvious disadvantage is that light can reach the retina from both sides, although only the reflected light is in focus.

The second example of reflection optics used in image formation occurs in the multiple eyes of shrimps and lobsters. These are superposition eyes, but without the graded index lens elements shown in Fig. 22.4(b). The multiple lens surface is replaced by an array of radially placed mirrors, which collect and focus light from many facets onto the curved retina (Fig. 22.5(b)). The mirror array is arranged in square facets rather than the hexagonal array ·of the

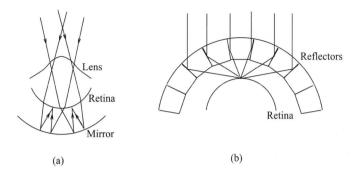

Fig. 22.5 Reflection optics in eyes: (a) the spherical mirror system of the scallop eye; (b) the radial mirror system of the lobster eye.

refracting multiple lenses; this means that the units act as corner reflectors, giving improved focusing.

This amazing application of reflection optics in the 'lobster eye' has provided the inspiration for a recent new design of X-ray telescope system. X-ray images can only be made using reflection at grazing incidence on a metallic surface, as in the Wolter system (Chapter 3) which uses large parabolic and hyperbolic reflectors. The new system, designed by Roger Angel, uses a focusing system consisting of an array of square box reflectors similar to those of the lobster eye.

22.6   THIN FILM REFLECTORS IN NATURE

Reflectors in biological organisms such as the shiny surfaces of fish scales, and the glint of reflected light in a cat's eyes, are familiar examples of remarkably good specular but non-metallic reflectors. Their high reflectivity derives from multilayers of materials with alternating high and low refractive indices. The layers are thin, and their reflectivity depends on the thin film interference phenomena described in Chapter 9. The reflectivity is therefore wavelength dependent, and the reflected light is coloured, often showing a marked variation in colour and intensity with angle.

The multiple layers of biological reflecting surfaces take many different forms. The reflector in the scallop's eye (Fig. 22.6) has stacks of cells in which the plates are of guanine (refractive index 1.83) separated by body cytoplasm. There are about five layers per micron, giving an optical thickness of a quarter wavelength at the centre of the visible spectrum. Where there are many layers, the reflectivity can exceed 90% over a restricted wavelength range. Different layer thicknesses give coloured reflections; such a selective reflectivity in multilayer structures is responsible, for example, for many of the colours of fish skin. A smaller step in the refractive index, or a difference in thickness of the two types of layer, decreases the wavelength range of high reflectivity, leading to brighter and more obvious colours.

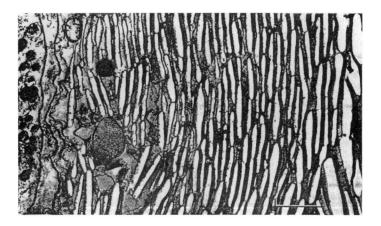

Fig. 22.6 A biological reflector; the multilayer reflector in the scallop's eye, in which layers of guanine ($n = 1.83$) alternate with body fluid ($n = 1.33$). The scale bar is 1 μm. (From M. Land, *Prog. Biophys. Molec. Biol.* **24**, 75–106, 1972.)

The iridescent colours in butterfly wings, and in particular the bright blue colours of many tropical butterflies, are derived from the structure of scales on the wing surfaces.[2] Here the thin film effect is achieved by the structure shown in Plate 22.3. The cells on the surface form an array of slots, which contain a serrated structure. The projections form a multilayer of chitin (the main structural material of insects) and air in which the refractive index alternates between 1.54 and 1.0. The maximum reflection is in blue light, where natural pigments are rarely encountered. The interference nature of this structural colour in the butterfly wing can be demonstrated by placing a drop of liquid such as acetone on the wing, thus displacing the air in the slots. This decreases the contrast in refractive index, and changes the colour of the wing from blue to green. The blue colour returns when the acetone evaporates.

22.7 BIOLOGICAL LIGHT DETECTORS

The process of detection of light is very similar over the whole of the animal world. As in the evolution of the geometric optics of eyes, several different evolutionary routes have converged on the same system, involving a protein molecule with a photonic absorption in the visual range and a signalling system to convey a nerve impulse to the brain. The photoreceptor molecule correspondingly has two parts, a *chromophore* and an *opsin*. Most proteins are colourless, with no specific absorption of light in the visual range; the chromophore must be a pigment with a resonant response to light. The basic chemical

---

[2] For a full description, see P. Vukusic, J. R. Sambles, C. R. Lawrence and R. J. Wootton, *Proc. Roy. Soc.*, B **266**, 1403, 1999.

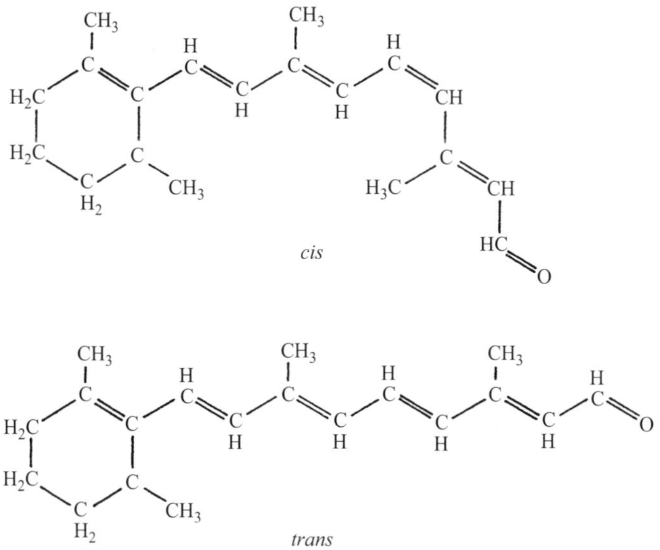

Fig. 22.7 The chromophore 11-*cis*-retinal, and the *trans* configuration to which it is converted when it absorbs a photon of light with a wavelength near 500 nm.

structure that achieves this both in eyes and in plants is a carbon chain with alternating single and double bonds, the *conjugated* chain. The universal animal chromophore is *retinal*, shown in Fig. 22.7. The combination with opsin forms the detector molecule *rhodopsin*.

Electrons in the conjugated carbon chain of retinal are held only loosely, and the absorption of a photon can temporarily disrupt a double bond allowing it to rotate before reforming in an *isomeric* form. Figure 22.7 shows the two spatial forms, 11-*cis*-retinal and its *trans* configuration to which it is converted when it absorbs a photon of light with a wavelength near 500 nm. The elongated rhodopsin is dipolar, and consequently the photon absorption is sensitive to the polarization of light. Parts of insect eyes contain aligned arrays of rhodopsin, allowing them to detect the angle of polarization of the sky, particularly in ultraviolet light.

The opsin signals a nerve impulse to the brain; the retinal then breaks from the opsin, recovers to the *cis* isomer and attaches again to form rhodopsin. The process takes a few milliseconds, after which the receptor is ready to receive the next photon. Prolonged exposure to bright light results in failure to recover, which is experienced in 'after images' of complementary colours. There are also random nerve impulses from the eye, the 'noise' of the detection process, and there is a very complex recognition process involved in distinguishing real signals from noise. Recognition of a light stimulus requires the simultaneous firing of several cells; in fact the most sensitive condition in the eye occurs when light falls on a considerable area of the retina, corresponding to at least one square degree in the field of view. The most sensitive detection corresponds to about 50 photons falling on this area of retina, which corresponds to about 100 photons on the cornea, since half of the light is lost on its way to the retina.

Evidently the DQE (detective quantum efficiency, see Chapter 21) of the human eye is only about 10^{-2} at best; however the astonishing adaptability of the eye is such that it preserves a DQE above 10^{-3} over a range of 10^7:1 in light intensity. The adaptation involves several processes, including a change of up to a factor of 10 in the area of the pupil; the most important is a change between two types of detecting cells, the rods and the cones. The cones are concerned with higher intensities, and in the human eye (as in other primates) they contain three forms of rhodopsin, absorbing at three different wavelengths: this provides for colour vision in sufficiently bright light. The rods are the most sensitive elements, but they are not colour selective; in consequence, colours cannot be seen in faint illumination, as for example in moonlight. Many stars are coloured, but except for the brightest they can only be detected by rods and appear colourless.

Rhodopsin occurs with an entirely different function in the *halobacteria*, which are microbes that live in very salty water. They actually need an external concentration of about 20% sodium chloride, and the protoplasm inside the cell contains about 5%. But their life style also needs potassium chloride, which must be present inside the cell at about 30% even though the environment may contain 1000 times less. The chemical engine that performs the necessary sorting of ions is powered by sunlight, which is absorbed by patches of rhodopsin in the cell membrane. In the process of generating ATP, the universal source of biological energy, ions are exchanged between the interior and exterior, leaving sodium outside and potassium inside. In halobacteria photon energy is used solely to drive an ion pump, and not to detect light or to use the energy to synthesize complex molecules.

22.8 PHOTOSYNTHESIS

Resonance in a conjugated carbon chain is also at the heart of *chlorophyll*, which is responsible for photosynthesis in plants. Chlorophyll *a*, found in all eukaryotes, has a resonance at 680 nm; chlorophyll *b*, found in vascular plants, has a very similar structure with a resonance at 650 nm. Figure 22.8(b) shows the simpler molecule β *carotene*, which is associated with chlorophyll in many plants and contributes to the photosynthetic process. Carotene, which gives colour to many plants including carrots, is probably a stage in the synthesis both of chlorophyll and of rhodopsin; it is one of a group of *xanthophylls* which contribute to the colours of a wide variety of proteins in diverse substances such as egg yolks, shrimps, corn and fruit.

Photosynthesis in plants, and the process of vision, both involve the movement of an electron in response to light. In chlorophyll the process is very complex. The free electron goes to a receptor molecule which uses the energy to

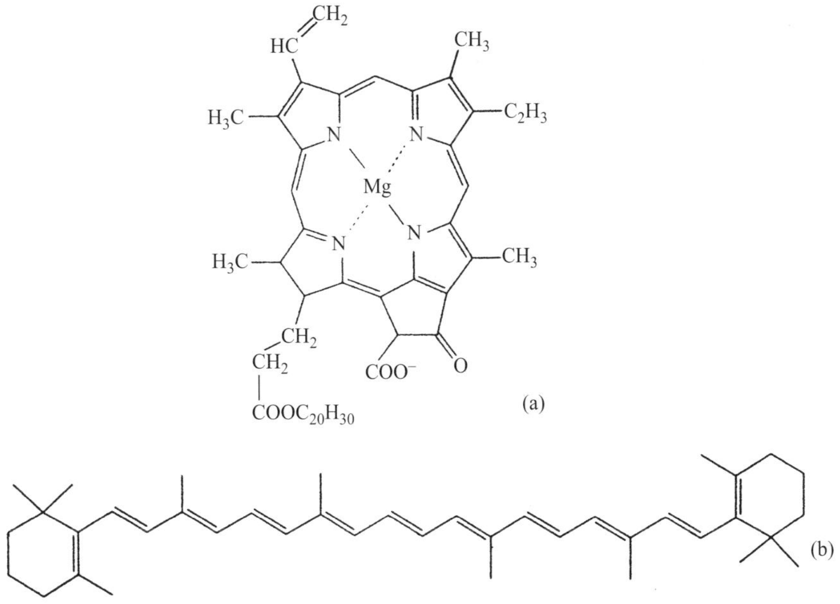

Fig. 22.8 The structures of (a) chlorophyll *a* and (b) β carotene, the two principal agents of photosynthesis.

synthesize ATP. The positively charged chlorophyll then removes an electron from other molecules; four such actions result in the breakdown of water molecules into oxygen.

One of the most remarkable aspects of all these processes is the complexity of the molecules. Chlorophyll has the chemical formula $C_{55}H_{72}MgN_4O_5$, 11-*cis*-retinal is $C_{20}H_{28}O$, and carotene is $C_{40}H_{56}$. Despite their complexity, it seems that each of these proteins has evolved along several different evolutionary routes, which have converged on the same complex but identical molecules. Given the diversity of the natural world, it is remarkable that the solutions of the problems of detecting light have been arrived at within such a small range of molecular structures. The evidence is all around us in the universal chlorophyll green of plants, with carotene and its relations contributing to the colours of leaves in autumn and in fruit.

22.9 FURTHER READING

I. Glynn, *An Anatomy of Thought: The Origin and Machinery of the Mind*. Weidenfield and Nicolson, 1999.

M. F. Land, The Optics of animal eyes, *Contemp. Phys.* **29**, 435, 1988.

D. K. Lynch and W. Livingston, *Color and Light in Nature*. Cambridge University Press, 1995.

M. G. J. Minnaert, *Light and Colour in the Outdoors*, Centenary edition. Springer-Verlag, 1993.

L. Stryer, The molecules of visual excitation, *Sci. Am.* **157**, 32, 1987.

R. Wehner, Polarized light navigation by insects, *Sci. Am.* July 106, 1976.

Appendix:
Radiometry and photometry

The science of measurement of energy in electromagnetic waves is *radiometry*, and its application to light is *photometry*. The system of units in radiometry is based on the SI system; for example, energy density is measured in units of joules per cubic metre. In photometry, however, there has been a plethora of units, arising from the need to define the illumination or visibility of a surface in terms which depend on the spectral characteristics of the human eye. We start with the basic concepts of radiometry.

The *radiant flux* Φ of a source is the rate of emission of energy, measured in watts. This relates to the total energy; there will be a dependence on direction, leading to the important definition of *radiant intensity* I, which is radiant flux per unit solid angle (steradian), measured in watts per steradian:

$$\text{radiant intensity} = I = \frac{d\Phi}{d\Omega}. \tag{A.1}$$

For an isotropic radiator $\Phi = 4\pi I$.

Irradiance S is the radiant flux across a surface per unit area perpendicular to a specified direction, measured in watts per square metre; *radiance* is irradiance per steradian. The term *radiant exitance* may be used for radiation leaving a surface.

The energy received from or falling on an illuminated surface depends on its angle θ to the direction of the radiation. A perfectly diffuse surface emits equally well in all directions, but the projected area reduces the radiance by the factor $\cos\theta$; this is *Lambert's cosine law*.

The total radiant flux from a flat surface, with area A and radiance (exitance) S, is the integral

$$2\pi A \int_0^{\pi/2} S \sin\theta \, d\theta. \tag{A.2}$$

For an illuminated surface obeying Lambert's law, where $S = S_0 \cos\theta$, the integral is $\pi A S_0$.

The irradiance of a flat surface under a hemispherical surface with radiance L is πL; it is useful to check that this is the same for a flat surface under an infinite parallel slab.

In radiometry we are often also concerned with the spectrum of the radiation, and each of the terms already defined may require the restriction 'per unit frequency band', or occasionally 'per unit wavelength range'. For example, the radiant intensity may be the integral of a spectrum of electromagnetic radiation with a *specific radiant intensity* $X(\nu)$ or $Y(\lambda)$:

$$I = \int X(\nu)\,d\nu \quad \text{or} \quad I = \int Y(\lambda)\,d\lambda. \tag{A.3}$$

Photometric units involve the response of the human eye, usually expressed in terms of wavelength. A standard response curve has been defined by the CIE (Commission International de l'Éclairage); this allows the relation between luminous flux. (including the response of the eye) to radiant flux to be plotted as in Fig. A.1. (The actual response of an individual eye may differ from this, especially at low levels of illumination where the peak is normally at 510 nm rather than 555 nm.) The photometric equivalents of the radiometric quantities already defined include a factor $V(\lambda)$, the *luminous efficiency*; for example, *luminous flux* is related to radiant flux by

$$\text{luminous flux} = k \int P(\lambda)V(\lambda)\,d\lambda \ \text{lm} \tag{A.4}$$

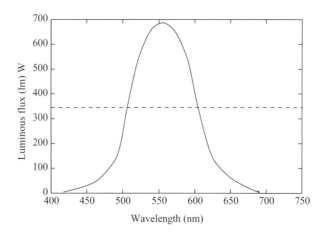

Fig. A.1   The standard relation between luminous flux and radiant flux, over the visible spectrum, as defined by the CIE. The peak sensitivity of the eye is at 555 nm, where 1 W is equivalent to 685 lm.

The most useful photometric units are shown in Table A.1. The word candela is a reminder that light was originally measured in terms of candlepower; it is now defined as the luminous intensity of $1/60\,\mathrm{cm}^{-2}$ of a blackbody radiator at the freezing temperature of platinum. In this system the constant in Eq. (A.4) is $k = 683\,\mathrm{lm}\ \mathrm{W}^{-1}$.

Table A.1

Radiometric system	Photometric system
Radiant flux (W)	Luminous flux (lumens, lm)
Radiant intensity (W sr^{-1})	Luminous intensity (candela, cd = lm sr^{-1})
Irradiance (W m^{-2})	Illuminance (lux = lm m^{-2})
Radiance (W sr^{-1}m^{-2})	Luminance (cd m^{-2})

A.1   REFERENCES

M. Born and E. Wolf, *Principles of Optics*, 6th edn. Pergamon Press, Oxford, 1980.
W. R. McCluney and R. McCluney, *Introduction to Radiometry and Photometry*. Artech House, 1994.

A.2   SOME EXAMPLES

1. Lambert's law: the Sun and the Moon. The disc of the Sun appears uniformly bright, while the brightness of the Moon's disc falls from the centre to the limb. If the surface obeyed Lambert's law the brightness would fall by the factor $\cos\theta$ in the illumination.
2. Solar photometry: luminous flux and illuminance (the photometric equivalents of radiant flux and irradiance). The illuminance of the Earth's surface by the Sun at normal incidence is 10^5 lux and the Sun subtends an angle of 30 arcmin $= 1/120\,\mathrm{rad}$. The luminous flux of the Sun is therefore $10^5(4/\pi)120^2 = 2\times10^9$ lm.
3. Irradiance: sunlight and moonlight. Assuming the Moon's surface obeys Lambert's law, we compare the radiant intensities in the direction of the Earth at full Moon.

   An elementary ring on the Moon's surface has area $a = 2\pi r^2 \sin\theta\,d\theta$, and its irradiance is $S_0\cos\theta$, where S_0 is the irradiance of full sunlight (assumed to be the same on the Earth and the Moon). The total radiant flux $d\Phi$ from the ring is $\pi^{-1}aS_0\cos\theta$; the total radiant intensity in the direction θ, back towards the Earth, is therefore

$$2r^2S_0 \int_0^{\pi/2} \cos^2\theta \sin\theta\,d\theta = \frac{2}{3}r^2 S_0.$$

At the distance d of the Earth the luminance is $\frac{2}{3}(r^2/d^2)S_0$.

The angular diameter of the Moon as seen from the Earth is $1/120$ rad. The reflectivity of the Moon's surface is about $1/5$: the final result is that the ratio of full moonlight to sunlight is 2.3×10^{-6}.

4. A burning glass: the luminous flux of an image in a simple converging lens. The angular width of the Sun ($\alpha = 1/120$ rad) is effectively increased by the lens. If the lens has diameter D and focal length f, and the normal illuminance of the Sun is S, then the illuminance is $\pi D^2/(4 f^2 \alpha^2)$. For a lens with diameter 10 cm and focal length 100 cm, an illuminance of 10^5 lux gives a luminous flux of 10^7 lm.

5. Radiance: comparison of a laser and a thermal emitter. A high-pressure mercury lamp may emit some hundreds of watts from an area of about 1 cm^2, giving a source irradiance of order 2×10^6 W m^{-2}sr^{-1}. A typical 5 mW He–Ne laser with a waist diameter of 0.5 mm concentrates light into a beam with angular diameter 1.6 mrad. The solid angle is 8×10^{-6} sr and the emitting area is 2×10^{-7} m^2, giving irradiance $= 3 \times 10^9$ W m^{-2}sr^{-1}, i.e. 1500 times larger than the thermal lamp.

Solutions to numerical examples

1.1 Electron volt $= 1.602 \times 10^{-19}$ J. Kinetic energy $\frac{1}{2}mv^2 = 10^{20}$ eV $= 10^{20} \times 1.6 \times 10^{-19}$. Velocity of 0.06 kg tennis ball is

$$v = \sqrt{\frac{2 \times 10^{20} \times 1.6 \times 10^{-19}}{60 \times 10^{-3}}} = 23\text{m s}^{-1} (= 83\,\text{km h}^{-1}).$$

1.2 In statistical physics each degree of freedom has an average energy of $\frac{1}{2}kT$. A hydrogen atom has 5 degrees of freedom (3 translational and 2 rotational). Hence thermal energy $= \frac{5}{2}kT =$ photon energy $h\nu = hc/\lambda$ and $T = \frac{2}{5}hc/k\lambda = 0.068$ K.

1.3 Photon energy $= h\nu = hc/\lambda = 100 \times 1.6 \times 10^{-19}$. So

$$\lambda = \frac{6.6 \times 10^{-34} \times 3 \times 10^8}{1.6 \times 10^{-17}} = 1.24 \times 10^{-8}\,\text{m}(= 12.4\,\text{nm}).$$

1.4
$$h\nu = \frac{hc}{\lambda} = \frac{6.6 \times 10^{-34} \times 3 \times 10^8}{1.5 \times 10^{-11}} = 1.32 \times 10^{-14}\,\text{J} = 8.25 \times 10^4\,\text{eV}.$$

1.5 Radiation approaching the top face at angle r to the normal will be refracted into an angle i outside the GaAs, where $\sin i = n \sin r$. This can happen for all angles of r smaller than the critical angle when $i = 90°$, or $r_{\text{crit}} = \sin^{-1}(1/3.6) = 16.1°$. For angles greater than 16.1°, the rays will be totally internally reflected. $R = 0.3 \tan r_{\text{crit}} = 0.087$ mm.

2.1 Using Eq. (2.8)
$$-\frac{1}{u} + \frac{1}{v} = \frac{1}{f},$$

we have $u = -2$m, $f = 2$ cm, hence $v = 2.02$ cm.

2.2 Using Eq. (2.2) for $v = u - 25$ gives

$$\frac{1}{u} + \frac{1}{u - 25} = \frac{1}{20}.$$

Solving this gives a quadratic with solutions $u = 8.9$ cm and $u = 56.1$ cm. The latter solution has a real image in front of the mirror and inverted; the former is the virtual upright image we seek. The linear magnification of the image is $-v/u = (25 - 8.9)/9.9 = 1.81$. The angle subtended by the pupil is thus $1.81 \times 0.008 = 0.0145$ rad.

2.3 A sphere is not a thin lens, so we use Eq. (2.15) for each surface in turn. Place the lamp on the left so that at the first surface $u = -9$ cm giving an image distance v_1 from

$$-\frac{1}{-9} + \frac{1.4}{v_1} = \frac{1.4 - 1}{-1}$$

and $v_1 = 4.85$ cm. This is positive and thus to the right of the first surface. For the second the object distance is $4.85 - 2 = 2.85$ cm, and the image distance v_2 is given by

$$-\frac{1.4}{2.85} + \frac{1}{v_2} = \frac{1 - 1.4}{-1}$$

giving $v_2 = 1.2$ cm. This is the distance of the image from the surface of the sphere.

3.1 Magnification $= f_1/f_2 = 15 = 30/f_2$. Hence $f_2 = 30$ cm. The field of view is determined by the second lens acting as a field stop. Hence $D/(f_1 + f_2) = 5\pi/180$ giving $D = 28$ mm.

3.2 The objective focuses the light at $0.01 \times 50 = 0.5$ cm from the axis, and the light reaches the eyepiece at $0.01 \times 52 = 0.52$ cm from the axis. The magnification is $50/2 = 25$, so the light leaves the eyepiece at an angle $0.01 \times 25 = 0.25$ rad and crosses the axis at $D = 0.52/0.25 = 2.08$ cm. (Note that this distance, although calculated for a specific angle of starlight, does not depend on angle. The general expression is $D = (f_1 + f_2)/M = f_2^2(1 + 1/M)$.)

4.1 See Fig. 1. Note that $2 \sin \omega t$ can be written as $2 \cos(\omega t - \pi/2)$. The sum has components $A_x = 5.828$, $A_y = 0.828$, giving $A = (A_x^2 + A_y^2)^{1/2} = 5.887$ with phase $\tan^{-1}(A_x/A_y) = 8.09°$.

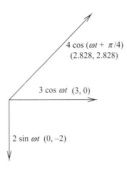

Fig. 1 Example 4.1.

5.1 Equations (5.42) give reflectance $R = 32\%$ and $T = 68\%$.

5.2 From Eq. (5.42) the two-layer reflectance is

$$\left(\frac{1.5 - 1.3}{1.5 + 1.3}\right)^2 = 5.1 \times 10^{-3}.$$

The reflectance at the first surface of the three-layer system is

$$\left(\frac{1.5 - 1.4}{1.5 + 1.4}\right)^2 = 1.2 \times 10^{-3};$$

at the second it is

$$\left(\frac{1.4 - 1.3}{1.4 + 1.3}\right)^2 = 1.4 \times 10^{-3}.$$

Thus the total fraction reflected is 2.6×10^{-3}, half that of the two-layer system.

5.3 Equation (5.32) gives $E = Bc/n$. Hence $B = 1.5 \times 10/3 \times 10^8 = 5 \times 10^{-8}$ T. Equation (5.26) gives $U = \frac{1}{2} \times 1.5^2 \times 4 \times \pi \times 10^{-7} \times 100 = 1.4 \times 10^{-4}$ J m^{-3}.

5.4 Equation (5.53) is Wien's law. A useful approximation is $\lambda_{\max} T = 0.3$ cm K, where T is in Kelvins. For 3 K, $\lambda_{\max} = 1$ mm (this is the cosmic microwave background); for 20°C $= 293$ K, $\lambda_{\max} = 10\,\mu$m, which is infrared; for 5800 K, $\lambda_{\max} = 500$ nm, in the visible (this is solar radiation).

5.5 Radiation pressure (section 5.6) at any wavelength equals the electromagnetic flux divided by the velocity of light. Hence the total force on the surface of the Earth is

$$\frac{1.4 \times 10^3 \times \pi \times \left(6.4 \times 10^6\right)^2}{3 \times 10^8} = 6 \times 10^8 \text{ N}$$

6.1 From Eq. (6.15); 0.85 μm.

7.1 Equation (7.7) gives maxima where $d \sin \theta / \lambda = n$, an integer. For small θ, $\theta = \lambda n / d = 7 \times 10^{-3} n$ rad for red light and $4.5 \times 10^{-3} n$ rad for blue light.

7.2 The diffraction pattern for each slit is given by the sinc function (Eq. (7.10)) which falls to zero at $\psi = \pi$ when $\lambda w \sin \theta$, or for small angles $\theta_{\text{diff}} = \lambda / w$. The Young's fringes are separated by an angle λ / d, so the number of fringes between the zeros at $\pm \theta_{\text{diff}}$ are $2(\lambda / w)/(\lambda / d) = 2d/w$ or $0.2/0.01 = 20$ fringes.

7.3 The angle between the central maximum and the first zero is $\lambda / d = 20$ arcmin. The double slit gives 20 interference fringes between the zeros of the single slit pattern.

7.4 Only the diameter of the objective matters. The Airey function, Eq. (7.32), gives the Fraunhofer pattern due to a circular aperture. The first zero is at an angle $1.22 \lambda / D = 1.8 \times 10^{-5}$ rad $= 4$ arcsec at $\lambda = 600$ nm.

8.1 From Eq. (8.1) the Rayleigh distance is 450 km.

8.2 Equation (8.2) gives the phase change $\phi = \pi h^2 / \lambda s$ due to the extra path length. Hence $h = \sqrt{5 \lambda s} = 224 \lambda$. On the Cornu diagram the phase will have rotated 2.5π times; the contribution is located just above the point $(0.5, 0.5)$. Equation (8.4) gives the parameter $v = \sqrt{10} = 3.2$.

8.3 The factor is $(\cos \chi_0 + \cos \chi)$. For $h = 0$ both angles are zero. For $h = 224 \lambda$, $\chi_0 = 0$ and $\chi = 224/10\,000 = 3.2 \times 10^{-2}$ rad. Hence the ratio of factors is $\frac{1}{2}(1 + \cos \chi)$ and the decrease is a factor of 1.25×10^{-4}. (Very small: it is often reasonable to ignore the inclination factor entirely.)

9.1 If the spacing is d, then $d\alpha = \lambda / 2$. Hence $\alpha = 3 \times 10^{-4}$ rad, or about 1 arcmin.

9.2 Use Eq. (9.1) to find the radii 0.54 mm and 0.94 mm.

9.3 Find N for $N\lambda_1 = (N - \frac{1}{2})\lambda_2$, giving $N = 500$.

10.1 Bragg's law, Eq. (10.30), gives the condition for reflection with adjacent planes reflecting in phase. For (i) we have $d = 2.8 \times 10^{-10}$ m giving $\theta = \sin^{-1}(N \times 1.54/2 \times 2.8) = 16.0°, 33.4°, 55.6°$; for (ii) $d = 1.98 \times 10^{-10}$ m, giving $\theta = 22.9°, 51.1°$.

10.2 Equation (10.9) gives the values of θ for the maxima, i.e. $\theta = N\lambda/d = 0.05$ and 0.1 rad. The adjacent zeros occur where phasors for the 10 lines form a closed polygon (see Fig. 10.2 for a five-line grating); the phase difference between adjacent lines is then $2\pi/10$ and $\lambda/10 = d \sin \delta\theta$, giving $\delta\theta = 0.005$ rad for both first and second orders. The spectrograph would distinguish a wavelength separation of 50 nm in the first order and 25 nm in the second order.

11.1 The mirror moves $\lambda/2$ for each fringe, so that $\lambda = (50 \times 2/185) \times 10^{-6} = 540$ nm.

11.2 The angular separation of the fringes is $\lambda/d \approx 6 \times 10^{-4}$ rad. The lamp must be more than $0.1/6 \times 10^{-4} \approx 170$ m away.

11.3 (i) angular separation of fringes $\lambda/d = 5 \times 10^{-4}$, giving linear separation 0.5 mm; (ii) the visibility is zero at the nth fringe either side of the centre, where $n = (\lambda/\delta\lambda) \cdot 2n = 25$ fringes will be seen; (iii) the angular width must be small compared with $\lambda/d = 5 \times 10^{-4}$; the linear width must be less than 2×10^{-2} mm.

12.1 The dispersion $\delta n/\delta\lambda = 3.3 \times 10^5$ nm^{-1}, giving resolving power $\lambda/\delta\lambda \approx 3 \times 10^4$. This easily resolves the sodium doublet but not the hydrogen doublet. The angular separation of the sodium lines, using Eq. (12.2), is 2×10^{-4} rad $= 0.01°$.

12.2 The total baseline is approximately $\pi D/2$. From Eq. (12.12) the resolving power is 15 000.

12.3 (i) (From Eq. (12.15)) 4400; (ii) (from Eq. (12.19)) 18 000; (iii) the resolution $\lambda/\delta\lambda$ is approximately $1.5p\sqrt{F}$, where $p = 2d/\lambda$. Hence for this spacing the resolution is 1.6×10^6.

13.1 For atomic mass number M' Eq. (13.18) gives

$$\frac{\delta\lambda}{\lambda} = 7.17 \times 10^{-7}\sqrt{\frac{T}{M'}}$$

and for neon ($M' = 20$) the Doppler line width is 1.66×10^{-3} nm. The frequency bandwidth is 1.2 GHz.

13.2 The half-width is 0.017 nm.

14.1 The coherence length for the Sun is approximately $120\lambda = 70\,\mu$m; for starlight it is approximately 20 cm.

14.2 The coherence length (approximately $c/\delta\nu$) is 5 m. For a coherence length of 10 km the bandwidth must be only 30 kHz; at wavelength 600 nm the linewidth would be $\delta\lambda = 3 \times 10^{-8}$ nm.

15.1 We require the rate of emission of photons, with energy $h\nu$. (i) 3×10^{18} s^{-1}; (ii) 5×10^{20} s^{-1}.

15.2 The stored energy is one photon per molecule. For (i) the available energy is 0.675 J and the peak power is 675 kW; for (ii) the available energy is 159 J and the peak power is 15.9 MW.

15.3 Equation (15.37) gives the mode separation 500 MHz; four modes lie within the gain curve.

17.1 Irradiance = power/area = $4/\pi \times 10^{-10} = 1.27 \times 10^{10}$ W m^{-2}. Note that this is a mean value, since the focal spot has a Gaussian spatial distribution of irradiance.

From Eq. (5.27) the Poynting vector $= N = \frac{1}{2}c\epsilon\epsilon_0 E_0^2$, where E_0 = the electric field amplitude. In free space $E_0^2 = 2N(\epsilon_0/\mu_0)^{1/2} = 2 \times 1.27 \times 10^{10}(\epsilon_0/\mu_0)^{1/2}$ giving $E_0 = 6 \times 10^6$ V m^{-2}.

17.2 The interval is $2nL/c$ which is the time for a pulse to make a double journey within the rod, i.e. 2.5 ns.

17.3 From Eq. (17.20), and assuming a Gaussian TEM$_{00}$ helium–neon laser beam, the beam radius w at a large distance z is $w = \lambda z/\pi w_0$. For surveying, the beam should be no larger than 10 cm radius, requiring a waist

$$w_0 \gg \frac{\lambda z}{\pi w} = \frac{632.8 \times 10^{-9} \times 10^4}{\pi \times 0.1} = 0.02 \text{ m}.$$

18.1 NA = 0.3, $\Delta = 2\%$.

18.2 1.56 μm.

18.3 The allowable loss is a factor of 750, or 28.7 dB. This is not quite sufficient for 50 km with 4 joints (29 dB); the answer is 49.6 km.

21.1 The photon rate is

$$\frac{10^{-9} \times 10^{-7} \times 10}{1.6 \times 10^{-19}} = 6.25 \times 10^3 \text{s}^{-1}.$$

A change of 1% requires a count of at least 10^4 to obtain S/N = 1 (Eq. (21.4)), obtained in 16 s integration time. A reasonably certain detection would need S/N = 3, obtained in 144 s.

21.2 The photon energy hc/λ for the photoemitter is $E_g + E_c$, and for the photo-conductor it is E_g, giving $E_g = 0.7$ eV and $E_c = 5.5$ eV.

Solutions to problems

1.1 At the liquid interface $n \sin 90° = N \sin r$; so $\sin r = n/N$. At emergence $\sin e = N \sin(90° - r) = N \cos r = N(1 - \sin^2 r)^{1/2}$. Hence $\sin^2 e = N^2(1 - n^2/N^2)$. Rearranging we get $n^2 = N^2 - \sin^2 e$.

1.3 In Fig. 2(a) the capillary has outer radius d and inner radius a. The edge of the inner appears to have radius b. Snell's law gives $\sin i = n \sin r$. For small angles $i \approx nr, a = dr, b = di$, giving $b = na$ independent of the outer radius.

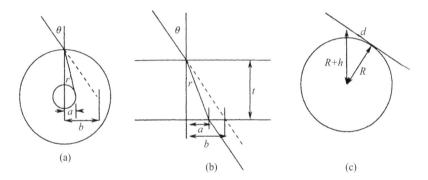

(a)

(b)

(c)

Fig. 2 (a) Problem 1.3. (b) Problem 1.4. (c) Problem 1.6

1.4 In Fig. 2(b) the displacement is $b - a$, where for small angles $a = tr$ and $b = ti$. Snell's law then gives displacement $ti(1 - n^{-1})$.

1.5 Consider a section of wavefront δy across. In time τ secondary waves from one end travel a distance $c\tau/n$; those from the other end travel a distance $c\tau/\{n + (dn/dy)\delta y\}$. If $\{(dn/dy)\delta y\}$ is small, then the extra distance is $-c\tau(dn/dy)\delta y\}$. Hence the angle turned is $c\tau/n\{1/n(dn/dy)\}$, and the radius of curvature is $n(dn/dy)^{-1}$.

1.6 In Fig. 2(c) the horizon is at a tangent to the circle, so that $(R + h)^2 = d^2 + R^2$. Expand and neglect $(h/R)^2$.

The radius of curvature of a ray is $h_0(n-1)^{-1}$, where h_0 is the scale height; in the example this is 36 000 km, giving an increase of 10% in the horizon distance.

1.7 If the ray is at an angle θ to the surface, then $\cos\theta \approx 1 - \theta^2/2 = n = 1 - \delta$. In the example $\theta \approx 12$ arcmin.

2.1 Use Eq. (2.22), noting that O must be closer than the focal point.

2.2 The ratio is $(n-1)/n$.

2.3 Adapting the analysis of section 2.4, the focal length is proportional to $n_1/(n_1 - n_2)$, where n_1 and n_2 are the refractive indices outside and inside the lens. Hence the change is a factor -7.7.

2.4 Use

$$\frac{n}{u} + \frac{1}{v} = \frac{n-1}{R},$$

giving

$$\frac{1}{v} = \frac{(n-1)}{R} - \frac{n}{u}.$$

Close to the front surface the second term dominates and $1/v \approx n/u$; the fish appears closer by a factor n. At the centre of the bowl where $u = R$ the equation gives $v = -R$, so that the fish appears in its correct place. For a distant observer, as the fish swims through a focal point at $u = R[n/(n-1)]$, the image will invert; but for water with $n = 1.4$ this point lies outside a spherical bowl.

2.5 As in Fig. 3 (similar to Fig. 2.9), first find the image distance v_1 for the first surface from

$$n/v_1 = (n-1)/R_1 = P_1.$$

This image becomes the object for the second surface at distance $(v_1 - d)$; the final image distance from the second surface v_2 is given by

$$-n/(v_1 - d) + 1/v_2 = (n-1)/R_2 = P_2.$$

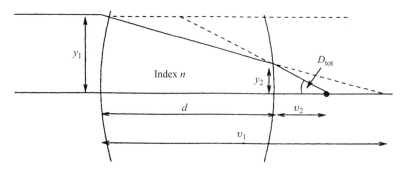

Fig. 3 Rays in a thick lens (Problem 2.5).

Now the total deviation of the ray (see section 2.7) is

$$D_{total} = y_2/v_2 = y_1 P.$$

From similar triangles $y_1/v_1 = y_2/(v_1 - d)$. Thus $P = (1/v_2)(v_1 - d)/v_1 = (P_2 + n)/(v_1 - d)/v_1$ and $P = P_1 + P_2 - dP_1 P_2/n$.

Note that the combined focal length is $f = f_1 f_2/(f_1 + f_2 - d/n)$.

2.6 From Problem 1.8 the radius of curvature of a ray in the disc is $n(dn/dr)^{-1}$. The lens requires

$$\frac{1}{n}\frac{dn}{dr} = \frac{r}{f}.$$

At radius r a ray must deviate by an angle r/f, so that for thickness t

$$t\left(\frac{1}{n}\frac{dn}{dr}\right)^{-1} = \frac{r}{f}.$$

This integrates to give the required variation of refractive index $n = 1.4 - 0.3r^2$.

2.7 With the origin of coordinates at the apex of the surface, and a focal length f the optical path through the point (x, y, z) is equal to the path fn_2 for an axial ray. Hence

$$xn_1 + ((f - x)^2 + y^2 + z^2)^{1/2}n_2 = fn_2,$$

and the equation for the surface is

$$(n_2^2 - n_1^2)x^2 + n_2^2(y^2 + z^2) - 2fn_2(n_2 - n_1)x = 0.$$

When $n_2^2 > n_1^2$ this is a prolate ellipsoid of revolution: when $n_2^2 < n_1^2$ it is a hyperboloid of revolution. The hyperboloid gives a virtual image. Note that if $n_1 = n_2$, then the aplanatic surface corresponds to a paraboloid of revolution; this is the perfect reflector.

2.8 The focusing property applies only to a single point, and no image can be formed of an extended object.

2.10 Use

$$\frac{1}{u} + \frac{1}{v} = \frac{1}{u+x} + \frac{1}{v-x} = \frac{1}{f},$$

giving $x = v - u$ and $f = uv/(u + v)$.

2.11 If the ray is incident at an angle θ, then it must be refracted at an angle $\theta/2$. Then $\alpha r = r \sin\theta$ and $\sin\theta = n \sin\theta/2$. Using $\sin\theta = 2\sin(\theta/2)\cos(\theta/2)$ the result $\alpha = n(1 - n^2/4)^{1/2}$ follows.

2.12 $1/f_1 = (n-1)(2/r)$. Hence for $n = 1.51$ the focal length is 216 mm. The aberration is given by

$$\frac{\delta f}{f} = \frac{\delta(n-1)}{n-1}$$

giving the aberration $\delta f = 3$ mm. For compensation $\delta n_1/f_1 = -\delta n_2/f_2$, giving $f_2 = 524$ mm. The combined focal length $1/f_1 + 1/f_2^{-1}$ is 272 mm.

2.14 Using Problem 2.13 with $1/u = 0$, find $b = 0.3$ mm and $c = 3$ mm.

2.15 The plate is as in Fig. 2.17. The extra thickness is $(1/n)a$, where a is found from Problem 2.13. The plate is thicker by 1.3 μm at the edge.

2.16 The image distance beyond the second surface is $[(2-n)/2(n-1)^2]r$. Differentiate to obtain the chromatic aberration $= 4$ mm.

3.1 Each surface of the sphere has power $P = (n-1)/r$. Use Eq. (2.30) for the combined power of the two surfaces separated by $2r$, giving

$$P = \frac{2}{r}(n-1) - \frac{2r}{n}\frac{(n-1)^2}{r^2},$$

and Eq. (3.5) for the magnification, giving $P \approx 800$. For glass with $n = 1.5$ the diameter would be $2r = 1.7$ mm.

3.3 The half-angle θ of the cone increases (as Snell's law) to $49°$, and the solid angle increases as $1 - \cos\theta$ giving nearly three times the total light (provided that the objective can accept the wider cone).

3.5 A point at u_1 will be imaged at $v = f + x$. The blurring at the focus will be $d = xD/f$. From

$$\frac{1}{u_1} + \frac{1}{f+x} = \frac{1}{f}$$

find $u_1 \approx f^2/Fd$. Hence $u_1 = 20$ m.

3.6 (i) ×8 (as specified); (ii) 1.9 cm (ratio of focal lengths= 8); (iii) 5 mm (= 40/8); (iv) $40/150 = 0.27$ rad $= 15°$.

4.1 Use the convenient expression $v_g = d\omega/dk$ with $v = c/n$, $\lambda\nu = c$, $\omega = kc/n$ and proceed as follows.

In terms of k, n:

$$\frac{d\omega}{dk} = \frac{c}{n} - \frac{kc}{n^2}\frac{dn}{dk}.$$

In terms of λ, v:

$$\frac{d\omega}{dk} = \frac{d\omega}{dv}\frac{d\lambda}{dv}\frac{d\omega}{d\lambda} = \frac{1}{k^2}\frac{dv}{d\lambda}\left(k + v\frac{dk}{dv}\right).$$

4.2 Group velocity $= v - dv/d\lambda = a = $ constant. *Any* pair of wavelengths have the same relative phase at the prescribed intervals of time and distance, so that a group made up of any number of component waves will reproduce exactly.

4.3 Group velocities are $V/2, 3V/2, 2V$ and c^2/V.

4.4

$$v_g = \frac{c}{n + \omega\frac{dn}{d\omega}};$$

$$\omega\frac{dn}{d\omega} = \frac{1}{n}(1 - n^2).$$

4.5 The coefficients a_n and b_n are found from integrating over one half-period:

$$a_n = E_0 \int_{-T/4}^{+T/4} \cos\omega t \cos(n\omega t)dt.$$

The sine terms are all zero since an odd parity function is being integrated over symmetric limits.

4.6 The Fourier transform is

$$F(v) = h \int_{-\infty}^{+\infty} \exp\left\{-\frac{t^2}{2\sigma^2}\exp(-2\pi ivt)\right\}dt.$$

Write

$$-\frac{t^2}{2\sigma^2}\exp(-2\pi ivt) = -\left(\frac{t}{\sigma\sqrt{2}} + i\sqrt{2}\pi v\sigma\right)^2 - 2\pi^2 v^2\sigma^2.$$

4.7 The Fourier transform is

$$F(v) = \int_{-b/2}^{+b/2} \left(h + \frac{2h}{bt}\right)dt.$$

4.8 The Fourier transform is

$$F(v) = a \int_{-\infty}^{+\infty} \exp\left(\frac{-1}{\tau} + i\omega_0 - i\omega\right)tdt.$$

4.9 The amplitude modulated wave has sidebands at $p \pm q$.

4.10 The frequency modulated wave has sidebands at $p \pm q, p \pm 2q, p \pm 3q$, etc.

4.11 Instantaneously the amplitudes must add, but over a time average and with random phases only the intensities add.

4.12 Expand the relativistic formula to

$$1 - \frac{V}{c} + \frac{1}{2}\frac{V^2}{c^2}.$$

The difference is 1.2 Hz.

4.13 24 days.

4.14 0.006 Hz.

4.15 Aberration 21 arcsec. Parallax 0.33 arcsec. Parallax and aberration are in quadrature.

4.16 Points on the wavefront δy apart in the plane of incidence arrive on a stationary mirror at times $(\delta y/c)\sin i$ apart. In this time the mirror moves $\delta y(v/c)\sin i$, and the secondary waves are retarded (or advanced) by $\delta y(2v/c)\sin i$.

5.1 The amplitude reflected is 20%; the intensity reflected is 4%. A coating with refractive index 1.225 is needed (see Numerical Example 5.2); a zero reflection coefficient can be achieved at one particular wavelength from a quarter-wavelength thick coating (see Chapter 9).

5.2 The gravitational force on the Earth is $M\omega^2 R = 6 \times 10^8$ N; the ratio is 7×10^{-15}. The ratio scales as mass/area, i.e. as radius. The particle radius must be 10^{-9} m.

5.3 For a 60 W light bulb, about 1 W goes into light, the rest into heat. If the light is emitted isotropically, with no reflector, then the irradiance at one metre from the bulb is $1/4\pi$ W m$^{-2} = \frac{1}{2}E^2/377$. The field $E = \sqrt{377 \times 2/4\pi} = 7.7$ V m^{-1}.

5.4 Suitable pairs of polarizations would be orthogonal linear, or circular with opposite hands.

5.5 If the waves are in phase everywhere, then the source amplitude must be doubled also.

6.1 Resolve the amplitude emerging from the first onto the direction of the second, and again onto the third, giving $\frac{1}{8}I_0$.

6.2 6.3 mm.

6.3 The plate produces a small lateral displacement, not an angular displacement.

6.4 There is dispersion in the difference between the two refractive indices.

6.6 Major axis at $52°$; axial ratio $= \cot 28° = 1.88$.

7.1 Derive the phasor diagrams from Fig. 7.8. On the first zero the phasor diagram is a closed circle; removing a central strip increases the amplitude to a maximum value of π^{-1}.

7.2 (i) In this region there are two crossing wide beams, giving Young's fringes at a spacing of 1 mm. (ii) Here the two pinholes are focused, giving images 5 mm apart, with Airy discs 0.5 mm across.

7.3 Each point of the source forms a pattern independently. The elongated source at an angle therefore forms fringes parallel to the line source, separated by an angle λ/D in a direction perpendicular to the diffracting slit.

7.4 (i) The beam width is usually quoted as the full width to half power (FWHP). This is approximately equal to the angular distance of the first zero in an Airy diffraction pattern $1.22\lambda/d = 10$ arcmin in this case. (ii) 12 arcsec.

7.5 In directions lying in a plane perpendicular to a side there are no zeros; the phasor diagram is a spiral that does not close. In planes parallel to the sides there are zeros where symmetrical halves of the triangle cancel. The angular distance to the zeros is very approximately $(2n + 1)\lambda/L$, where L is the side of the triangle; the angular distance to the first zero is 0.05 rad.

7.6 The zeros occur where $\tan z = 1$.

7.7 Interference fringes are formed between the transmitter and its image, which has a reversed phase. The minimum value of h is therefore 5 wavelengths, i.e. 4.5 m.

7.8 At each position the slit spreads the rays by an angle λ/D. The widths of the standard diffraction patterns in the three cases are: (i) 2 mm; (ii) 1 mm; (iii) 2 mm.

7.9 (i) 2 arcmin; (ii) 0.5 arcmin; (iii) 0.03 arcsec; (iv) 10^{-3} arcsec; (v) 3×10^{-3} arcsec.

8.1 The Fresnel zone is about 500 m across, comparable with the height of surface features. The angular width is about 250 microarcsec, about ten times the angular diameter of the nearest star.

8.2 The particles are about 35 μm in diameter. There must be a spread in diameter as the diffraction pattern is spread into an indistinct halo.

8.3 The phasor diagram for each zone is a semicircle, and alternate zones are missing, giving an amplitude ratio of $\frac{1}{2} \times 2/\pi$, and an intensity ratio of π^{-2}. Reversing the phase of alternate zones quadruples the intensity. The remaining energy is diffracted around the focal point.

8.4 The two polarized sectors act independently, and the two edge patterns add in intensity giving shadow fringes on both sides of the geometric shadow edge.

8.5 The pattern is a superposition of a Fresnel shadow from a half-amplitude wavefront on an unobstructed half-amplitude wavefront. Shadow fringes will be seen on both sides of the geometric shadow edge, as can be shown by using the Cornu spiral.

9.1 The interference fringes are concentric circles.

9.2 The reflected waves from the front and back of the film are in opposite phase. The film thickness must therefore be considerably less than one wavelength.

9.3 A sideways dispersive shift by the prism moves the white light fringe away from the centre.

9.4 At the centre the glass surfaces are in contact and there is no reflection. The oil will reduce the reflection coefficient at both surfaces, giving fringes with reduced intensity and with smaller diameters.

9.5 The reflection coefficient takes the form $(r + \exp i\theta)/(1 + r \exp i\theta)$, whose modulus is unity.

9.6 The spacing reduces to d/n. The reflection coefficient becomes small, and the fringes become faint. The *visibility* does not fall.

9.7 The order is highest in the centre. Let this be $n + \epsilon$, where $\epsilon < 1$. Then

$$n + \epsilon = \frac{2d}{\lambda}.$$

For 600.000 nm we find $n_1 + \epsilon_1 = 33\,333.33$. For 600.010 nm, $n + \epsilon$ is decreased by 1 part in 6×10^4 and $n_2 + \epsilon_2 = 33\,332.788$. Hence $n_1 = 33\,333$ and $n_2 = 33\,332$.

Fringes appear at θ where $\cos\theta = n\lambda/2d$. Putting $\cos\theta = 1 - \theta^2/2$, the first fringe appears at $\theta^2 = 2\epsilon/n$. Hence $\theta_1 = 4.5$ mrad and $\theta_2 = 6.9$ mrad.

9.8 Thickness $d = n(\lambda/2)$. If a change in $\delta\lambda$ changes n by δ/n, then

$$\frac{\delta n}{n} = -\frac{\delta\lambda}{\lambda}.$$

Here $\delta n = 1$, so that

$$n = \frac{475}{0.043} = 1.10 \times 10^5$$

and

$$d = \frac{1}{2} \times 1.10 \times 10^5 \times 475 \times 10^{-9}\,\mathrm{m} = 2.62\,\mathrm{cm}.$$

9.9 The parameter $F = (4 \times 0.6)/(0.4)^2 = 15$. The ratio required is $(1 + F) = 16$.

9.10 The spectrum consists of a regular sequence of bright bands at wavelengths $\lambda = 2t/n$, where n is an integer. If $n = 2t/\lambda_1$ and $n + 200 = 2t/\lambda_2$, then $t = 0.216\,\mathrm{mm}$.

10.1 (i) The effect is similar to the diffraction from a circular aperture as compared with a rectangular aperture. The diffraction maxima become wider, and the subsidiary maxima become smaller.

(ii) The grating spacing becomes $2d$, so that the maxima of all orders are closer spaced. The intensity in the direction of the original maxima is reduced by a factor of four. The widths of the maxima are unchanged.

(iii) A superposition of two grating patterns, one with spacing d and amplitude α, the other with spacing $2d$ and amplitude $1 - \alpha$, should be considered. Then the original maxima will be reduced in intensity by a factor $[(1 + \alpha)/2]^2$, and subsidiary maxima appear at a half spacing in sin θ, with intensity α^2.

10.2 (i) The pattern is shifted by an angle θ, where $\sin\theta = \lambda/2d$. The envelope pattern, determined by the width of a single slit, remains unchanged.

(ii) Superpose two patterns, one from a grating with spacing d and unit amplitude, the other from a grating with spacing $3d$ and amplitude -2. Then subsidiary maxima appear at one-third intervals of sin θ, with intensity reduced in intensity to $(1/3)^2$.

(iii) Each half should be considered separately, then the two halves added to give an interference pattern, repeating at intervals of $2\lambda/D$ in sin θ. This is half the total width of a diffraction maximum, so that each maximum breaks up into two halves, with a zero in the centre.

10.5 Consider two pieces of grating δx apart, with rays emerging at angles θ and $\theta + \delta\theta$. They converge at a distance D, where $D\delta\theta = \delta x \cos\theta$. Differentiate the grating equation to obtain $\delta x/\delta\theta$.

10.6 The modulation on the grating is represented by the product of an ideal grating function and a sinusoid. The Fourier transform is the convolution of the two individual transforms. Such a modulation gives ghosts on either side of a spectral line, apparently from light with a wavelength difference $\delta\lambda = \lambda/q$.

11.1 Each half prism deviates by an angle $(n_1 - 1)\alpha$. Immersion changes this to $(n_1 - n_2)\alpha$; also the wavelength reduces to λ/n_2. The fringe spacing changes by a factor $(n_1 - 1)/n_2(n_1 - n_2)$.

11.2 Approximately 10 cm.

11.3 $n - 1 = 1.7 \times 10^{-4}$. The visibility is at a minimum when the path difference corresponds to 500 fringes, which requires 2.7 atm.

11.4 Differentiating Eq. (9.9) gives

$$\frac{d\phi}{d\theta} = \frac{4\pi nh}{\lambda} \sin r \frac{dr}{d\theta};$$

differentiating Snell's law gives

$$\cos\theta = n\cos r \frac{dr}{d\theta}.$$

11.5 5000630 ± 1 nm.

12.1 (i) $m = 15.3$; (ii) 100 arcsec; (iii) resolving power $= 8 \times 10^5$.

14.1 The reception pattern is the product of the two amplitude diffraction patterns. The angular resolution is approximately λ/D.

14.2 (ii) The microscope resolving power is $1.2\lambda(\sin\theta)^{-1}$, giving a resolution of 0.1 nm.

14.3 $\tau = 20\,\mu s$, $N = 100$.

15.1 Rates are equal when $\exp(h\nu/kT) - 1 = 1$. For $\lambda = 10\,\mu m$, $T = 2078$K.

15.2 Irradiance $I = B\omega$ and $I = cP$. For a black body the solid angle $\omega = 4\pi$.

15.3 The density of modes $d(\nu) = 8\pi\nu^2/c^3 = A_{21}/B_{21}$. The rate of stimulated emission is

$$\left(\frac{dN_2}{dt}\right)_{stim} = -B_{12}N_2\rho(\nu) = -\frac{A_{12}N_2}{d(\nu)h\nu}\rho(\nu).$$

If one photon is excited per mode, i.e. $d(\nu) = 1$, then $\rho(\nu) = h\nu$, and $(dN_2/dt)_{stim} = -A_{21}N_2$, which is the rate of spontaneous emission.

15.4 For a helium–neon laser, $\Delta\nu_D = 1.52$ GHz. For Ar^{+r}, $\Delta\nu_D = 4.9$ GHz.

15.5 Use Eqs. (15.27) and (15.34). The minimum length is 0.136 m.

15.6 (i) 0.67 mrad $= 0.038°$. (ii) $B = I/\omega = P/A\omega = 2.50 \times 10^9$ W m^{-2}sr^{-1}.
(iii) $T \approx B\lambda^2/2k = 3.62 \times 10^{19}$ K.

15.7 (i) Threshold gain coefficient $\gamma_{\text{thres}} = 0.71$ m^{-1}; (b) inversion $(N_2 - (g_2/g_1)N_1)$
$= 8.27 \times 10^{22}$ m^{-3}.

15.8 Power within the waist is $\int_0^{w_0} I_0 \exp(-2r^2/w^2)\, dA$, compared with the same
integral from 0 to infinity. The ratio is $1 - e^{-2} = 86.5\%$.

15.9 (i) The transition is homogeneously broadened. (Doppler broadening (inho-
mogeneous) is $\Delta\nu_D = 61$ MHz. Pressure broadening (homogeneous) is
$\Delta\nu_P = 790$ MHz.) (ii) The contributions are equal at a pressure of
7.84×10^3 Pa.

15.10 Doppler broadening is 4.66 GHz. Mode spacing is 187.5 MHz. So 24 modes
are present. To ensure oscillation in a single mode the cavity must be less than
6 cm long.

15.11 The resonance condition is $m\lambda = L$, where L is the circumference of the ring;
compare $m(\lambda/2) = L$ for a standing wave cavity.

17.1 From Eq. (17.7), the laser linewidth $= \Delta\nu_L = (2\pi h\nu/P)(\Delta\nu_c)^2$, where
$P =$ laser power and $\Delta\nu_c = 1/2\pi\tau_c =$ width of cavity resonance. Then

$$\Delta\nu_L = \frac{2\pi h\nu}{P}\left(\frac{1}{2\pi\tau_c}\right)^2 = \frac{2\pi \times 6.6 \times 10^{-34} \times 3 \times 10^8}{10^{-3} \times 632.8 \times 10^{-9}}\left(\frac{1}{2\pi \times 10^{-7}}\right)^2$$

$$= 4.97 \times 10^{-3}\text{ Hz}.$$

The theoretical minimum linewidth is not attainable since it is dominated by
cavity vibrations and thermal drift.

For a length change ΔL the line frequency changes by $\Delta\nu$ where

$$\frac{\Delta\nu}{\nu} = -\frac{\Delta L}{L} = \frac{4.97 \times 10^{-3} \times 632.8 \times 10^{-9}}{3 \times 10^8} = 1.05 \times 10^{-17}.$$

The equivalent temperature change for a coefficient of 10^{-6} K^{-1} is 10^{-11} K.

17.2 A useful general expression for coherence time is $= \tau_c \approx 1/\Delta\nu$, where $\Delta\nu =$
bandwidth. (A more precise relation for quasi-monochromatic radiation is
$\tau_c = 1/2\pi\Delta\nu$.) The coherence length $= l_c = c\tau_c$.

(i)

$$\Delta\nu \approx \frac{c}{\lambda^2}\Delta\lambda = \frac{3 \times 10^8 \times 300 \times 10^{-9}}{(550)^2 \times 10^{-18}} = 2.97 \times 10^{15} \text{ Hz.}$$

$$\tau_c \approx \frac{1}{\Delta\nu} = \frac{1}{2.97 \times 10^{15}} = 3.36 \times 10^{-16}\text{s.}$$

$$l_c = c\tau_c = 10^{-7}\text{ m.}$$

(ii) $\tau_c = 5 \times 10^{-5}\text{s}, l_c = 1.5 \times 10^4\text{ m.}$

(iii) For the laser longitudinal modes the mode spacing $\Delta\nu = c/2L = 3 \times 10^8/2 \times 0.3 = 5 \times 10^8$ Hz.   For   three   modes $\Delta\nu = 10^9$ Hz, $\tau_c \approx 1/\Delta\nu = 10^{-9}\text{s}$, and $l_c = 3 \times 10^8 \times 10^{-9} = 0.3\,\text{m.}$

18.1 $\tau = 0.25\mu\text{s}$. $B_T = 2 \times 10^6$ bits s^{-1} = 2 Mbs^{-1}.

18.2 See Problem 1.8. Consider the turning point where the ray is parallel with the axis. When the ray has turned through an angle θ the refractive index has increased in proportion to cos θ, and so has the optical path.

18.3 2 ns km^{-1} and 0.1 ns km^{-1}.

Index